Lecture Notes in Mathematics

Editors:
J.-M. Morel, Cachan
B. Teissier, Paris

For further volumes:
http://www.springer.com/series/304

Tadahito Harima • Toshiaki Maeno
Hideaki Morita • Yasuhide Numata
Akihito Wachi • Junzo Watanabe

The Lefschetz Properties

Springer

Tadahito Harima
Department of Mathematics Education
Niigata University
Nishi-Ku, Niigata, Japan

Toshiaki Maeno
Department of Mathematics
Meijo University
Nagoya, Japan

Hideaki Morita
Muroran Institute of Technology
Muroran, Japan

Yasuhide Numata
Department of Mathematical Sciences
Shinshu University
Matsumoto, Japan

Akihito Wachi
Department of Mathematics
Hokkaido University of Education
Kushiro, Japan

Junzo Watanabe
Department of Mathematics
Tokai University
Hiratsuka, Japan

ISBN 978-3-642-38205-5   ISBN 978-3-642-38206-2 (eBook)
DOI 10.1007/978-3-642-38206-2
Springer Heidelberg New York Dordrecht London

Lecture Notes in Mathematics ISSN print edition: 0075-8434
                    ISSN electronic edition: 1617-9692

Library of Congress Control Number: 2013942530

Mathematics Subject Classification (2010): 13-02, 13A02, 13A50, 06A07, 06A11, 14M15, 14F99,
                    14M10, 14L30, 17B10

Printed on acid-free paper

Springer is part of Springer Science+Business Media (www.springer.com)

Introduction and Historical Note

An Artinian local ring is geometrically a point with multiplicity. The coordinate ring for such a point is a quotient of a regular local ring by an ideal of finite colength. Those ideals are, in contrast to ideals of low codimension, very complicated. It could be said that the study of Artinian rings is at the outset made difficult by their apparent simplicity and therefore lack of obvious invariants. Nonetheless, and precisely for this reason, there are good grounds to be interested in Artinian rings.

For example, (1) many problems of local rings often "reduce" to problems of Artinian rings, (2) standard graded Artinian Gorenstein algebras are parameterized by the n-ary r-ics, which were central objects of study during the early development of abstract algebra in the nineteenth century, their classification was the main problem of classical invariant theory, (3) a strong parallelism can be observed between Artinian rings and finite posets and this aids understanding of both, (4) there is no harm in looking at graded Artinian rings as cohomology rings of some (probably nonexistent) algebraic varieties, and if we do so, it gives us a new way of looking at Artinian rings, and (5) it helps us understand the Schur–Weyl duality which plays a very basic role in representation theory. All commutative algebraists would most likely agree with us if we said that the best way to understand the Schur–Weyl duality was through commutative algebra. (The interested reader may wish to see [103].)

The viewpoint (1) above is traditional and well understood. What we try to do in this book is to emphasize less known aspects of the theory of Artinian rings encompassed by (2)–(5).

A major theorem in the theory of finite sets is Dilworth's theorem, which states that the maximum size of antichains in a poset is equal to the minimum number of chains into which the poset can be disjointly decomposed. This common number is called the Dilworth number of the poset. This result is a typical minimax theorem and it is a powerful tool to determine the number for a poset. In fact, if one finds a disjoint chain decomposition and an antichain whose size is equal to the number of the chains, then this number is the Dilworth number. This idea works practically unchanged for Artinian rings. A minimal generating set of an ideal in an Artinian local ring is almost exactly an antichain in a poset. The counterpart of a chain decomposition of a poset is the Jordan block decomposition of a linear form in the regular representation of a local ring. With this interpretation, we may attach new meanings to Artinian rings and moreover provide a new method for proving Sperner type theorems in the theory of Artinian local rings. In his pioneering work [136], Stanley showed that the hard Lefschetz theorem can be used to prove that certain posets have the Sperner property.

The investigation of the Lefschetz properties of Artinian local rings was first suggested by one of the authors in the Japan–US joint seminar on combinatorics and commutative algebra held in Kyoto, Japan in the summer 1985, where R. Stanley gave an expository lecture, among other things, on $\mathfrak{sl}_2$-posets and J. Watanabe announced the results of [146] on the Sperner number of Artinian rings. However, for the first 10–15 years few results were obtained in this direction; there were many natural problems but they seemed too difficult to deal with.

The first breakthrough was made by J. Migliore and U. Nagel in [55], and after that many results were obtained and many connections to other areas of mathematics began emerging. These results are now known as Lefschetz properties of graded Artinian rings.

By now there are many results and applications on or of the Lefschetz properties and it would be a demanding task for any authors of a book to include all major results on the Lefschetz properties. In this monograph, the authors intend to concentrate on presenting the results from what has become a vast literature with which they are most familiar, many of which were discovered by the authors themselves in the last 10 years. Even this selection seems more than enough to show the diversity of applications of the Lefschetz properties. The authors have tried to put the emphasis on the diversity of applications and not on the completeness of the results presented here.

There are many topics on the Lefschetz properties that are being developed and investigated. For example, very recently the Lefschetz properties have been found to have surprising new connections with enumerative combinatorics (with plane partitions and lattice path enumeration, in particular). This was discovered algebraically by D. Cook II and U. Nagel [21, 22, 26] and independently by J. Li and F. Zanello [80] and more importantly was then explained combinatorically by Chen, Guo, Jin and Liu [15]. A few of their results are collected in the Appendix without proof. There is also a surprising relation to Laplace equations discovered by E. Mezzetti, M. -R. Miro–Róig and G. Ottaviani [8, 100]. Regretfully these are not included in this book.

There is a very good survey paper by Migliore and Nagel "A Tour of the Weak and Strong Lefschetz Properties" [101]. It is amazing that there is little overlapping between the topics dealt with in their survey and those in this monograph. The fact is an evidence for the diversity of the Lefschetz properties.

In Chap. 1 of this book we have collected some basic definitions and theorems of the Sperner theory of finite sets as are found in the books [2, 9, 33] and [34]. Particularly the treatment of Dilworth's Theorem and the Sperner theory using the language of "set systems" as in [9] is of help to a newcomer to this area. The purpose of this section is to introduce the reader to the theory. Particularly important for us are the four examples of posets with rank function that we introduce here and are companions throughout the book. They have a direct translation in terms of Artinian rings.

In Chap. 2 we provide basic definitions and theorems of Artinian local rings with emphasis on Gorenstein algebras. Except for a few cases we restrict ourselves to Artinian rings and do not treat commutative rings in general, since Artinian Gorenstein algebras can be understood without knowledge of commutative algebra as a whole. This section is intended for those readers who wish to acquire knowledge of Artinian Gorenstein algebras as quickly as possible.

In Chaps. 3–6 we present recent results that were obtained and developed by the authors over the last 10 years.

The motivation in proving these results was a desire to prove that Artinian complete intersections have the strong Lefschetz property over a field of characteristic zero. We show some special cases where this holds. In Chap. 7 we relate the Lefschetz property of Artinian Gorenstein algebras with the hard Lefschetz theorem in algebraic geometry and show some ring-theoretic methods for proving certain cohomology rings have the strong Lefschetz property. In Chaps. 8 and 9 we show that the Lefschetz elements have a special meaning in representation theory.

m-Full Ideals and Dilworth Number of Artinian Rings

The number of generators of ideals in a ring has been one of the central topics in the theory of commutative rings. In fact some theorems on the number of generators have fundamental importance in the theory of commutative rings, for example, the Hilbert basis theorem, Krull's principal ideal theorem, and Serre's theorem which states that height two Gorenstein ideals are complete intersections. More results can be found in Sally's book [122].

In 1983, D. Rees raised the following problem: for which ideals I (say, in a local ring R with maximal ideal $\mathfrak{m}$), is it true that $\mu(I) \geq \mu(J)$ for any $J \supset I$. (Here μ denotes the number of generators.) To approach this problem, one has to ask oneself: what is the number

$$\max\{\,\mu(J) \mid J \supset I\,\},\tag{1}$$

for a given I, where J runs over all ideals containing I. Set $A = R/\mathfrak{m}I$. Then this number is equal to

$$\max\{\,\mu(\mathfrak{a}) \mid \mathfrak{a} \subset A\,\}.\tag{2}$$

If this is finite, then it follows that Krull dim $R/I \leq 1$. Let us confine ourselves to the case the Krull dimension is zero or in other words, R/I is Artinian. If R is a polynomial ring over a field and I can be generated by monomials, then finding the number (1) is a combinatorial problem. This can be explained as follows.

Let $R = K[x_1, x_2, \ldots, x_n]$ be the polynomial ring over a field K and let $\mathrm{m} = (x_1, x_2, \ldots, x_n)$ be the maximal ideal. Further, let S be the set of all monomials in R. Then S is a set partially ordered by divisibility. Let I be an ideal in R generated by monomials such that R/I is Artinian. Consider the set

$$P := S \setminus \mathrm{m}I.$$

The set P, being a subset of S, is a poset. Notice that P is a vector space basis for $R/\mathrm{m}I$. If we want to determine the number (2) for $A = R/\mathrm{m}I$, we need to find a set of monomials in P of maximum size whose elements are totally disordered. The maximum number of elements in antichains (totally disordered sets) in a poset is known as the Dilworth number. The name came from the theorem of Dilworth, as mentioned earlier.

The definition of "Dilworth number" can be extended to all commutative rings of Krull dimension at most one, but the case of Artinian local rings is essential. The investigation made in [146] was motivated by a desire to determine the numbers (1) and (2) for m-primary ideals and Artinian local rings in general.

To answer Rees's question let $y \in R$ be any non-unit element and let J be any ideal of R which contains I. Then it is not difficult to prove that

$$\mu(J) \leq \mathrm{length}(R/(\mathrm{m}I + yR)).$$

(See Proposition 2.30.) For $y \in \mathrm{m}$, the numerical value $\nu_y(I)$ for the ideal $I \subset R$ is defined by

$$\nu_y(I) = \mathrm{length}(R/(\mathrm{m}I + yR)).$$

Then obviously we have

$$I_1 \subset I_2 \implies \nu_y(I_1) \geq \nu_y(I_2).$$

Thus, for a given ideal I, if there exists an element $y \in \mathrm{m}$ such that

$$\mu(I) = \nu_y(I), \tag{3}$$

then I has the property which Rees's question asks for. It is Rees himself who defined the notion of an m-full ideal. Namely, I is m-**full** if there exists an element y such that $\mathrm{m}I : y = I$. This is equivalent to the claim that the equality (3) $\mu(I) = \nu_y(I)$ holds. Hence, we have the proposition stating that an m-primary m-full ideal has the Rees property. (See Proposition 2.55.)

It is easy to see that the intersection of two m-primary m-full ideals is m-full. Hence an m-primary ideal has an m-full closure. Given an m-primary ideal I, it seems natural to hope that, for any m-primary ideal I, we have

$$\mu(I^*) = \max\{\mu(J) \mid J \supset I\},$$

where I^* is the m-full closure of I. This equality seems too good to be true, but at the very least there should be no harm trying to prove it. We have not yet found a counter example to this equality in a polynomial ring in characteristic zero. Another interesting question is under what conditions does the m-full closure of I have the form $I + \mathfrak{m}^k$ for some k. It is clear that this is not always true, but at least for complete intersections, there are many hints that suggest this should be true.

Finite Posets Versus Artinian Rings

As was mentioned in the preceding section there can be observed a strong parallelism between the theory of posets and that of commutative rings. The set S of monomials in the variables

$$x_1, x_2, \ldots, x_n$$

is a partially ordered set with divisibility as the order. It is a basis for the polynomial ring

$$R = K[x_1, x_2, \ldots, x_n]$$

as a linear space over a field K. The set of square-free monomials in S is the standard basis for the Artinian algebra

$$K[x_1, x_2, \ldots, x_n]/(x_1^2, x_2^2, \ldots, x_n^2).$$

An antichain in S is a minimal generating set of a monomial ideal of R. Let I be an ideal in R generated by monomials, and let P be the standard basis for R/I. Then P is a poset with rank function. This way of translation continues and a lot of ring-theoretic notions can find their counterpart in the theory of posets and conversely many combinatorial notions have ring-theoretic interpretations. Table 1 is a small dictionary of terms in the theory of Artinian rings and terms in the theory of finite sets, which interested readers may wish to expand.

One of the topics of this book is the Dilworth number of the Artinian rings of the form $A = R/I$, where the Dilworth number is defined to be the supremum of the number of generators of ideals in A. The definition of the Dilworth number for Artinian rings was introduced in [146] based on the theorem of Dilworth, as the definition for Artinian local rings is natural and is necessary to solve the problem of Rees. If I is generated by monomials, then the Dilworth number of R/I as an Artinian ring coincides with the Dilworth number of the poset $S \setminus I$ in the original sense. We want to know the number $\max \{ \mu(J) \mid J \supset I \}$, which is the same as the Dilworth number of $A := R/\mathfrak{m}I$. The two Artinian rings A and $A' := R/I$ differ only by $I/\mathfrak{m}I$, which may be regarded as a set of generators for I and those

Table 1 A ring-poset dictionary of terms

Ring	Poset	Description
$R = K[x_1, \ldots, x_n]$	$S = \{x_1^{d_1} \cdots x_n^{d_n}\}$	S : a basis of R
Artinian ring	Finite poset	
Canonical module	Dual poset	
Hilbert function	Whitney numbers	
Degree	Rank	
Ideal	Filter	
R^*-ideal	ideal	$R^* = K[\frac{\partial}{\partial x_1}, \ldots, \frac{\partial}{\partial x_n}]$
Minimal generating set	Antichain	
$\mathfrak{m}I$	Neighbor	$\mathfrak{m} = (x_1, \ldots, x_n)$
Multiplication by l	Hasse diagram	$l = x_1 + \cdots + x_n$
Multiplication by $l + l^1 + l^2 + \cdots$	Reachability matrix	$l = x_1 + \cdots + x_n$
Standard basis for R/I	$S \setminus I$	I : an ideal of R
Dilworth number of R/I	Dilworth number of $S \setminus I$	I : an ideal of R
#1	Disjoint chain decomposition	
$R/(x_1^2, \ldots, x_n^2)$	Boolean lattice	
$R/(x_1^{d_1}, \ldots, x_n^{d_n})$	Divisor lattice	
#2	Young diagram lattice	
#3	Vector space lattice	

See Sect. 3.5 (p. 126) for #1, Sect. 1.4.3 (p. 22) for #2, Sect. 1.4.4 (p. 29) for #3.

elements are the socle elements of A. So it is often the case that the Dilworth number of A is known from that of A'. (See Remark 2.44.)

Very fortunately the combinatorialists have developed a theory to determine the Dilworth number for various posets (e.g., [33, 34, 42]). In many important cases the Dilworth number turns out to be equal to the maximum of the Whitney numbers. (For Whitney numbers see Definition 1.24.) In such cases the poset is said to have the Sperner property. The following are basic examples of posets with rank function:

1. The Boolean lattice ($= 2^{[n]}$, all subsets of $[n] = \{1, 2, \ldots, n\}$, see Example 1.2)
2. The divisor lattice or the finite chain product (see Sect. 1.4.2 for details)
3. The vector space lattice (see Sect. 1.4.4 for details)
4. The lattice of the Young diagrams contained in a rectangle (see Sect. 1.4.3)
5. The lattice of partitions of a set (see [42])

Among these five examples, (1)–(4) have the Sperner property. It was a long standing conjecture that (5) has the Sperner property [42], but in 1978 it was disproved by Canfield [14].

In 1928, Sperner [132] proved that the Boolean lattice has the Sperner property. More specifically, he proved that if F is a family of subsets of $\{1, 2, \ldots, n\}$, in which no two members are related by inclusion, then $|F| \leq \binom{n}{[n/2]}$, and the equality occurs only in the case all sets have the same size.

This result was the origin of the Sperner theory. Once this was proved, it should have been easy to realize that the divisor lattice has the Sperner property. In spite of simple nature of this assertion, no proof was known for 20 years. In 1949, the Dutch Mathematical Society posed it as a prize problem, and in response three Dutch mathematicians gave a proof [11]. The idea of their proof is to decompose a divisor lattice into a disjoint union of "symmetric chains." (For details see, e.g., Aigner [2], Greene–Kleitman [42].)

Their method called "symmetric chain decomposition" resembles the Jordan block decomposition of a general element in the regular representation of an Artinian monomial complete intersection, and this seems to be a good evidence for the most natural method for proving the Sperner property for divisor lattice being the use of the theory of the special linear Lie algebra $\mathfrak{sl}_2$. This will be elucidated in Chap. 3.

The Hard Lefschetz Theorem and Sperner Property

Stanley [136] used the hard Lefschetz theorem to prove that certain partially ordered sets arising from a class of algebraic varieties have the Sperner property. To see an example, put

$$X = \mathbb{P}_{\mathbb{C}}^{d_1-1} \times \mathbb{P}_{\mathbb{C}}^{d_2-1} \times \cdots \times \mathbb{P}_{\mathbb{C}}^{d_n-1}.$$

The cohomology ring $H^*(X, \mathbb{C})$ of X is isomorphic to

$$\mathbb{C}[x_1, x_2, \ldots, x_n]/(x_1^{d_1}, x_2^{d_2}, \ldots, x_n^{d_n}).$$

The assertion of the hard Lefschetz theorem proves that the divisor lattice has the Sperner property. (In fact the hard Lefschetz theorem is a stronger statement than the Sperner property.) L. Reid, L. Roberts and M. Roitman [119] gave a purely algebraic proof for the equivalent statement to the hard Lefschetz theorem for the monomial complete intersection. One of the main objectives of this book is to provide a systematic method to prove the Sperner property for various posets including the four posets (1)–(4) above.

Schur–Weyl Duality and the Strong Lefschetz Property

The theory of the strong Lefschetz property for Artinian rings gives us a new way to look at representation theory of classical groups. A very basic fact of the representation of the general linear group $GL(d, K)$ is that all irreducible representations are obtained by decomposing the tensor representation

$$GL(d) \to GL(nd) = GL\left((K^d)^{\otimes n}\right)$$

into irreducible representations, and to decompose $(K^d)^{\otimes n}$ into irreducible $GL(d)$-modules is the "dual" to decomposing it into S_n-modules. To explain this in more detail, let the symmetric group S_n act on the tensor space $(K^d)^{\otimes n}$ by permutation of the components. A basic fact is that the actions of the two groups commute with each other. So the tensor space is in fact an $S_n \times GL(d)$-module, and an irreducible $S_n \times GL(d)$-module is of the form

$$F \otimes_K G,$$

where F is an irreducible S_n-module and G an irreducible $GL(d)$-module.

It is possible to identify the tensor space $(K^d)^{\otimes n}$ with the Artinian algebra:

$$A = K[x_1, x_2, \ldots, x_n]/(x_1^d, x_2^d, \ldots, x_n^d).$$

In fact, the isomorphism

$$(K^d)^{\otimes n} \to A = K[x_1, x_2, \ldots, x_n]/(x_1^d, x_2^d, \ldots, x_n^d)$$

is given by

$$e_{i_1} \otimes e_{i_2} \otimes \cdots \otimes e_{i_n} \mapsto x_1^{i_1-1} x_2^{i_2-1} \cdots x_n^{i_n-1},$$

where $\{ e_1, \ldots, e_d \}$ is the standard basis of K^d. The tensor space $(K^d)^{\otimes n}$ is acted on by the symmetric group S_n as permutation of components and an isotypic component of it as an S_n-module *is* an isotypic component of it as a $GL(d)$-module.

So, what is the advantage of considering the algebra A rather than the tensor space $(K^d)^{\otimes n}$? First, in the obvious sense A has a ring structure, so more information is available than just the tensor space $(K^d)^{\otimes n}$ offers. At least from the viewpoint of commutative algebraists, the algebra A should be easier to deal with than the tensor space.

Second, the action of the symmetric group S_n has a clear ring-theoretic meaning, namely, it is the permutation of the variables, where this action preserves the grading of A. One may realize that decomposing A into irreducible S_n-modules is more or less same as decomposing the polynomial ring $K[x_1, x_2, \ldots, x_n]$ into irreducible S_n-modules, which is well understood. For example, the ring of invariants R^{S_n} is the algebra generated by the elementary symmetric polynomials.

Third, the linear form $l := x_1 + x_2 + \cdots + x_n$ is an invariant of S_n and the multiplication map $\times l : A \to A$ as well as $\times l : R \to R$ preserves S_n-module structure. Thus, when the irreducible decomposition of R/lR is known, it is also known for R. The same is true for A/lA and A.

We have seen what the action of S_n means on A. On the other hand, one might ask what is the ring-theoretic meaning of the tensor representation,

$$\Phi : GL(d) \to GL\left((K^d)^{\otimes n}\right),$$

if we replace the tensor space by the Artinian ring $A = K[x_1, \ldots, x_n]/(x_1^d, \ldots, x_n^d)$. This can be explained as follows. Put $B = K[x]/(x^d)$ and let B^* be the multiplicative group of B. Similarly, let A^* be the multiplicative group of $A = K[x_1, \ldots, x_n]/(x_1^d, \ldots, x_n^d)$. Then we may define a group homomorphism

$$\Phi' : B^* \to A^*$$

by

$$f(x) \mapsto f(x_1)f(x_2) \cdots f(x_n).$$

With this notation we have the following commutative diagram of algebraic groups:

$$
\begin{array}{ccc}
GL(B) & \xrightarrow{\;\Phi\;} & GL(A) \\[4pt]
\uparrow & & \uparrow \\[4pt]
B^* & \xrightarrow{\;\Phi'\;} & A^*
\end{array}
\qquad (4)
$$

where the vertical maps are the group homomorphism induced by the regular representation of the algebras B and A. For example, the first vertical map is written as

$$
f(x) \mapsto \begin{pmatrix}
a_0 & & & & & \\
a_1 & a_0 & & & & \\
a_2 & a_1 & a_0 & & & \\
\vdots & \vdots & \vdots & \ddots & & \\
a_{d-2} & \vdots & \vdots & & \ddots & \\
a_{d-1} & a_{d-2} & a_{d-3} & \cdots & \cdots & a_0
\end{pmatrix},
$$

where

$$f = a_0 + a_1 x + a_2 x^2 + \cdots + a_{d-1}x^{d-1} \in B.$$

We have explained the group homomorphism $\Phi' : B^* \to A^*$ and the vertical maps. So this explains ring-theoretically a part of

$$GL(d) \to GL(dn),$$

in the diagram (4) and it should give us a clear picture of the lower half of the tensor representation. As well as the multiplication by elements of A, the differential operator

$$a_0 + a_1 \frac{\partial}{\partial x} + \cdots + a_{d-1} \frac{\partial^{d-1}}{\partial x^{d-1}}$$

induces an automorphism of A (provided that $a_0 \neq 0$) and it explains the upper half of the tensor representation.

Historical Note by J. Watanabe

I learned the definition of "m-full ideal" as well as the associated problem (mentioned earlier) from Prof. D. Rees himself through a private conversation with him while he was visiting the Department of Mathematics of Nagoya University in 1983. He also showed me how one could define a "general element" for Artinian rings and showed me a proof for the statement in the Appendix of [147]. What it asserts is only too natural but the proof is very difficult. The proof written in Appendix of [147] and Theorem 5.1 and Proposition 5.2 of this book is due to Rees. Without the clear definition of general elements, the theory of Artinian rings could not start. Theorem 5.1 of this book is the basis of the theory of the SLP.

In [146], I used the term "strong Stanley property" for an Artinian Gorenstein graded algebra $A = \bigoplus_{i=0}^{c} A_i$ if there exists a linear form $l \in A$ such that the multiplication map $\times l^{c-2i} : A_i \to A_{c-i}$ is bijective for all $0 \leq i \leq [c/2]$, and other authors followed [66, 123]. It is an abstraction of the hard Lefschetz theorem and I came to this definition because it seemed to be the quickest way to prove that the maximum number $\mu(I)$ for ideals I in A is attained by a power of the maximal ideal. (It is a ring-theoretic interpretation of the Sperner property.) In 1990s the term "strong/weak Lefschetz property" began to be commonly used by many authors.

I would like to thank A. Iarrobino who showed his interest in the theory of Dilworth number of Artinian local rings at the earliest stage. Particularly, I would like to thank him for many discussions and encouragement. Special thanks are due to J. Migliore and U. Nagel for their invitation to let me share their idea to use the Grauert–Mülich theorem to prove that any complete intersection in embedding codimension three has the weak Lefschetz property [55]. It was the first breakthrough in an attempt to prove the Lefschetz property for complete intersections and it encouraged me greatly to continue my work on the strong and weak Lefschetz properties. H. Ikeda has contributed greatly to this theory. She constructed various examples of Gorenstein algebras without the SLP or WLP and gave a new proof for the SLP of the Boolean lattice [123], from which the Sperner property of monomial complete intersections follows. This is not well known but it preceded the proof of L. Reid, L. Roberts and M. Roitman [119].

Hiratsuka, Japan   J. Watanabe

Acknowledgements The authors are extremely grateful to L. Smith and A. Iarrobino for their invaluable suggestions and encouragement. Prof. Larry Smith read the first manuscript of this monograph and made various comments from the viewpoint of algebraic topology and made corrections for improvement of English usage. The authors learned that there are many problems of Lefschetz properties in positive characteristic to be investigated in connection with algebraic topology. Regrettably the authors were unable to include most of them because of the short space and time. They are also grateful to S. Murai, who gave many helpful suggestions after careful reading of the manuscript of this monograph. This book grew out of seminars in Tokai University starting in 2003. The authors express thanks to the participants of the seminars.

T. H. is supported by Grant-in-Aid for Scientific Research (C) (20540035 and 23540052). T. M. is supported by Grant-in-Aid for Scientific Research (C) (22540015). H. M. is supported by Grant-in-Aid for Scientific Research (C) (22540003). Y. N. is supported by JST CREST. A. W. is supported by Grant-in-Aid for Scientific Research (C) (23540179). J. W. is supported by Grant-in-Aid for Scientific Research (C) (19540052).

Contents

Chapter 1
Poset Theory

This chapter was written to furnish a starting point for the study of Artinian rings in commutative algebra. We are primarily interested in the Sperner theory of finite posets. The basic examples of finite posets to keep in mind are (1) the **Boolean lattice** $2^{[n]}$, (2) the **divisor lattice** $\mathscr{L}(n)$ or the **finite chain product**, (3) the **order ideal** $\mathscr{I}(s^r)$ **of Young diagrams** contained in a rectangle (s^r), and (4) the **vector space lattice** $\mathscr{V}(n,q)$ of finite dimensional vector spaces over a finite field. We are interested in the methods used to prove the Sperner property for these lattices, and how they can be translated into terms of commutative rings, which together with the theory of Artinian rings can be applied to prove that certain classes of such rings have the Sperner property.

1.1 Poset and Dilworth Number

We assume that the reader is familiar with the definition of a partially ordered set (poset) and a lattice. In this book it is enough to consider finite posets, but sometimes an infinite poset may appear. The purpose of this section is to introduce to the reader the Dilworth number and the Dilworth theorem of posets. As an application of the Dilworth theorem, we prove Hall's marriage theorem. For completeness we start with the definition of a poset. (For a definition of a lattice see the first paragraph of Sect. 1.3.)

Definition 1.1. A **partially ordered set** (**poset**) is a set P together with a binary relation $\leq$ satisfying

1. For all $x \in P$, $x \leq x$.
2. If $x \leq y$ and $y \leq x$, then $x = y$.
3. If $x \leq y$ and $y \leq z$, then $x \leq z$.

T. Harima et al., *The Lefschetz Properties*, Lecture Notes in Mathematics 2080, DOI 10.1007/978-3-642-38206-2_1, © Springer-Verlag Berlin Heidelberg 2013

The binary relation $\leq$ is called a **partial order** or simply an **order**. The notation $x < y$ is used in the obvious sense: $x \leq y$ and $x \neq y$. Two posets are regarded as isomorphic if there exists an order preserving bijection.

We say that two elements $x, y \in P$ are **comparable**, if either $x \leq y$ or $y \leq x$, and otherwise they are **incomparable**. A subset of a poset is automatically a poset with the induced order. It is called a **subposet**.

Example 1.2. Let $[n] = \{1, 2, \ldots, n\}$, and $2^{[n]}$ the power set of $[n]$, i.e., the set of subsets of $[n]$. Then $2^{[n]}$ is a poset ordered by inclusion. The elements $\{1, 2\}, \{1, 2, 3\} \in 2^{[4]}$ are comparable. The elements $\{1, 2\}, \{2, 3\} \in 2^{[4]}$ are incomparable. The empty set $\emptyset$ is a unique minimum element and $\{1, 2, \ldots, n\}$ a unique maximum element.

A product of two posets is defined as follows:

Example 1.3. Let P and Q be posets with orders $\leq_P$ and $\leq_Q$, respectively. Then $P \times Q = \{(x, y) \mid x \in P, y \in Q\}$ is ordered by the binary relation $\leq$:

$$(x, y) \leq (x', y') \iff x \leq_P x' \text{ and } y \leq_Q y'$$

for $(x, y), (x', y') \in P \times Q$. The poset $P \times Q$ is the **product** of posets P and Q.

Example 1.4. Let P be a poset. For $x, y \in P$ with $x \leq y$, we define the **interval** $[x, y]$ between x and y to be the set $\{z \mid x \leq z \leq y\}$. Each interval is automatically a poset with the induced order.

Example 1.5. Let $P = 2^{[n]}$. For $X, Y \in P$ with $X \subset Y$, the interval $[X, Y]$ is isomorphic to the Boolean lattice $2^{Y \setminus X}$ as posets via the map φ defined by

$$\varphi: [X, Y] \ni Z \mapsto Z \setminus X \in 2^{Y \setminus X}.$$

Chains and antichains play a vital role in the Sperner theory. The definitions are as follows:

Definition 1.6. A poset P is a **chain** if any two elements are comparable. A poset P is an **antichain** if no two elements are comparable. The definition applies to a subset of P. Thus a subposet of P is a chain if any two elements in it are comparable, and an antichain if any two elements in it are incomparable.

Remark 1.7. For a poset P, the empty set $\emptyset$ is a chain and an antichain.

Example 1.8. We denote by $C(d)$ the **chain with $d+1$ elements**. Thus it is possible to write $C(d) = \{x^0, x^1, \ldots, x^d\}$, where

$$x^0 < x^1 < \cdots < x^d.$$

Definition 1.9. For a poset P, we denote by $\mathscr{A}(P)$ the **set of antichains in P**.

Definition 1.10. For a poset $(P, \leq_P)$, the set $\mathscr{A}(P)$ is a poset with the order $\leq$:

$$A \leq A' \iff \forall a \in A, \quad \exists a' \in A', \quad a \leq_P a',$$

for $A, A' \in \mathscr{A}(P)$.

Remark 1.11. Let P be a poset. For $P' \subset P$, $\mathscr{A}(P')$ is a subposet of $\mathscr{A}(P)$.

Definition 1.12 (Dilworth number). For a poset P, we define the **Dilworth number** $d(P)$ of P as

$$d(P) = \max_{A \in \mathscr{A}(P)} |A|,$$

where $|A|$ is the cardinality of A. So the Dilworth number of P is the maximum size of an antichain of P.

Example 1.13. Let $P = 2^{[3]}$. Following is a list of all maximal antichains in P. Any element of $\mathscr{A}(P)$ is a subset of one of the antichains listed below:

$$\{\emptyset\},$$
$$\{\{1,2,3\}\},$$
$$\{\{1\},\{2,3\}\},$$
$$\{\{2\},\{1,3\}\},$$
$$\{\{3\},\{1,2\}\},$$
$$\{\{1\},\{2\},\{3\}\},$$
$$\{\{1,2\},\{2,3\},\{1,3\}\}.$$

Hence $d(P) = 3$. $\mathscr{A}(P)$ consists of all subsets of these sets. In fact there are 20 antichains in all.

Next we state the famous theorem of Dilworth.

Theorem 1.14 (Dilworth [29]). *For a finite poset P, the Dilworth number $d(P)$ equals the minimum number of disjoint chains into which P can be decomposed.*

The original proof given in [29] was seven pages long. Perles' proof [115] is known to be the shortest.

Proof of Theorem 1.14 (Perles). No two elements of an antichain can appear in a chain. So the size of any antichain cannot exceed the number of chains in a disjoint chain decomposition. So the claim is clear in one direction. To prove the converse, let n be the maximum size of antichains, i.e., $n = d(P)$. We have to show that P can be decomposed into n chains. We induct on $|P|$. If $|P| = 1$, then there is nothing to prove. Assume that $|P| > 1$. Suppose first that there exists an antichain U of maximum size, $|U| = n$, which contains neither all maximal elements nor all minimal elements of P. Define P^+ and P^- by

$$P^+ = \{ p \in P \mid p \geq u \text{ for some } u \in U \},$$

$$P^- = \{ p \in P \mid p \leq u \text{ for some } u \in U \}.$$

Notice that $P^+ = P$ (resp. $P^- = P$) would imply that U is the set of minimal (resp. maximal) elements. Since we have assumed that U contains neither all maximal elements nor all minimal elements, it follows that P^+ and P^- are not all of P. Hence by the induction hypothesis P^+ and P^- can be decomposed into n chains and we may glue these chains at U to obtain a decomposition of P into n chains. Next suppose that there is no such U with the above property. This means that an antichain of maximum size is either the set of minimal elements or the set of maximal elements. (Thus in fact there are at most two antichains with the maximal cardinality.) Let a be a maximal element and b a minimal element such that $b \leq a$. Put $P' := P \setminus \{a, b\}$. One notices that the maximum size of antichains in P' is $n - 1$. By induction hypothesis we can decompose P' into $n - 1$ disjoint chains. We add to it the chain $\{a, b\}$, which gives us a decomposition of P into n disjoint chains.

Example 1.15. Let $P = 2^{[3]}$. Then we have $d(P) = 3$ and a minimal chain decomposition $P = C_1 \sqcup C_2 \sqcup C_3$ with $C_1 = \{\emptyset, \{1\}, \{1, 2\}, \{1, 2, 3\}\}$, $C_2 = \{\{2\}, \{2, 3\}\}$, $C_3 = \{\{3\}, \{1, 3\}\}$.

As was pointed out by Dilworth himself in [29], Hall's marriage theorem is a consequence of Theorem 1.14.

A "matching" for an incidence structure may have many interpretations and can be defined in several ways depending on the situation involved, and the definition can have variants. Here we define it as follows:

Definition 1.16. Let X, Y be sets. A **matching** in $X \times Y$ is a subset M of the Cartesian product $X \times Y$, satisfying

$$(x, y) \neq (x', y') \implies x \neq x', y \neq y',$$

for any $(x, y), (x', y') \in M$. A matching M is **full** if $|M| = \min\{|X|, |Y|\}$.

Example 1.17. Let $X = \{1, 2, 3\}$, $Y = \{4, 5, 6, 7\}$.

1. $M = \{(1, 5), (3, 7)\}$ is a matching in $X \times Y$.
2. $M' = \{(1, 4), (1, 5), (3, 7)\}$ is not a matching in $X \times Y$.

Example 1.18. Let $X = Y = \{1, 2, 3, 4, 5, 6, 7, 8\}$.

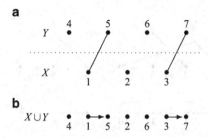

Fig. 1.1 M in Example 1.17 (**a**) M as a bipartite graph (**b**) M as a directed graph

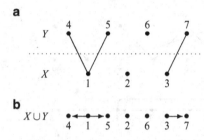

Fig. 1.2 M' in Example 1.17 (**a**) M' as a bipartite graph (**b**) M' as a directed graph

1. $M = \{(1,2),(2,3),(3,4),(4,1),(5,6),(7,7)\}$ is a matching in $X \times Y$.
2. $M' = \{(1,2),(3,2),(3,4),(4,1),(5,6),(7,7)\}$ is not a matching in $X \times Y$.

Remark 1.19. 1. A subset M in $X \times Y$ may be expressed as a 01-matrix with rows and columns indexed by X, Y. In this case a 01-matrix is a matching if there exists at most one 1 in each row and in each column.
2. If X and Y are distinct sets, then a matching $M \subset X \times Y$ may be thought of as a pairing between members of X and Y, in such a way that each member of X has at most one partner in Y and no two members in X have the same partner.
3. If $X = Y$, any subset $M \subset X \times Y$ may be thought of as defining a directed graph on X as a vertex set. This means that M is considered as a set of edges. An edge $(x_i, x_j) \in M$ may be expressed as $x_i \rightarrow x_j$. Some vertices may not belong to any edges while loops like $x_i \rightarrow x_i$ may exist. With this interpretation a matching $M \subset X \times X$ may be thought of as a directed graph consisting of disjoint circuits and chains. Note that a circuit with one vertex is a loop and a chain with one vertex is a vertex without edges. By abuse of notation suppose that we use the same M to denote the matrix $M = (m_{ij})$ such that $m_{ij} = 1$ if $x_i \rightarrow x_j$ and $m_{ij} = 0$ otherwise. If $M^r = O$ for some integer r, then the directed graph consists of disjoint chains with at most r vertices.
4. In Example 1.17 both M and M' define directed graphs on the vertex set $X \sqcup Y = \{1,2,3,4,5,6,7\}$. These are illustrated as Figs. 1.1b and 1.2b.

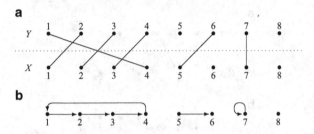

Fig. 1.3 M in Example 1.18 (**a**) M as a bipartite graph (**b**) M as a directed graph

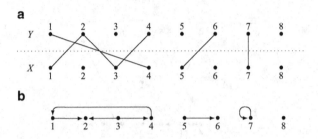

Fig. 1.4 M' in Example 1.18 (**a**) M' as a bipartite graph (**b**) M' as a directed graph

5. In Example 1.18, both M and M' define directed graphs on the vertex set $X = \{1, 2, 3, 4, 5, 6, 7, 8\}$. These are illustrated as Figs. 1.3b and 1.4b. The directed graph of Fig. 1.3b consists of two circuits and two chains.

Theorem 1.20 (P. Hall [46]). *Let $E \subset X \times Y$, where X and Y are finite sets with $|X| \le |Y|$. Then E contains a full matching if and only if*

$$|S| \le |N(S)|,$$

for any subset $S \subset X$, where

$$N(S) = \{\, y \in Y \mid (x, y) \in E \text{ for some } x \in S \,\}.$$

Proof. If E contains a full matching, then we may construct an injection f from X to Y such that $(x, f(x)) \in E$ for each $x \in X$. Since $f(S) \subset N(S)$ for any $S \subset X$, we have

$$|S| = |f(S)| \le |N(S)|.$$

Next we show the other implication. We may assume that $X \cap Y = \emptyset$. Then $P = X \cup Y$ becomes a poset with the following order:

$$x < y \iff (x, y) \in E.$$

Note that a chain in P is either a set $\{x, y\}$ of two elements such that $(x, y) \in E$ or a singleton. Let $U \subset P$ be an antichain of maximum size. Let $U_1 = U \cap X$ and $U_2 = U \cap Y$. Then $N(U_1) \cup U_2$ is an antichain with size at least the size of U. This means that Y is an antichain of the maximum size. By the Theorem of Dilworth (Theorem 1.14) there exists a chain decomposition $X \cup Y = C_1 \sqcup C_2 \sqcup \cdots \sqcup C_n$ such that $n = |Y|$. Each C_i contains one and only one element of Y and every $x_i \in X$ is contained in exactly one of C_j, which means there exists a full matching.   □

1.2   Ranked Posets and the Sperner Property

In many finite posets it is often the case that two maximal chains between two elements have the same length. Such a poset, with a little stronger condition, is called a ranked poset. (Shortly we define it precisely.) A ranked poset P has subsets called rank sets and it has the Sperner property if $d(P)$ is attained by the size of a rank set. After the definition we prove that the Boolean lattice has the Sperner property.

Definition 1.21. Let $(P, \leq)$ be a poset. For $x, y \in P$, we say that y **covers** x if $x < y$ and there are no elements properly between x and y. Namely y covers x if and only if $x < y$ and $[x, y] = \{x, y\}$. The **covering relation** is denoted by $x \lessdot y$. (Note the "·" over the <.)

Definition 1.22. A poset P is said to be **ranked** if there exists a **rank function** $\rho: P \to \mathbb{N}$ such that $\rho(x) = 0$ if x is a minimal element, and $\rho(y) > \rho(x)$ if $y > x$, and $\rho(y) = \rho(x) + 1$ if $y \lessdot x$. ($y \lessdot x$ of course means that y covers x.)

Remark 1.23. Suppose P is a poset with rank function $\rho : P \to \mathbb{N}$. Put

$$P_i = \{x \in P \mid \rho(x) = i\}.$$

Then we have a decomposition $P = \bigsqcup_{i \geq 0} P_i$, which satisfies:

1. P_0 is the set of minimal elements of P,
2. $P_{i+1} = \{y \in P \mid \text{there exists } x \in P_i \text{ such that } x \lessdot y\}$ for $i \geq 0$.

If P is finite, then $P = \bigsqcup_{i=0}^{s} P_i$ for some $s \in \mathbb{N}$ such that $P_s \neq \emptyset$. Obviously P_k is an antichain. Thus, for any k,

$$|P_k| \leq d(P).$$

Definition 1.24. In the notation of Remark 1.23, we call P_k a **rank set**. The sizes $|P_k|$ of P_k are called the **Whitney numbers** or also **rank numbers**.

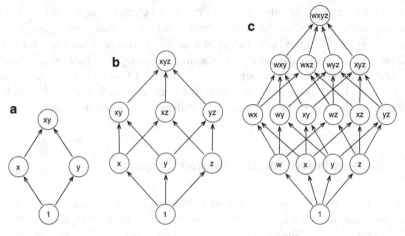

Fig. 1.5 Hasse diagrams of Boolean lattices. (a) $2^{[2]}$, (b) $2^{[3]}$, (c) $2^{[4]}$

Definition 1.25. Let P be a finite ranked poset with rank decomposition $P = \bigsqcup_k P_k$. We say that P has the **Sperner property** if

$$d(P) = \max_k |P_k|.$$

Definition 1.26. For a poset P ordered by $\leq$, we define the **Hasse diagram** $B(P)$ of P by $B(P) = \{ (x, y) \in P \times P \mid x \lessdot y \}$.

Example 1.27. The Boolean lattice $2^{[n]}$ is a ranked poset. Its rank decomposition is $2^{[n]} = \bigsqcup_{k=0}^{n} P_k$, where $P_k = \{ X \subset [n] \mid |X| = k \}$. The rank function is given by $\rho(X) = |X|$ for $X \in 2^{[n]}$. Figure 1.5 shows the Hasse diagrams of the Boolean lattice $2^{[n]}$ for $n = 2, 3, 4$. In the figure, a small circle denotes an element in the lattice. An arrow from x to y denotes the covering relation $x \lessdot y$.

Definition 1.28. For a ranked poset $P = \bigsqcup_i P_i$, we define $B_i(P)$ to be the set

$$\{ (x, y) \in P_{i-1} \times P_i \mid x \lessdot y \} = (P_{i-1} \times P_i) \cap B(P) = B(P_{i-1} \cup P_i).$$

Example 1.29. The posets whose Hasse diagrams are Fig. 1.6a, b have the Sperner property. Figure 1.6c shows a poset that does not have the Sperner property.

Definition 1.30. A sequence of positive integers $(h_0, h_1, \ldots, h_n)$ is said to be **unimodal** if there exists j such that

$$\begin{cases} h_i \leq h_{i+1} & (i < j) \\ h_i \geq h_{i+1} & (j \leq i). \end{cases}$$

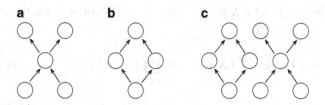

Fig. 1.6 Examples of ranked posets

A sequence of positive integers $(h_0, h_1, \ldots, h_n)$ is said to be **symmetric** if $h_i = h_{n-i}$ for $i = 0, 1, 2, \ldots, n$.

Theorem 1.31. *Let P be a ranked poset with the rank decomposition $P = \bigsqcup_{i=0}^{s} P_i$. Assume that $(|P_0|, |P_1|, \ldots, |P_s|)$ is unimodal. If $B_i(P)$ contains a full matching for each i, then P has the Sperner property.*

Proof. Let j be an integer such that $h_0 \leq h_1 \leq \cdots \leq h_j \geq h_{j+1} \geq \cdots \geq h_s$. Choose a full matching of $B_i(P)$ for every $i = 1, 2, \ldots, j$. It gives us a decomposition of $\bigsqcup_{i=0}^{j} P_i$ into k chains, where $k = |P_j|$. Similarly choosing a full matching for $B_i(P)$ for every $i = j + 1, j + 2, \ldots, s$, we have a chain decomposition for $\bigsqcup_{i=j}^{s} P_i$. Glue these chains at P_j giving a decomposition of P into k disjoint chains. Thus we obtain $k \geq d(P)$. Since $d(P) \geq |P_j|$, we obtain $d(P) = k$. □

By now many methods are known to prove that the Boolean lattice has the Sperner property. Here we give a proof which uses Theorem 1.31 above along the lines of Sperner's original idea. It seems not to be a coincidence that Hall's marriage theorem [46] appeared soon after Sperner's paper.

Theorem 1.32. *The Boolean lattice $2^{[n]}$ has the Sperner property.*

Proof. Let $P = \bigsqcup_{k=0}^{n} P_k$ be the Boolean lattice $2^{[n]}$. Since $|P_k| = \binom{n}{k}$, $|P_i| \leq |P_{i+1}|$ for $i < n/2$ and $|P_{i-1}| \geq |P_i|$ for $i > n/2$. Hence the sequence $(|P_0|, |P_1|, \ldots, |P_n|)$ is unimodal. First consider the case when $2i < n$. Let S be a subset of P_i, and T the set $\{Y \in P_{i+1} \mid X \subset Y \text{ for some } X \in S\}$. Then it follows that

$$\{(X, Y) \in S \times T \mid X \subset Y\} = \bigsqcup_{X \in S} \{(X, Y) \mid X \subset Y \in T\}$$

$$= \bigsqcup_{X \in S} \left\{(X, X \cup \{x'\}) \,\middle|\, x' \in [n] \setminus X\right\},$$

which implies $\left|\{ (X,Y) \in S \times T \mid X \subset Y \}\right| = |S| \cdot (n - i)$. On the other hand, we have

$$\{ (X,Y) \in P_i \times T \mid X \subset Y \} = \bigsqcup_{Y \in T} \{ (X,Y) \mid X \subset Y, X \in P_i \}$$

$$= \bigsqcup_{Y \in T} \{ (Y \setminus \{y\}, Y) \mid y \in Y \},$$

which implies $\left|\{ (X,Y) \in P_i \times T \mid X \subset Y \}\right| = |T| \cdot (i + 1)$. Since

$$\{ (X,Y) \in S \times T \mid X \subset Y \} \subset \{ (X,Y) \in P_i \times T \mid X \subset Y \},$$

$|S| \cdot (n - i) \leq |T| \cdot (i + 1)$. Since $2i < n$, $i + 1 \leq n - i$, which implies $|S| \leq |T|$. By the marriage theorem (Theorem 1.20), $B_{i+1}(2^{[n]})$ contains a full matching for $2i < n$. Similar proof works for the case when $2i > n$. Hence, by Theorem 1.31, the Boolean lattice $2^{[n]}$ has the Sperner property. $\square$

1.3   The Dilworth Lattice

In Definition 1.10 the ordering was given on the set $\mathscr{A}(P)$ of antichains in a poset P, with which $\mathscr{A}(P)$ is a poset. It is in fact a lattice as we will see below.

Recall that a **lattice** is a poset in which any two elements have a least upper bound and a greatest lower bound. The least upper bound is sometimes called **join** and is denoted by $x \vee y$ and the greatest lower bound is called **meet** and is denoted by $x \wedge y$. Let P be a lattice and P' a subposet of P. Then P' is a sublattice of P if $x \wedge y \in P'$ and $x \vee y \in P'$. It is an immediate consequence of Dilworth's theorem that the set of antichains of maximum size in a poset form a lattice. We call this lattice the *Dilworth lattice* of the poset (see Definition 1.44). We are interested in it, because it has an immediate generalization to an Artinian local ring, which we discuss in Sect. 2.3.5.

It is stated in Stanley [138] that the most important class of lattices from the combinatorial point of view are the **distributive lattice**. These are defined by the distributive laws:

$$x \wedge (y \vee z) = (x \wedge y) \vee (x \wedge z)$$
$$x \vee (y \wedge z) = (x \vee y) \wedge (x \vee z)$$

(One can prove that either of these laws implies the other.)

Example 1.33. The Boolean lattice $P = 2^{[n]}$ is a distributive lattice. The meet and join are given by $X \vee Y = X \cup Y$ and $X \wedge Y = X \cap Y$ for each $X, Y \in P$.

Example 1.34. The poset consisting of five elements $\hat{0}, x_1, x_2, x_3, \hat{1}$ with ordering defined by $\hat{0} \leq x_i \leq \hat{1}$ for all i is a lattice. Since $x_1 \wedge (x_2 \vee x_3) = x_1$ does not equal $(x_1 \wedge x_2) \vee (x_1 \wedge x_3) = \hat{0}$, the lattice is not distributive.

Example 1.35. Let P be a Boolean lattice $2^{[3]}$. Consider the subposet

$$P' = \{ \emptyset, \{1\}, \{2\}, \{3\}, \{1,2,3\} \}.$$

Then P' is a lattice, but P' is not a sublattice of P. For example, the join $\{1\} \vee_P \{2\} = \{1,2\}$ in P is not equal to the join $\{1\} \vee_{P'} \{2\} = \{1,2,3\}$ in P'. Moreover P is distributive, but P' is not distributive.

Before we state the fundamental theorem of distributive lattices (Theorem 1.42), we give a definition for **order ideal** and **filter** of a poset for the reader's convenience:

Definition 1.36. Let P be a poset.

1. A subset $I \subset P$ is an **ideal** or **order ideal** if

$$y \in P, x \in I \text{ and } y \leq x \implies y \in I.$$

2. A subset $I \subset P$ is a **dual order ideal** or **filter** if

$$y \in P, x \in I \text{ and } x \leq y \implies y \in I.$$

Remark 1.37. For a poset P, P itself is an ideal and a filter of P. The empty set is also an ideal and a filter of P.

Example 1.38. Let $P = 2^{[2]}$. All order ideals of $2^{[2]}$ are:

$$\emptyset = \{ \},$$

$$\{\emptyset\},$$

$$\{\{1\}, \emptyset\},$$

$$\{\{2\}, \emptyset\},$$

$$\{\{1\}, \{2\}, \emptyset\},$$

$$P = \{\{1,2\}, \{1\}, \{2\}, \emptyset\}.$$

Example 1.39. For a ranked poset $P = \bigsqcup_{i=0}^c P_i$, the set $\bigcup_{i=0}^k P_i$ (resp. $\bigcup_{i=k}^c P_i$) is an order ideal (resp. filter) for each $k = 0, 1, \ldots, c$.

Notation 1.40. *For a poset P, we denote by $J(P)$ the set of finite order ideals in P.*

Remark 1.41. For a poset P, the set $J(P)$ ordered by the inclusion $\subset$ is a distributive lattice. The meet and join are the intersection and union, respectively. The FTFDL, which we state below, says that the converse is also true.

Theorem 1.42 (Fundamental theorem of finite distributive lattice). *A finite distributive lattice L is isomorphic to J(P) for a poset P that is unique up to isomorphism.*

We refer the reader to [138, §3.4] for details.

Proposition 1.43. *For a finite poset P, J(P) and $\mathscr{A}(P)$ are isomorphic as posets.*

Proof. For $A \in \mathscr{A}(P)$, define $\varphi(A)$ to be the order ideal generated by A, i.e., $\{x \in P \mid x \le a$ for some $a \in A\}$. Since P is finite, φ is a map from $\mathscr{A}(P)$ to $J(P)$. Let $A, A' \in \mathscr{A}(P)$ satisfy $A \le A'$. By definition, for each $a \in A$, there exists $a' \in A'$ such that $a \le a'$. Since $A \subset \varphi(A')$, $\varphi(A) \subset \varphi(A')$.

For $I \in J(P)$, define $\psi(I)$ to be the set of maximal elements in I. Since a set of maximal elements of a subset is an antichain, ψ is a map from $J(P)$ to $\mathscr{A}(P)$. Let $I, I' \in J(P)$ satisfy $I \subset I'$. Then $\psi(I) \subset I'$. Hence, for each $a \in \psi(I)$, there exists $a' \in \psi(I')$ such that $a \le a'$. Thus $\psi(I) \le \psi(I')$.

Finally we prove that ψ is the inverse map of φ. First we show that $\varphi \circ \psi(I) = I$ for each $I \in J(P)$. For each $x \in I \in J(P)$, there exists $a \in \psi(I)$ such that $x \le a$. Hence $I \subset \varphi \circ \psi(I)$. On the other hand, if x satisfies $x \le a$ for some $a \in \psi(I)$, then x must be an element in I. Hence $\varphi \circ \psi(I) \subset I$. Thus $\varphi \circ \psi(I) = I$ for $I \in J(P)$. Next we show that $\psi \circ \varphi(A) = A$ for $A \in \mathscr{A}(P)$. Let $a \in A \in \mathscr{A}(P)$. If there does not exist $x \in \varphi(A)$ such that $a < x$, then a is a maximal element in the order ideal $\varphi(A)$, so that $a \in \psi \circ \varphi(A)$. Let $x \in \varphi(A)$ satisfy $a \le x$. For each $x \in \varphi(A)$, there exists $a' \in A$ such that $x \le a'$. Hence we have $a, a' \in A$ and $a \le x \le a'$. Since A is an antichain, we obtain $a = a'$, which implies $a \in \psi \circ \varphi(A)$. Hence $A \subset \psi \circ \varphi(A)$. On the other hand, let $a \in \psi \circ \varphi(A)$. Then, since a is an element of the order ideal $\varphi(A)$ generated by A, there exists $a' \in A$ such that $a \le a'$. Since a is a maximal element in $\varphi(A)$, we obtain $a' = a$, which implies $a \in A$. Hence $\psi \circ \varphi(A) \subset A$. Thus $\psi \circ \varphi(A) = A$ for $A \in \mathscr{A}(P)$. □

As we have seen, the set $\mathscr{A}(P)$ of antichains of a poset is a distributive lattice. Dilworth showed that the set of all antichains with maximum size is also a distributive lattice [30]. In [36], Freese showed that the lattice is a sublattice of the lattice $\mathscr{A}(P)$.

Definition 1.44. For a finite poset P, we define the **Dilworth lattice** $\mathscr{F}(P)$ of P to be the family of antichains of P of maximum size, i.e., $\mathscr{F}(P) = \{A \in \mathscr{A}(P) \mid |A| = d(P)\}$.

Theorem 1.45. *For a finite poset P, $\mathscr{F}(P)$ is a sublattice of $\mathscr{A}(P)$. Hence $\mathscr{F}(P)$ is a distributive lattice.*

Proof. Let A and A' be antichains of P. It follows from Proposition 1.43 that the join $A \vee_{\mathscr{A}(P)} A'$ (resp. the meet $A \wedge_{\mathscr{A}(P)} A'$) in $\mathscr{A}(P)$ equals the set of maximal (resp. minimal) elements in $A \cup A'$. If $a \in A$ is not maximal nor minimal in $A \cup A'$, then there exist $a', a'' \in A'$ such that $a' < a < a''$, which contradicts the assumption that A' is an antichain. Hence each $a \in A$ is a maximal and/or minimal element. Similarly it follows that each $a \in A'$ is a maximal and/or minimal element. The set

$(A \vee_{\mathscr{A}(P)} A') \cap (A \wedge_{\mathscr{A}(P)} A')$ is the set of elements $a \in A \cup A'$ such that a is a maximal and minimal element. Assume that $|A| = |A'| = d(P)$. It is easy to show that a is an element of $(A \vee_{\mathscr{A}(P)} A') \cap (A \wedge_{\mathscr{A}(P)} A')$ for $a \in A \cap A'$. For $a \in (A \vee_{\mathscr{A}(P)} A') \cap (A \wedge_{\mathscr{A}(P)} A')$, there does not exist $a' \in A \cap A'$ such that $a' < a$ or $a < a'$, which implies $A \cup \{a\}$ and $A' \cup \{a\}$ are also antichains. Hence we have $d(P) = |A| \leq |A \cup \{a\}| \leq d(P)$ and $d(P) = |A'| \leq |A' \cup \{a\}| \leq d(P)$. This implies that $a \in A \cap A'$. Therefore we have $(A \vee_{\mathscr{A}(P)} A') \cap (A \wedge_{\mathscr{A}(P)} A') = A \cap A'$. Hence it follows that

$$\left| (A \vee_{\mathscr{A}(P)} A') \right| + \left| (A \wedge_{\mathscr{A}(P)} A') \right| = |A \cup A'| + \left| (A \vee_{\mathscr{A}(P)} A') \cap (A \wedge_{\mathscr{A}(P)} A') \right|$$

$$= |A \cup A'| + |A \cap A'|$$

$$= |A| + |A'| = 2d(P).$$

If $\left| A \wedge_{\mathscr{A}(P)} A' \right| < d(P)$, then $\left| A \vee_{\mathscr{A}(P)} A' \right| > d(P)$, which is a contradiction. Hence we have $\left| A \wedge_{\mathscr{A}(P)} A' \right| = d(P)$ and $A \wedge_{\mathscr{A}(P)} A' \in \mathscr{F}(P)$. Thus $A \wedge_{\mathscr{F}(P)} A' = A \wedge_{\mathscr{A}(P)} A'$. Similarly we obtain $A \vee_{\mathscr{F}(P)} A' = A \vee_{\mathscr{A}(P)} A'$.   □

Proposition 1.46 (Sperner [132]). *For $P = 2^{[n]}$, we have*

$$\mathscr{F}(P) = \begin{cases} \{ P_l \}, & (n = 2l), \\ \{ P_l, P_{l+1} \}, & (n = 2l + 1), \end{cases}$$

where P_k is the set of elements of P of rank k, so here subsets of $[n]$ of size k.

Proof. In the proof of Theorem 1.32, we proved that

$$|S| \leq |T| \tag{1.1}$$

for any subset $S \subset P_i$, where T is the set $\{ Y \in P_{i+1} \mid X \subset Y \text{ for some } X \in S \}$, for all $0 \leq i < [n/2]$. Sperner [132] in fact proved that, in addition to this, the equality occurs only if n is odd and $S = P_{(n-1)/2}$, in which case $T = P_{(n+1)/2}$. This means that if $S \subsetneqq P_j$ for any $j < n/2$, then $|S| < |T|$. For a short stretch we take this for granted, and proceed to the proof of proposition.

   We assume that n is even. (The same argument applies to the case n is odd.) Suppose that $S \subset P$ is an antichain of maximum size. Put $S_i = S \cap P_i$ and let j be the least integer i such that $S_i \neq \emptyset$ and j' be the maximum integer i such that $S_i \neq \emptyset$. We want to show that $j = j'$. By way of contradiction assume that $j < j'$. Then we have either $j < n/2$ or $n/2 < j'$. Assume $j < n/2$. Obviously $S_j \subsetneqq P_j$. Put $S' = (S \setminus S_j) \cup T$, where

$$T = \{ Y \in P_{j+1} \mid Y \supset X \text{ for some } X \in S \}.$$

Then, by the assumption $|S_j| < |T|$, we have $|S| < |S'|$, a contradiction. If $n/2 < j'$, instead of $j < n/2$, we can use duality to get a contradiction. So we have proved that S is a rank set.

It remains to prove that the equality

$$|S| \leq |T|$$

occurs for $S \subset P_i$ with $i < n/2$ only if n is odd and $S = P_{(n-1)/2}$. By way of contradiction assume that

$$S = \{ A_1, \ldots, A_m \}, \qquad T = \{ B_1, \ldots, B_m \}$$

for a subset $S \subsetneq P_i$ with $i < n/2$. By definition of T, any B_i is obtained by adding an element to some A_j. We claim that there exist $B \in P_{i+1} \setminus T$ and some $B_k \in T$ such that B and B_k differ only by one element. Indeed, $B \in P_{i+1} \setminus T$ exists anyway. Starting with B, one may reach one of $B_1, \ldots, B_m$ by repeating the process of replacing an element in the set with another element. One sees easily that in this way one may find such subsets B and B_k in P_{i+1} as claimed. Put $A = B \cap B_k$. We have $A \in P_i$. If $A \in S$, then B must be in T, since $B = A \cup \{z\}$ for some $z \in [n]$, a contradiction. If $A \notin S$, then B_k can not be in T, since $B_j \in T$ with an element deleted must be in S. This contradiction proves that

$$|S| < |T|$$

as long as $S \subsetneq P_i$ with $i < n/2$. Furthermore, as long as $i < n/2$, we have $|S| < |T|$ even for $S = P_i$ except $i = (n-1)/2$. Thus the assertion follows. □

Example 1.47. Consider the poset

$$P = C(r) \times C(s) = \{ (x^i, y^j) \mid 0 \leq i \leq r, 0 \leq j \leq s \}$$

for $r \leq s$. In this case, $\mathscr{F}(P)$ is the set

$$\left\{ \{ (x^0, y^{j_0}), (x^1, y^{j_1}), \ldots, (x^r, y^{j_r}) \} \mid s \geq j_0 > j_1 > \cdots > j_r \geq 0 \right\}.$$

1.4 Examples of Posets with the Sperner Property

In this section we give some examples of ranked posets with the Sperner property. We introduce some methods to verify the Sperner property, and apply them to the examples.

1.4.1 Boolean Lattice

Let P be the Boolean lattice $2^{[n]}$. Then P ordered by the inclusion $\subset$ is a ranked poset. For each $X \in P$, the rank $\rho(X)$ equals the number $|X|$ of elements in X. Hence the rank decomposition $P = \bigsqcup_{k=0}^{n} P_k$ is given by $P_k = \{ X \subset [n] \mid |X| = k \}$, which implies $|P_k| = \binom{n}{k}$. Let $h_k = |P_k|$. Then $(h_0, h_1, \ldots, h_n)$ is unimodal. For $v \le v'$ in P, the interval $[v, v']$ is isomorphic to the Boolean lattice $2^{[\rho(v') - \rho(v)]}$. We have already shown that the Boolean lattice P has the Sperner property (Theorem 1.32). Here we give another proof, because it has wider applications.

To give this argument, we will first prove some lemmas not only for Boolean lattices but also any finite ranked poset.

Definition 1.48. Let P be a finite ranked poset with rank decomposition $P = \bigsqcup_{k=0}^{s} P_k$, and $P_k = \left\{ v_1^{(k)}, v_2^{(k)}, \ldots, v_{h_k}^{(k)} \right\}$. For $k \ge 1$, we define the $(h_k \times h_{k-1})$ matrix $M^{(k)} = (M_{ij}^{(k)})$ by

$$M_{ij}^{(k)} = \begin{cases} 1 & (v_j^{(k-1)} \le v_i^{(k)}), \\ 0 & (\text{otherwise}). \end{cases}$$

We call $M^{(k)}$ the **incidence matrices** of $B_k(P)$. (For the meaning of $B_k(P)$ see Definition 1.28.) We do not fix any linear ordering on P. The incidence matrices are defined only upto permutation of the elements of P.

Lemma 1.49. *Let P be a finite ranked poset with the rank decomposition $P = \bigsqcup_{k=0}^{s} P_k$ such that $P_k = \left\{ v_1^{(k)}, v_2^{(k)}, \ldots, v_{h_k}^{(k)} \right\}$ and the sequence $(h_0, h_1, \ldots, h_s)$ is unimodal. Let $L^{(k)} = (L_{ij}^{(k)})$ be an $(h_k \times h_{k-1})$ matrix satisfying*

$$L_{ij}^{(k)} \ne 0 \implies v_j^{(k-1)} \le v_i^{(k)}.$$

If $L^{(k)}$ has full rank for each k, then P has the Sperner property.

Proof. First we consider the case when k satisfies $h_{k-1} = |P_{k-1}| \le h_k = |P_k|$. Let $\{l_1, l_2, \ldots, l_n\} \subset \{1, 2, \ldots, h_{k-1}\}$. Since $L^{(k)}$ has full rank and $h_k \ge h_{k-1} \ge n$, the $(h_k \times n)$-matrix $L' = (L_{i,l_j}^{(k)})_{i,j}$ also has full rank. Hence the number of nonzero row vectors in L' is at least n. This means the number of elements v in P_k satisfying $v' < v$ for some $v \in P'_{k-1} = \{ v_{l_1}^{(k-1)}, v_{l_2}^{(k-1)}, \ldots, v_{l_n}^{(k-1)} \} \subset P_{k-1}$ is greater than or equal to n. Hence, by Marriage Theorem (Theorem 1.20), $B_k(P) = \{ (x, y) \in P_{k-1} \times P_k \mid x \le y \}$ contains a full matching. Similarly, we can also show that $B_k(P)$ contains a full matching for k such that $h_{k-1} \ge h_k$. Therefore we have proved the lemma by Theorem 1.31. $\qquad\square$

Lemma 1.50. *Let P be a finite ranked poset with the rank decomposition $P = \bigsqcup_{k=0}^{s} P_k$ such that $P_k = \left\{ v_1^{(k)}, v_2^{(k)}, \ldots, v_{h_k}^{(k)} \right\}$ and the sequence $(h_0, h_1, \ldots, h_s)$ is unimodal and symmetric. Suppose that matrices $L^{(k)} = (L_{ij}^{(k)})$ of size $(h_k \times h_{k-1})$ satisfy*

$$L_{ij}^{(k)} \neq 0 \implies v_j^{(k-1)} < v_i^{(k)}.$$

Let $\tilde{L}^{(k)} = L^{(s-k)} L^{(s-k-1)} \cdots L^{(k)}$ for $k < s/2$. If $\tilde{L}^{(k)}$ has an inverse for each k, then P has the Sperner property.

Proof. Since $\tilde{L}^{(k)}$ and $\tilde{L}^{(k+1)}$ have inverses, both $L^{(k)}$ and $L^{(s-k)}$ have full rank. Hence, by Lemma 1.49, we have the lemma. □

The incidence matrices $M^{(k)}$ satisfy the condition

$$M_{ij}^{(k)} \neq 0 \implies v_j^{(k-1)} < v_i^{(k)}.$$

Hence we have Lemmas 1.51 and 1.52.

Lemma 1.51. *Let P be a finite ranked poset with the rank decomposition $P = \bigsqcup_{k=0}^{s} P_k$ such that the sequence $(|P_0|, |P_1|, \ldots, |P_s|)$ is unimodal. If the incidence matrix $M^{(k)}$ of $B_k(P)$ has full rank for each k, then P has the Sperner property.*

Lemma 1.52. *Let P be a finite ranked poset with the rank decomposition $P = \bigsqcup_{k=0}^{s} P_k$ such that the sequence $(|P_0|, |P_1|, \ldots, |P_s|)$ is unimodal and symmetric. Let $M^{(k)}$ be the incidence matrix of $B_k(P)$, and $\tilde{M}^{(k)}$ the product $M^{(s-k)} M^{(s-k-1)} \cdots M^{(k+1)}$ with $k < s/2$. If $\tilde{M}^{(k)}$ has the inverse for each k, then P has the Sperner property.*

Remark 1.53. Even if we assume that $(|P_0|, |P_1|, \ldots, |P_s|)$ is unimodal and symmetric, the following condition is not sufficient to prove that a ranked poset P with the rank decomposition $P = \bigsqcup_{i=0}^{s} P_i$ has the Sperner property: $\{ (x, y) \in P_k \times P_{s-k} \mid x < y \}$ contains a full matching for each $k < s/2$. For example, the ranked poset (c) in Fig. 1.6 does not have the Sperner property. However, $\{ (x, y) \in P_0 \times P_2 \mid x < y \}$ contains a full matching.

For the Boolean lattice $P = 2^{[n]}$, Hara and Watanabe proved the following Lemma by an elementary method.

Lemma 1.54 (Hara–Watanabe [48]). *Define $d(n, k)$ by $d(n, k) = |\det \tilde{M}^{(n,k)}|$, where $\tilde{M}^{(n,k)}$ is the product $M^{(n,n-k)} M^{(n,n-k-1)} \cdots M^{(n,k)}$ of incidence matrices $M^{(n,j)}$ of $B_j(2^{[n]})$. Then*

$$d(n, k) = \begin{cases} 1 & (k = 0 \text{ or } k = \frac{n}{2}), \\ \left(1 + \frac{1}{n-2k}\right)^{\binom{n-1}{k-1}} d(n-1, k) d(n-1, k-1) & (\text{otherwise}). \end{cases}$$

By Lemma 1.54, for each n, the product $\tilde{M}^{(n,k)}$ of the incidence matrices is invertible and this shows that M_i not only has a full matching but actually it has full rank. Hence we have the Sperner property of the Boolean lattice as a corollary to Lemma 1.54. Thus Lemma 1.54 gives another proof of Theorem 1.32.

For the rest of this section, we consider the matching number of the incidence matrix for posets. Recall that in Definition 1.16 we defined a matching for the Cartesian product $X \times Y$ for any sets X, Y Here we define the matching number of a 01-matrix.

Definition 1.55. Let $M = (m_{ij})$ be a 01-matrix of size $n \times l$. The **matching number** $\beta(M)$ of the matrix M is defined by

$$\beta(M) = \max \left\{ \sum_{i,j} m'_{ij} \,\middle|\, M' \in \mathscr{M}(M) \right\},$$

where $\mathscr{M}(M)$ denotes the set of 01-matrices $M' = (m'_{ij})$ of size $n \times l$ satisfying

1. $m_{ij} = 0 \implies m'_{ij} = 0$,
2. $\sum_{i=1}^{n} m'_{ij} \leq 1$,
3. $\sum_{j=1}^{l} m'_{ij} \leq 1$.

Note that $\sum_{i,j} m'_{ij}$ is the number of non-zero entries of M'. A matrix $M' \in \mathscr{M}(M)$ may be interpreted as a selection of non-zero entries of M subject to the condition no two entries should lie in a row or column. (Cf. Remark 1.19.)

Example 1.56. Consider the matrices:

$$M = \begin{pmatrix} 0\,0\,0\,0\,0\,0 \\ 0\,1\,1\,1\,1\,1 \\ 1\,1\,1\,1\,1\,1 \\ 0\,1\,1\,1\,1\,1 \\ 0\,0\,1\,1\,1\,0 \\ 0\,0\,0\,0\,0\,0 \end{pmatrix}, \quad \text{and} \quad M' = \begin{pmatrix} 0\,0\,0\,0\,0\,0 \\ 0\,0\,1\,0\,0\,0 \\ 0\,0\,0\,1\,0\,0 \\ 0\,0\,0\,0\,1\,0 \\ 0\,0\,0\,0\,0\,0 \\ 0\,0\,0\,0\,0\,0 \end{pmatrix}.$$

Let $X = \{x_1, x_2, x_3, x_4, x_5, x_6\}$, $Y = \{y_1, y_2, y_3, y_4, y_5, y_6\}$. Since M' contains at most a 1 in each row and column, M' gives a matching in $X \times Y$ in the sense of Definition 1.16. In fact the matching M' gives is the subset of $X \times Y$:

$$\{(x_2, y_3), (x_3, y_4), (x_4, y_5)\}.$$

Since 1's in M' are chosen from among those of M, we have $M' \in \mathscr{M}(M)$. Since M' contains three 1's, we have

$$\sum_{i,j} m'_{ij} = 3$$

in the notation of Definition 1.55. One sees easily that the 01-matrix M' cannot be expanded to a bigger matching by selecting another 1 from M. However, we have $\beta(M) = 4$, since, for example,

$$M'' = \begin{pmatrix} 0\,0\,0\,0\,0\,0 \\ 0\,0\,0\,0\,1\,0 \\ 0\,0\,0\,0\,0\,1 \\ 0\,0\,1\,0\,0\,0 \\ 0\,0\,0\,1\,0\,0 \\ 0\,0\,0\,0\,0\,0 \end{pmatrix}$$

gives a matching

$$\{\,(x_2, y_5), (x_3, y_6), (x_4, y_3), (x_5, y_4)\,\}$$

of $X \times Y$, which is obviously a maximal matching of $X \times Y$, or by abuse of language, of M. If we assume $X \cap Y = \emptyset$, then M'' gives four "matchings" of elements of X and Y. On the other hand, if we consider $X = Y$, $x_i = y_i$, $i = 1, 2, \ldots, 6$, then M may be considered as defining a digraph. In this case M'' gives a chain of length three:

$$x_2 \to x_5 \to x_4 \to x_3 \to x_6.$$

Similarly M' gives a chain of length two:

$$x_2 \to x_3 \to x_4.$$

Consider the matrix:

$$M''' = \begin{pmatrix} 0\,0\,0\,0\,0\,0 \\ 0\,0\,0\,0\,0\,1 \\ 1\,0\,0\,0\,0\,0 \\ 0\,0\,1\,0\,0\,0 \\ 0\,0\,0\,1\,0\,0 \\ 0\,0\,0\,0\,0\,0 \end{pmatrix}$$

Then, again $M''' \in \mathcal{M}(M)$, since 1's of M''' are chosen from among those of M. If we regard M as a digraph on $X = Y$, then M''' gives a decomposition of X into two chains

$$x_2 \to x_6, \text{ and } x_5 \to x_4 \to x_3 \to x_1.$$

(See Remark 1.19.)

The next theorem is another interpretation of Dilworth's theorem.

Theorem 1.57. *Let* $P = \{ v_1, v_2, \ldots, v_n \}$ *be a finite poset, and* $M = (m_{ij})$ *the* $n \times n$ *matrix defined by*

$$m_{ij} = \begin{cases} 1 & (v_i < v_j), \\ 0 & (otherwise). \end{cases}$$

Then $d(P) = |P| - \beta(M)$.

Proof. Generally speaking, let $I \subset P \times P$ be a matching as defined in Definition 1.16, and let M' be the 01-matrix representing the matching I. (Cf. Remark 1.19.) Then with a renumbering of elements of P, the matrix M' decomposes into blocks of circuits and chains. Here a circuit is a block

$$\begin{pmatrix} 0 & 1 & 0 & & 0 \\ 0 & 0 & 1 & & 0 \\ 0 & 0 & & \ddots & \vdots \\ \vdots & \vdots & & & 1 \\ 1 & 0 & & & 0 \end{pmatrix}, \quad \text{and a chain} \quad \begin{pmatrix} 0 & 1 & 0 & & 0 \\ 0 & 0 & 1 & & 0 \\ 0 & 0 & & \ddots & \vdots \\ \vdots & \vdots & & & 1 \\ 0 & 0 & & & 0 \end{pmatrix}.$$

Now let M be the matrix in consideration of this theorem, let $\mathcal{M}(M)$ be the set as defined in Definition 1.55 and let $M' \in \mathcal{M}(M)$. Since P is a poset, P does not contain circuits. Thus M' decomposes only into chains. In other words the set $\mathcal{M}(M)$ and the set of disjoint chain decompositions of P are in one-to-one correspondence. Notice that

$$|P| - \text{number of non-zero entries of } M' = \text{number of blocks of } M'.$$

Hence $|P| - \beta(M)$ is equal to the minimum number of chains into which P is decomposed. $\qquad\square$

1.4.2 The Divisor Lattice and Finite Chain Product

Here we consider the poset $\mathscr{L}(n)$ of divisors of a given positive integer n and show it has the Sperner property. We give a ring theoretical proof based on facts which will be proved in subsequent chapters. The idea of the proof is the following: First we define a graded algebra $A = \bigoplus_{i=0}^{n} A_i$ such that a linear basis for the homogeneous component of degree i is parameterized by the elements of the poset of rank i. We find an element $l \in A_1$ of degree one such that the representation matrix of the multiplication map $\times l \colon A_{i-1} \to A_i$ with respect to the basis is the incidence matrix

$M^{(i)}$ of $B_i(P)$ for each i. The Sperner property for the poset $\mathscr{L}(n)$ is obtained from the non-degeneracy of the multiplication map $\times l: A_{i-1} \to A_i$ for each i.

For a positive integer n, we define the **divisor lattice** $\mathscr{L}(n)$ to be the set of divisors of n. Then $P = \mathscr{L}(n)$ ordered by the divisibility is a ranked poset. For $m, m' \in \mathscr{L}(n)$, m' is covered by m if and only if m/m' is a prime number. Let n be decomposed into the product of distinct primes: $n = \prod_{p: \text{prime}} p^{e_p}$. Then it is not difficult to see that the integers $h_k = |P_k|$ are given as coefficients of t^k in the polynomial

$$\sum_k h_k t^k = \prod_{p: \text{prime}} (1 + t + \cdots + t^{e_p}).$$

For a positive integer n, $\mathscr{L}(n)$ is a lattice. The join (resp. meet) is given as the LCM (resp. GCD) for each pair of elements in $\mathscr{L}(n)$.

Remark 1.58. Note that, for $m, n'm \in \mathscr{L}(n)$, the interval $[m, n'm]$ is isomorphic to $\mathscr{L}(n')$.

Remark 1.59. The finite chain product

$$C(d_1) \times C(d_2) \times \cdots \times C(d_k)$$
$$= \left\{ (x_1^{i_1}, x_2^{i_2}, \ldots, x_k^{i_k}) \;\middle|\; 0 \le i_1 \le d_1, 0 \le i_2 \le d_2, \ldots, 0 \le i_k \le d_k \right\}$$

is a poset, whose order is defined by

$$(x_1^{i_1}, x_2^{i_2}, \ldots, x_k^{i_k}) \le (y_1^{j_1}, y_2^{j_2}, \ldots, y_k^{j_k}) \iff i_l \le j_l \text{ for all } l.$$

The rank function ρ is given by $\rho((x_1^{i_1}, x_2^{i_2}, \ldots, x_k^{i_k})) = i_1 + i_2 + \cdots + i_k$. Let n be a positive integer such that $n = p_1^{d_1} p_2^{d_2} \cdots p_k^{d_k}$ with distinct primes $p_1, p_2, \ldots, p_k$. Then the finite chain product $C(d_1) \times C(d_2) \times \cdots \times C(d_k)$ is isomorphic to the divisor lattice $\mathscr{L}(n)$.

Remark 1.60. Let $n_1, n_2, \ldots, n_r$ be positive integers and $P = \mathscr{L}(n_1) \cup \mathscr{L}(n_2) \cup \cdots \cup \mathscr{L}(n_r) \subset \mathbb{N}$. It is a ranked finite poset. The rank function ρ of P is defined as follows: Let $z \in \mathscr{L}(n_j)$ for some j and $z = p_1^{i_1} p_2^{i_2} \cdots p_k^{i_k}$ be the prime factorization of z. Then

$$\rho(z) = i_1 + i_2 + \cdots + i_k.$$

So $\rho(z)$ is the "number of prime factors" that occur with counting multiplicity in the prime factorization of an integer in P. The sequence of rank numbers

$$|P_0| = 1, |P_1|, |P_2|, \ldots, |P_s|$$

is known as an O-sequence in commutative algebra. Moreover any O-sequence appears as the rank function of some P. (See Sect. 6.2 for the definition of O-sequence.)

Theorem 1.61. *For a positive integer n, the divisor lattice $\mathscr{L}(n)$ has the Sperner property.*

Proof. We show that the finite chain product $P = C(d_1) \times C(d_2) \times \cdots \times C(d_k)$ has the Sperner property. Consider the polynomial ring $K[x_1, x_2, \ldots, x_k]$ over a field K of characteristic zero and identify the poset P with the set

$$\left\{ x_1^{i_1} x_2^{i_2} \cdots x_k^{i_k} \;\middle|\; 0 \le i_1 \le d_1, 0 \le i_2 \le d_2, \ldots, 0 \le i_k \le d_k \right\}$$

of monomials. The set is a K-linear basis for the graded ring

$$A = K[x_1, x_2, \ldots, x_k]/(x_1^{d_1+1}, x_2^{d_2+1}, \ldots, x_k^{d_k+1}).$$

Let $l = x_1 + x_2 + \cdots + x_k$ and $c = d_1 + d_2 + \cdots + d_k$. Consider the K-linear map

$$\times l: A_{i-1} \ni f \mapsto fl \in A_i.$$

The representation matrix of $\times l: A_{i-1} \to A_i$ is the incidence matrix $M^{(i)}$. On the other hand, it follows from Corollary 3.35, which appears later, that

$$\times l^{c-2i}: A_i \ni f \mapsto fl^{c-2i} \in A_{c-i}$$

is bijective. Hence, by Lemma 1.52, P has the Sperner property.   □

Remark 1.62. An element $l \in A_1$ satisfying the following condition plays an important role in the proof of Theorem 1.61:

$$\times l^{c-2i}: A_i \to A_{c-i}$$

is bijective for each i. We call an element with the above property a **strong Lefschetz element (in the narrow sense)**. In general, for a graded algebra $A = \bigoplus_i A_i$, we call $l \in A_1$ a strong Lefschetz element of A if $\times l^k: A_i \to A_{i+k}$ has full rank for each k and each i. The ring is said to have the strong Lefschetz property (SLP) if there exists a strong Lefschetz element. See Sect. 3.2 for the details. An element is called a **weak Lefschetz element** of A if $\times l: A_{i-1} \to A_i$ has full rank for each i. If A has a weak Lefschetz element, we say that A has the weak Lefschetz property (WLP). See Sect. 3.1 for details.

Remark 1.63. Let $p_1, p_2, \ldots, p_m$ be distinct primes, and $n = p_1 p_2 \cdots p_m$. Then the divisor lattice $\mathscr{L}(n)$ is isomorphic to the Boolean lattice $2^{[m]}$ as posets. Hence the proof of Proposition 1.61 is the third proof of the Sperner property for the Boolean lattice, which is a ring-theoretical proof. (See also Theorem 1.32 and Sect. 1.4.1 for the other proofs.)

1.4.3   Partitions of Integers

Here we consider partitions of integers or Young diagrams. The set of Young diagrams with the containment order is a poset. The order ideal $\mathscr{I}(s^r)$ of the poset generated by a rectangle has the Sperner property. A ring-theoretic argument to prove this works well also for $\mathscr{I}(s^r)$. The basic idea is the same as the case of divisor lattices. In this case, the ring of symmetric functions plays an important role. Hence we review some basic facts about symmetric functions here.

A non-increasing sequence $\lambda = (\lambda_1, \lambda_2, \ldots)$ of nonnegative integers such that $\sum_i \lambda_i = l$ is called a **partition** of l. We use the notation $\lambda \vdash l$ to mean that λ is a partition of l. We often omit some zeroes and regard $\lambda = (\lambda_1, \lambda_2, \ldots, \lambda_l, 0, \ldots)$ as $(\lambda_1, \lambda_2, \ldots, \lambda_l)$. Note that we do not exclude $\lambda_l = 0$ even if we write a partition as a finite sequence $(\lambda_1, \lambda_2, \ldots, \lambda_l)$. We identify a partition λ with a **Young diagram** $\{ (i, j) \mid 1 \le j \le \lambda_i \}$. We often write an array of boxes with left-justified rows such that the i-th row contains λ_i boxes, e.g.,

$$(4, 3, 3) = \quad \boxed{} \quad .$$

For partitions λ and μ, we define the **containment ordering** $\subset$ as follows:

$$(\lambda_1, \lambda_2, \ldots) \subset (\mu_1, \mu_2, \ldots) \iff \lambda_i \le \mu_i \text{ for all } i.$$

Then the set of partitions is a ranked poset with the rank function ρ given as

$$\rho((\lambda_1, \lambda_2, \ldots, \lambda_r)) = \sum_{i=1}^{r} \lambda_i,$$

the number of boxes in the Young diagram. The minimum element of the poset is the unique partition of 0, which is denoted by (0).

Let $\mathscr{I}(s^r)$ be the set of partitions defined by

$$\mathscr{I}(s^r) = \{ (\lambda_1, \lambda_2, \ldots, \lambda_r) \mid s \ge \lambda_1 \ge \lambda_2 \ge \cdots \ge \lambda_r \ge 0 \}.$$

In other words, $\mathscr{I}(s^r)$ is the order ideal generated by the rectangle $(s^r) := \underbrace{(s,\ldots,s)}_{r}$. Regarding partitions as Young diagrams, $\mathscr{I}(s^r)$ is isomorphic to the poset

$$J(C(r-1) \times C(s-1))$$

of order ideals as posets. In other words, the map

$$\mathscr{I}(s^r) \ni \lambda \mapsto \{(i,j) \mid 1 \le j \le \lambda_i\} \in J(C(r-1) \times C(s-1))$$

is an order-preserving bijection, where we realize the chain $C(l-1)$ as the set $\{1,2,\ldots,l\}$ ordered by $\le$.

To describe the number of elements in $\mathscr{I}(s^r)$ at each rank, we define the q-integers.

Definition 1.64. For $n > 0$ and $n \ge k \ge 0$, we define the q-**integer** $[n]_q$, the q-**factorial** $[n]_q!$ and the q-**binomial coefficient** $\begin{bmatrix} n \\ k \end{bmatrix}_q$ by

$$[n]_q = \frac{q^n - 1}{q - 1} = q^{n-1} + q^{n-2} + \cdots + q + 1,$$

$$[n]_q! = [n]_q [n-1]_q \cdots [2]_q [1]_q \text{ and } [0]_q! = 1,$$

$$\begin{bmatrix} n \\ k \end{bmatrix}_q = \frac{[n]_q!}{[k]_q! [n-k]_q!}.$$

If q is a power of a prime number, the q-binomial coefficient $\begin{bmatrix} n \\ k \end{bmatrix}_q$ equals the number of the linear subspaces of dimension k in the n-dimensional vector space $V = (\mathbb{F}_q)^n$ over the finite field with q elements. Generally q need not be an integer.

Remark 1.65. If we take the limit $q \to 1$, then

$$[n]_q \to n, \qquad [n]_q! \to n!, \qquad \begin{bmatrix} n \\ k \end{bmatrix}_q \to \binom{n}{k}.$$

Lemma 1.66. *For $n > k > 1$,*

$$\begin{bmatrix} n \\ k \end{bmatrix}_q = \begin{bmatrix} n-1 \\ k-1 \end{bmatrix}_q + q^k \begin{bmatrix} n-1 \\ k \end{bmatrix}_q.$$

Proof. Let $n > k > 1$. Then $[n]_q = [k]_q + q^k [n-k]_q$. Hence we prove the lemma by the following direct calculation

$$\begin{bmatrix} n-1 \\ k-1 \end{bmatrix}_q + q^k \begin{bmatrix} n-1 \\ k \end{bmatrix}_q = \frac{[n-1]_q!}{[k-1]_q! [n-k]_q!} + q^k \frac{[n-1]_q!}{[k]_q! [n-1-k]_q!}$$

$$= \frac{([k]_q + q^k [n-k]_q) [n-1]_q!}{[k]_q! [n-k]_q!}$$

$$= \frac{[n]_q [n-1]_q!}{[k]_q! [n-k]_q!} = \begin{bmatrix} n \\ k \end{bmatrix}_q. \qquad \square$$

Proposition 1.67. *Let* $\mathscr{I}(s^r) = \bigsqcup_{k=0}^{rs} \mathscr{I}(s^r)_k$ *be the rank decomposition. Then* $|\mathscr{I}(s^r)_k|$ *is equal to the coefficient of* q^k *in the q-binomial coefficient* $\begin{bmatrix} r+s \\ s \end{bmatrix}_q$, *or equivalently,*

$$\sum_{k=0}^{rs} |\mathscr{I}(s^r)_k| t^k = \begin{bmatrix} r+s \\ s \end{bmatrix}_t.$$

Proof. Let $h_{s,r}(t) = \sum_{k=0}^{rs} |\mathscr{I}(s^r)_k| t^k$. We verify the equation by induction on r and s. Consider the first case when $r = 1$. In this case, since $\mathscr{I}(s^1) = \{ (0), (1), \ldots, (s) \}$, we have $h_{s,1}(t) = \sum_{k=0}^{s} t^k$. On the other hand, since $\begin{bmatrix} s+1 \\ 1 \end{bmatrix}_t = [s+1]_t$, we have $h_{s,1}(t) = \begin{bmatrix} s+1 \\ 1 \end{bmatrix}_t$. Consider the next case where $s = 1$. In this case, since $\mathscr{I}(1^r) = \{ (0), (1), (1^2), \ldots, (1^r) \}$, we have $h_{1,r}(t) = \sum_{k=0}^{r} t^k = \begin{bmatrix} r+1 \\ 1 \end{bmatrix}_t$. Finally consider the case of $s > 1$ and $r > 1$. We divide the set $\mathscr{I}(s^r) = \mathscr{I}' \sqcup \mathscr{I}''$, where

$$\mathscr{I}' = \{ (\alpha_1, \alpha_2, \ldots, \alpha_r) \mid \alpha_1 < s \},$$
$$\mathscr{I}'' = \{ (\alpha_1, \alpha_2, \ldots, \alpha_r) \mid \alpha_1 = s \}.$$

Then $\mathscr{I}' = \mathscr{I}((s-1)^r)$. On the other hand, the correspondence $(\alpha_1, \alpha_2, \ldots, \alpha_r) \mapsto (\alpha_2, \ldots, \alpha_r)$ gives us a bijection $\mathscr{I}'' \to \mathscr{I}(s^{r-1})$. Thus $h_{s,r}(t)$ satisfies

$$h_{s,r}(t) = h_{s-1,r}(t) + t^s h_{s,r-1}(t).$$

Hence, by our induction hypothesis and Lemma 1.66, we have the proposition. $\square$

Example 1.68. Figure 1.7 is the Hasse diagram for $\mathscr{I}(4^2)$. For each k, $|\mathscr{I}(4^2)_k|$ is the coefficient of t^k in the polynomial

$$\begin{bmatrix} 6 \\ 2 \end{bmatrix}_t = 1 + t + 2t^2 + 2t^3 + 3t^4 + 2t^5 + 2t^6 + t^7 + t^8.$$

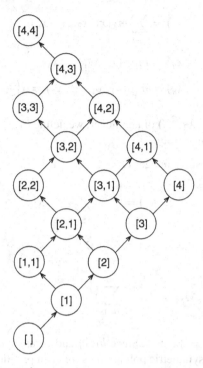

Fig. 1.7 Hasse diagram of $\mathscr{I}(4^2)$

The poset of Young diagrams is closely linked with the ring of symmetric polynomials. Here we summarize some facts concerning the ring of symmetric polynomials.

Fix a positive integer n and consider the ring of polynomials in n variables. The symmetric group on n objects S_n naturally acts on it by $\sigma f(x_1, x_2, \ldots, x_n) = f(x_{\sigma(1)}, x_{\sigma(2)}, \ldots, x_{\sigma(n)})$. We define the ring Λ_n of **symmetric polynomials** in n variables to be the subring $(\mathbb{Z}[x_1, x_2, \ldots, x_n])^{S_n}$ of polynomials fixed by this action. First we recall some important families of symmetric polynomials.

Definition 1.69. For a partition $\lambda = (\lambda_1, \lambda_2, \ldots, \lambda_n)$ of an integer, we define the **monomial symmetric polynomials** m_λ by

$$m_\lambda(x_1, x_2, \ldots, x_n) = \sum_{\alpha \in \left\{ (\lambda_{\sigma(1)}, \lambda_{\sigma(2)}, \ldots, \lambda_{\sigma(n)}) \mid \sigma \in S_n \right\}} x^\alpha,$$

where x^α denotes the monomial $x_1^{\alpha_1} x_2^{\alpha_2} \cdots x_n^{\alpha_n} = \prod_i x_i^{\alpha_i}$.

Definition 1.70. For a nonnegative integer l, we define the l-th **homogeneous complete symmetric polynomials** h_l, the l-th **elementary symmetric polynomials** e_l and the l-th Newton **power sums** p_l by

$$h_l(x_1, x_2, \ldots, x_n) = \sum_{\lambda \vdash l} m_\lambda(x_1, x_2, \ldots, x_n),$$

$$e_l(x_1, x_2, \ldots, x_n) = m_{(1^l)}(x_1, x_2, \ldots, x_n),$$

$$p_l(x_1, x_2, \ldots, x_n) = m_{(l)}(x_1, x_2, \ldots, x_n) = x_1^l + \cdots + x_n^l.$$

For a partition $\lambda = (\lambda_1, \lambda_2, \ldots)$ of an integer, we define h_λ, e_λ and p_λ to be $\prod_i h_{\lambda_i}$, $\prod_i e_{\lambda_i}$ and $\prod_i p_{\lambda_i}$, respectively.

Remark 1.71. For a positive integer l,

$$h_l(x_1, x_2, \ldots, x_n) = \sum_{1 \leq i_1 \leq i_2 \leq \cdots \leq i_l \leq n} x_{i_1} x_{i_2} \cdots x_{i_l},$$

$$e_l(x_1, x_2, \ldots, x_n) = \sum_{1 \leq i_1 < i_2 < \cdots < i_l \leq n} x_{i_1} x_{i_2} \cdots x_{i_l},$$

$$p_l(x_1, x_2, \ldots, x_n) = \sum_{i=1}^{n} x_i^l.$$

For $l = 0$, $h_l = e_l = p_l = 1$. Since h_l, e_l and p_l are symmetric polynomials, h_λ, e_λ and p_λ are also symmetric polynomials for each partition λ.

Remark 1.72. The elementary symmetric polynomials e_l and power sums p_l satisfy the following equation known as Newton formula:

$$p_m = -\sum_{j=1}^{n} (-1)^j e_j p_{m-j}$$

for $n < m$. The complete symmetric polynomials h_l and power sums p_l satisfy the equations

$$nh_n = \sum_{l=1}^{n} p_l h_{n-l},$$

$$ne_n = \sum_{l=1}^{n} (-1)^{l-1} p_l e_{n-l}.$$

See [85] for the details.

We also introduce another family of symmetric polynomials parameterized by partitions. Let $\delta = (n-1, n-2, \ldots, 1, 0)$. For a partition λ of an integer, the determinant $a_{\lambda+\delta} = \det((x_j^{\lambda_i+n-i})_{1 \leq i, j \leq n})$ is skew-symmetric, or equivalently, $(i, j)(a_{\lambda+\delta}) = -a_{\lambda+\delta}$ for a transposition $(i, j) \in S_n$. Hence the polynomial can be divided by $x_i - x_j$ for $i \neq j$. In the other words, $a_{\lambda+\delta}/D$ is a polynomial, where D is the Vandermonde determinant $\prod_{1 \le i < j \le n}(x_i - x_j) = \det((x_j^{n-i})_{1 \le i,j \le n})$. Moreover it is a symmetric polynomial since D is also skew-symmetric.

Definition 1.73. For a partition $\lambda = (\lambda_1, \lambda_2, \ldots, \lambda_n)$ of an integer, we define the **Schur polynomial** s_λ by

$$s_\lambda(x_1, x_2, \ldots, x_n) = \frac{\det((x_j^{\lambda_i + n - i})_{1 \le i,j \le n})}{\det((x_j^{n-i})_{1 \le i,j \le n})}.$$

Remark 1.74. The Schur polynomial has another combinatorial description. Let us regard a partition of an integer as a Young diagram. A map $T : \lambda \ni (i, j) \mapsto T_{ij} \in \mathbb{Z}_+$ is called a **semistandard Young tableau** if the following hold:

$$T_{i,j} \le T_{i,j+1},$$
$$T_{i,j} < T_{i+1,j}.$$

For $\lambda \vdash l$, a semistandard Young tableau $T : \lambda \to \{1, 2, \ldots, l\}$ is called a **standard Young tableau** if T is a bijection. We often write a diagram obtained by filling in the boxes of the Young diagram with corresponding numbers to denote a semistandard Young tableau T. For example,

$$\begin{array}{|c|c|} \hline 1 & 2 \\ \hline 3 \\ \cline{1-1} \end{array}$$

stands for the standard Young tableau T such that $T_{1,1} = 1$, $T_{1,2} = 2$, $T_{2,1} = 3$. The Schur polynomials s_λ can also be defined as the weighted generating functions of semistandard Young tableaux:

$$s_\lambda(x_1, x_2, \ldots, x_n) = \sum_T x^T,$$

where x^T denotes the monomial $\prod_{(i,j) \in \lambda} x_{T_{ij}}$, and the sum is over all semistandard Young tableaux $T : \lambda \to \{1, 2, \ldots, n\}$. From this definition it is easy to show that $s_{(l)} = h_l$ and $s_{(1^l)} = e_l$.

Proposition 1.75. *The following are $\mathbb{Z}$-bases of the homogeneous component of degree m of Λ_n:*

$\{\, m_\lambda(x_1, x_2, \ldots, x_n) \mid \lambda$ *is a partition of* m *such that* $\lambda_{n+1} = 0. \,\}$,

$\{\, s_\lambda(x_1, x_2, \ldots, x_n) \mid \lambda$ *is a partition of* m *such that* $\lambda_{n+1} = 0. \,\}$,

$\{\, h_\lambda(x_1, x_2, \ldots, x_n) \mid \lambda$ *is a partition of* m *such that* $\lambda_{n+1} = 0. \,\}$,

$\{\, h_\lambda(x_1, x_2, \ldots, x_n) \mid \lambda$ *is a partition of* m *such that* $\lambda_1 \le n. \,\}$,

$\{\, e_\lambda(x_1, x_2, \ldots, x_n) \mid \lambda$ *is a partition of* m *such that* $\lambda_1 \le n. \,\}$.

Proposition 1.76 is known as the Pieri formula, please see [37, 85] for the proof:

Proposition 1.76 (Pieri formula and dual Pieri formula). *For a partition* λ,

$$s_\lambda(x_1, x_2, \ldots, x_n) \cdot h_l(x_1, x_2, \ldots, x_n) = \sum_\mu s_\mu(x_1, x_2, \ldots, x_n),$$

$$s_\lambda(x_1, x_2, \ldots, x_n) \cdot e_l(x_1, x_2, \ldots, x_n) = \sum_\nu s_\nu(x_1, x_2, \ldots, x_n),$$

where the first sum is over all Young diagrams μ *obtained from* λ *by adding* l *boxes, at most one box for each column, and the second sum is over all Young diagrams* ν *obtained from* λ *by adding* l *boxes, at most one box for each row.*

Remark 1.77. We often consider the projective limit of the ring of symmetric polynomials. Let Λ_n^i be the homogeneous component of degree i of Λ_n. Then $\Lambda_n = \bigoplus_{i \ge 0} \Lambda_n^i$ has a structure of a graded ring. For $m \ge n$, define the map $\rho_{m,n}^i \colon \Lambda_m^i \to \Lambda_n^i$ by

$$\rho_{m,n}^i(f) = f(x_1, x_2, \ldots, x_n, 0, 0, \ldots, 0)$$

for $f(x_1, x_2, \ldots, x_n, x_{n+1}, x_{n+2}, \ldots, x_m) \in \Lambda_n^i$. The map $\rho_{m,n}^i \colon \Lambda_m^i \to \Lambda_n^i$ is a $\mathbb{Z}$-module homomorphism, and induces a graded ring homomorphism $\rho_{m,n} \colon \Lambda_m \to \Lambda_n$. These homomorphisms $\rho_{m,n}^i$ and $\rho_{m,n}$ are surjective. Moreover, for $m \ge n \ge i$, $\rho_{m,n}^i$ is bijective. For each i, $(\{\, \Lambda_n^i \,\}, \rho_{m,n}^i)$ is a projective system. We define Λ^i as the projective limit $\varprojlim \Lambda_n^i$. Then the direct sum $\Lambda = \bigoplus_i \Lambda^i$ is a graded ring, called the ring of symmetric functions. The element in Λ whose image in Λ_n is the Schur polynomial $s_\lambda(x_1, x_2, \ldots, x_n)$ for each n is called a Schur function. The set $\{\, p_\lambda \,\}$ forms a $\mathbb{Q}$-basis of $\mathbb{Q} \otimes_\mathbb{Z} \Lambda$, where p_λ is the element in Λ whose image in Λ_n is $p_\lambda(x_1, x_2, \ldots, x_n)$ for each n. On the other hand, the sets $\{\, s_\lambda \,\}, \{\, m_\lambda \,\}, \{\, h_\lambda \,\}$ and $\{\, e_\lambda \,\}$ form $\mathbb{Z}$-bases of Λ.

Theorem 1.78. *The poset* $\mathscr{I}(s^r)$ *has the Sperner property.*

Proof. Let Λ be the ring of symmetric functions. The set $\{\, s_\lambda \,\}$ of Schur functions is a $\mathbb{Z}$-basis for Λ. By the Pieri formula, $s_\lambda \cdot h_1$ equals the sum $\sum_\mu s_\mu$ over all Young diagrams μ obtained from λ by adding one box. Let

$$A = \Lambda / (\, s_\lambda \mid \lambda \notin \mathscr{I}(s^r) \,).$$

Then $\{\, s_\lambda \mid \lambda \in \mathscr{I}(s^r)\,\}$ is a $\mathbb{Z}$-basis for A. Hence the representation matrix for the basis of the linear map

$$\times h_1 \colon A_{i-1} \to A_i$$

gives the incidence matrix $M^{(i)}$ of $B_i(\mathscr{I}(s^r))$. On the other hand, h_1 is a strong Lefschetz element, or equivalently,

$$\times h_1^{rs-2k} \colon A_k \to A_{rs-k}$$

is bijective. See Chap. 7 for the proof. Hence, by Lemma 1.52, the poset has the Sperner property. $\qquad\qquad\qquad\qquad\qquad\qquad\qquad\qquad\qquad\qquad\qquad\qquad\qquad\square$

Remark 1.79. We can also prove Theorem 1.78 using the fact that the ring $A = \mathbb{Q}[e_1, e_2, \ldots, e_n]/(p_{s+1}, p_{s+2}, \ldots, p_{s+n})$ has the strong Lefschetz property, where e_i denotes the elementary symmetric polynomial in n variables, and p_i denotes the power sum in n variables (see Lemma 4.22). The ring A is the subring of $\mathbb{Q}[x_1, x_2 \ldots, x_n]/(x_1^{s+1}, x_2^{s+1}, \ldots, x_n^{s+1})$ fixed by the action of the n-th symmetric group S_n. It follows from direct calculation that

$$(\, p_l(x_1, x_2, \ldots, x_n) \mid l = s+1, s+2, \ldots, s+n)\,)$$
$$= (\, m_\lambda(x_1, x_2, \ldots, x_n) \mid \lambda \notin \mathscr{I}(s^n)\,).$$

Hence $\{\, m_\lambda(x_1, x_2, \ldots, x_n) \mid \lambda \in \mathscr{I}(s^n)\,\}$ forms a $\mathbb{Q}$-basis of the ring A. If μ is not obtained from λ by adding one box, then the coefficient of $m_\mu(x_1, x_2, \ldots, x_n)$ in the product $m_\lambda(x_1, x_2, \ldots, x_n) \cdot m_{(1)}(x_1, x_2, \ldots, x_n)$ is zero. Since the representation matrices of $\times(x_1 + x_2 + \cdots + x_n) \colon A_i \to A_{i+1}$ with respect to the basis $\{\, m_\lambda(x_1, x_2, \ldots, x_n)\,\}$ satisfy the condition in Lemma 1.50, so $\mathscr{I}(s^n)$ has the Sperner property.

We do not know if this can be proved purely poset-theoretically.

Problem 1.80. Give a purely poset-theoretical proof for the theorem.

1.4.4 The Vector Space Lattices

Here we consider the lattice of subspaces in a finite dimensional vector space over a finite field. First we recall the result of Kantor, which implies the Sperner property for the lattice. We also give another proof by ring-theoretic machinery. In this case, the corresponding ring has the strong Lefschetz property. To prove it, we use the criterion of higher Hessians, which we will prove in Sect. 3.6.

We define the **vector space lattice** $\mathscr{V}(n, q)$ to be the set of subspaces of the n-dimensional vector space $(\mathbb{F}_q)^n$ over the finite field $\mathbb{F}_q$ with q elements. Then $\mathscr{V}(n, q)$ is naturally ordered by the inclusion $\subset$. The rank function ρ is given by $\rho(W) = \dim(W)$ for $W \in \mathscr{V}(n, q)$. The following proposition is well-known.

Proposition 1.81. *For $n \geq k$,*

$$|\mathscr{V}(n,q)_k| = \begin{bmatrix} n \\ k \end{bmatrix}_q.$$

Proof. Since

$$\frac{(q^n - 1)(q^n - q^1)\cdots(q^n - q^{k-1})}{(q^k - 1)(q^k - q^1)\cdots(q^k - q^{k-1})} = \frac{(q^n - 1)}{(q^k - 1)} \frac{q(q^{n-1} - 1)}{q(q^{k-1} - 1)} \cdots \frac{q^{k-1}(q^{n-k+1} - 1)}{q^{k-1}(q - 1)}$$

$$= \begin{bmatrix} n \\ k \end{bmatrix}_q,$$

we will show that $|\mathscr{V}(n,q)_k|$ is given by

$$|\mathscr{V}(n,q)_k| = \frac{(q^n - 1)(q^n - q^1)\cdots(q^n - q^{k-1})}{(q^k - 1)(q^k - q^1)\cdots(q^k - q^{k-1})}.$$

To verify the equation, we consider the number of ordered sequences of k linearly independent vectors $v_1, v_2, \ldots, v_k$ in $V = \mathbb{F}_q^n$. The subspace $\langle v_1, v_2, \ldots, v_i \rangle$ spanned by the first i vectors contains q^i elements. A vector $v \in V$ is independent of $v_1, v_2, \ldots, v_i$ if and only if $v \in V \setminus \langle v_1, v_2, \ldots, v_i \rangle$. Thus, once the first i vectors are chosen, there are $q^n - q^i$ ways to choose a vector v_{i+1} such that it is linearly independent of $v_1, v_2, \ldots, v_i$. Hence the number of sequences of k linearly independent vectors is $(q^n - 1)(q^n - q^1)\cdots(q^n - q^{k-1})$.

Two sequences of such vectors $v_1, v_2, \ldots, v_k$ and $v_1', v_2', \ldots, v_k'$ span the same subspace if and only if there is an invertible $k \times k$ matrix T with entries in $\mathbb{F}_q$ such that $(v_1', v_2', \ldots, v_k') = (v_1, v_2, \ldots, v_k) \cdot T$.

Since T is nothing but a sequence of k linearly independent vectors in $\mathbb{F}_q^k$, the number of such matrices T is

$$(q^k - 1)(q^k - q^1)\cdots(q^k - q^{k-1}).$$

Thus the claimed equality follows.   □

As a special case of the theorem due to Kantor, we have Proposition 1.82.

Proposition 1.82 (Kantor [70]). *The incidence matrix $M^{(k)}$ of $B_k(\mathscr{V}(n,q))$ has full rank for each $k \leq \frac{n}{2}$.*

For $W \in \mathscr{V}(n,q)$, define $W^{\perp} \in \mathscr{V}(n,q)$ by

$$W^{\perp} := \left\{ w \in (\mathbb{F}_q)^n \ \middle| \ \sum_{i=1}^{n} w_i v_i = 0 \text{ for all } v \in W \right\}.$$

Then the map $\mathscr{V}(n,q) \ni W \mapsto W^{\perp} \in \mathscr{V}(n,q)$ is a bijection satisfying the following condition: $U \subset W$ if and only if $W^{\perp} \subset U^{\perp}$. In other words, the map is an order-reversing bijection. Hence the incidence matrix $M^{(k)}$ of $B_k(\mathscr{V}(n,q))$ has full rank for each k. This implies the following theorem by using Lemma 1.51 and Proposition 1.82.

Theorem 1.83. *The lattice $\mathscr{V}(n,q)$ has the Sperner property.*

The Sperner property for the $\mathscr{V}(n,q)$ follows from Kantor's theorem. However, we are interested in a ring theoretic proof for Theorem 1.83, because we can expect generalizations and by-products. For example our ring theoretic method will prove that the sequence

$$\begin{bmatrix} n \\ 0 \end{bmatrix}_q, \begin{bmatrix} n \\ 1 \end{bmatrix}_q, \begin{bmatrix} n \\ 2 \end{bmatrix}_q, \dots, \begin{bmatrix} n \\ n \end{bmatrix}_q$$

is an O-sequence. The proof is interesting in its own right, but it involves a lot of heavy computations. So we make a sketch of proof first.

Here is an outline of the proof.

1. Put $V = (\mathbb{F}_q)^n$ and $\mathscr{V} = \mathscr{V}(n,q)$, $N := |\mathscr{V}_1|$. So $N = [n]_q$. We work on the polynomial ring in N variables $R = \mathbb{Q}[x_1, x_2, \dots, x_N]$ over the field of rational numbers $\mathbb{Q}$.

2. Choose an injection

$$\phi : \{x_1, \dots, x_N\} \to V \setminus \{0\},$$

such that $\phi(x_i)$ and $\phi(x_j)$ are linearly independent if $i \neq j$. This is possible since $N = |\mathscr{V}_1|$. Once and for all we fix such an injection ϕ.

3. Define a polynomial $F \in R$ of degree n as follows:

$$F = \sum_{i_1, i_2, \dots, i_n} x_{i_1} x_{i_2} \cdots x_{i_n},$$

where the summation runs over the indices $(i_1, i_2, \dots, i_n)$ such that the images of the variables $x_{i_1}, x_{x_2}, \dots, x_{i_n}$ in V under ϕ are linearly independent.

4. Let $\operatorname{Ann}_R F := \left\{ f \in R \ \middle| \ f(\frac{\partial}{\partial x_1}, \frac{\partial}{\partial x_2}, \dots, \frac{\partial}{\partial x_N}) F = 0 \right\}$. Note that $x_j^2 \in \operatorname{Ann} F$ for any j, since every monomial that appears in F is square free. Put $A = R/\operatorname{Ann}_R F$. Then A is a graded Gorenstein $\mathbb{Q}$-algebra and for a graded

component A_k of A, we prove that $\dim A_k = |\mathcal{V}_k|$. In fact there is a one-to-one correspondence between the set of monomials in A_k and the set $\mathcal{V}_k$; the correspondence is give by

$$x_{i_1} x_{i_2} \cdots x_{i_k} \leftrightarrow \langle \phi(x_{i_1}), \phi(x_{i_2}), \ldots, \phi(x_{i_k}) \rangle .$$

This is the hardest part of proof, for which Lemma 1.84 is needed.

5. Let $B_k(\mathcal{V})$ be the incidence matrix for $\mathcal{V}_{k-1} \times \mathcal{V}_k$ as defined in Definitions 1.28 and 1.48. Put $l = x_1 + x_2 + \cdots + x_n$. Then one sees that, for each $k > 0$, the linear map :

$$\times l : A_{k-1} \to A_k$$

is represented in the monomial bases by a matrix which differs from $B_k(\mathcal{V})$ only by a scalar multiple.

6. The homomorphism $\times l^{n-2k} : A_k \to A_{n-k}$ is bijective. This can be proved by evaluating the higher Hessian of F at $x_1 = x_2 = \cdots = x_N = 1$. For the meaning of higher Hessians see Sect. 3.6. In the proof that follows we confine ourselves to showing this only for the first Hessian. (For the non-vanishing of higher Hessians we refer to [88].)

7. In the course of proof we introduce an extra ring Q, which is $Q = \mathbb{Q}[\partial_1, \ldots, \partial_N]$, where $\partial_i = \frac{\partial}{\partial x_i}$. Of course $Q \cong R$ and $Q/\operatorname{Ann}_Q F \cong R/\operatorname{Ann}_R F$. Furthermore, extra ideals $I_0, I_1, I_2 \subset R$ are introduced. These are used to establish a bijection between the monomial basis of A_k and $\mathcal{V}_k$. Once this is understood, the meaning of these ideals should be clear. It will be shown that $Q/\operatorname{Ann}_Q F \cong R/(I_0 + I_1 + I_2)$.

8. For the proof of the non-vanishing of Hessian, one should keep in mind the fact that the determinant of the square matrix for the linear map given by $(a_1 \partial_1 + \cdots + a_n \partial_N)^{n-2} : (Q_1 F) \to Q_{n-1} F$ is equal to $(n-2)!$ times the Hessian of F evaluated at $(x_1, \ldots, x_N) = (a_1, \ldots, a_N)$. For this see Theorem 3.76.

Let $V = (\mathbb{F}_q)^n$. Fix a maximal subset $\mathbb{P}V$ of V such that any distinct two elements in $\mathbb{P}V$ are $\mathbb{F}_q$-linearly independent. In other words, $\mathbb{P}V$ is a complete representatives for lines in V. It is easy to show that $|\mathbb{P}V| = [n]_q$. Let R be the polynomial ring $\mathbb{Q}[x_v | v \in \mathbb{P}V]$ in $[n]_q$ variables which correspond to vectors in $\mathbb{P}V$. For $B \subset \mathbb{P}V$, we define x_B to be the monomial $\prod_{v \in B} x_v$. Let I be the set of $\mathbb{F}_q$-linearly independent subsets of $\mathbb{P}V$. For $m = 0, 1, \ldots, n$, we define I_m to be $\{ B \in I \mid |B| = m \}$. We consider ideals I_0, I_1 and I_2 defined by

$$I_0 = \left(x_v^2 \mid v \in \mathbb{P}V \right),$$

$$I_1 = \left(x_B \mid B \notin I \right),$$

$$I_2 = \left(x_B - x_{B'} \mid B, B' \in I, \langle B \rangle = \langle B' \rangle \right),$$

where $\langle B \rangle$ denotes the vector space spanned by B. For each subspace $W \subset V$, fix $B_W \in I$ such that $\langle B_W \rangle = W$. Let $B = \{ B_W \mid W \in \mathcal{V}(n, q) \}$ and $B_i = B \cap I_i$. Then it is easy to show that the set $\{ x_B \mid B \in B \}$ is a basis of the ring $R/(I_0 + I_1 + I_2)$ as a $\mathbb{Q}$-vector space.

Let Q be the ring $\mathbb{Q}[\partial_v \mid v \in \mathbb{P}V]$ generated by derivations $\partial_v = \frac{\partial}{\partial x_v}$. The rings R and Q are isomorphic by the isomorphism φ defined by $\varphi(\partial_v) = x_v$. The ring Q naturally acts on R. For $B \subset \mathbb{P}V$, we define ∂_B to be the monomial $\prod_{v \in B} \partial_v$, respectively. Let F be the polynomial $\sum_{B \in I_n} x_B$. We define $\text{Ann}_Q(F)$ by $\text{Ann}_Q(F) = \{ P \in Q \mid PF = 0 \}$. It is easy to show that $\text{Ann}_Q(F)$ is an ideal of Q containing $\varphi^{-1}(I_0 + I_1 + I_2)$. Hence φ^{-1} induces a surjection $R/(I_0 + I_1 + I_2) \to Q/\text{Ann}_Q(F)$. We shall show that they are isomorphic to each other by proving that the dimension of the i-th homogeneous components $R/(I_0 + I_1 + I_2)$ as a $\mathbb{Q}$-vector space equal those of $Q/\text{Ann}_Q(F)$.

For a subspace $U \subset V$, we define $F_m(U)$ by (here read "|" as "such that")

$$F_m(U) = \sum_{B \in I_m \mid B \subset U} x_B.$$

Note that $F = F_n(V)$. Let $D_0 = \sum_{u \in \mathbb{P}V} \partial_u$. For $v \in \mathbb{P}V$, we define D_v and P_v by

$$D_v = \sum_{u \in \langle v \rangle^\perp \cap \mathbb{P}V} \partial_u,$$

$$P_v = D_0 - q D_v.$$

Then we have the following formula.

Lemma 1.84. *For $X \in I_k$ and $m \leq n - k$,*

$$P_v F_m(\langle X \rangle^\perp) = \begin{cases} q^{m-1} F_{m-1}(\langle X \rangle^\perp \cap \langle v \rangle^\perp) & (v \notin \langle X \rangle) \\ q^{m-1}(-q^{n-k-m+1} + 1) F_{m-1}(\langle X \rangle^\perp) & (v \in \langle X \rangle). \end{cases}$$

Proof. (Throughout proof, "|" in a subscript for $\sum$ should be read as "such that.") It follows by definition that

$$D_0 F_m(\langle X \rangle^\perp) = \sum_{u \in \mathbb{P}V} \partial_u \sum_{B \in I_m \mid B \subset \langle X \rangle^\perp} x_B = \sum_{\substack{(u,B) \in \mathbb{P}V \times I_m \mid \\ B \subset \langle X \rangle^\perp}} \partial_u x_B = \sum_{\substack{(u,B) \in \mathbb{P}V \times I_m \mid \\ u \in B \subset \langle X \rangle^\perp}} x_{B \setminus \{u\}}.$$

It is easy to show that

$$\sum_{(u,B) \subset \mathbb{P}V \times I_m \mid u \in B \subset \langle X \rangle^\perp} x_{B \setminus \{u\}} = \sum_{(u,B') \in I} x_{B'},$$

where

$$I = \left\{ (u, B') \in \mathbb{P}V \times \boldsymbol{I}_{m-1} \ \middle| \ \begin{array}{c} u \in \langle X \rangle^{\perp} \\ B' \subset \langle X \rangle^{\perp} \\ \{u\} \cup B \in \boldsymbol{I}_m \end{array} \right\}.$$

Since $u \in B \in \boldsymbol{I}_m$ if and only if $B \setminus \{u\} \in \boldsymbol{I}_{m-1}$ and $u \notin \langle B \rangle$, we have

$$I = \left\{ (u, B') \in \mathbb{P}V \times \boldsymbol{I}_{m-1} \ \middle| \ \begin{array}{c} u \in \langle X \rangle^{\perp}, \\ B \subset \langle X \rangle^{\perp}, \\ \langle B' \rangle \cap \langle u \rangle = 0 \end{array} \right\}$$

$$= \left\{ (u, B') \in \mathbb{P}V \times \boldsymbol{I}_{m-1} \ \middle| \ \begin{array}{c} u \in \langle X \rangle^{\perp} \setminus \langle B' \rangle, \\ B \subset \langle X \rangle^{\perp}. \end{array} \right\}.$$

Since the monomials depend only on B', we have

$$D_0 F_m(\langle X \rangle^{\perp}) = \sum_{B' \in \boldsymbol{I}_{m-1} | B \subset \langle X \rangle^{\perp}} \left| (\mathbb{P}V \cap \langle X \rangle^{\perp}) \setminus \langle B' \rangle \right| x_{B'}.$$

Similarly, for each $v \in \mathbb{P}V$, we obtain

$$D_v F_m(\langle X \rangle^{\perp}) = \sum_{B' \in \boldsymbol{I}_{m-1} | B \subset \langle X \rangle^{\perp}} \left| (\mathbb{P}V \cap \langle v \rangle^{\perp} \cap \langle X \rangle^{\perp}) \setminus \langle B' \rangle \right| x_{B'}.$$

Hence

$$P_v F_m(\langle X \rangle^{\perp})$$

$$= \sum_{B' \in \boldsymbol{I}_{m-1} | B \subset \langle X \rangle^{\perp}} \left(\left| (\mathbb{P}V \cap \langle X \rangle^{\perp}) \setminus \langle B' \rangle \right| - q \left| (\mathbb{P}V \cap \langle v \rangle^{\perp} \cap \langle X \rangle^{\perp}) \setminus \langle B' \rangle \right| \right) x_{B'}.$$

Let us show that

$$\left| (\mathbb{P}V \cap \langle X \rangle^{\perp}) \setminus \langle B' \rangle \right| = q \left| (\mathbb{P}V \cap \langle v \rangle^{\perp} \cap \langle X \rangle^{\perp}) \setminus \langle B' \rangle \right|$$

for $B' \not\subset (\mathbb{P}V \cap \langle v \rangle^{\perp} \cap \langle X \rangle^{\perp})$ such that $B' \subset \mathbb{P}V \cap \langle X \rangle^{\perp}$. Define L_i by

$$L_i = \left\{ u \in (\mathbb{P}V \cap \langle X \rangle^{\perp}) \setminus \langle B' \rangle \ \middle| \ \sum_k u_k v_k = i \right\}$$

for $i \in \mathbb{F}_q$. Then $L_0 = (\mathbb{P}V \cap \langle v \rangle^{\perp} \cap \langle X \rangle^{\perp}) \setminus \langle B' \rangle$. We will show that $|L_i| = |L_j|$ for all i, j, which implies the equation. Since B' satisfies $B' \not\subset (\mathbb{P}V \cap \langle v \rangle^{\perp} \cap \langle X \rangle^{\perp})$ and $B' \subset \mathbb{P}V \cap \langle X \rangle^{\perp}$, $B' \not\subset \langle v \rangle^{\perp}$. Hence there exists $x \in B'$ such that $\sum_i x_i v_i = c$ for some $c \in \mathbb{F}_q^{\times}$. Let $\bar{x} = \frac{1}{c} x$. Then $\bar{x}$ satisfies $\sum_i \bar{x}_i v_i = 1$ and $\bar{x} \in \langle B' \rangle$. Hence the map $+\bar{x} \colon L_i \ni u \mapsto u + \bar{x} \in L_{i+1}$ is a bijection with the inverse map $-\bar{x} \colon L_{i+1} \ni u \mapsto u - \bar{x} \in L_i$, which implies $|L_i| = |L_j|$. So we have
$$\left| (\mathbb{P}V \cap \langle X \rangle^{\perp}) \setminus \langle B' \rangle \right| = q \left| (\mathbb{P}V \cap \langle v \rangle^{\perp} \cap \langle X \rangle^{\perp}) \setminus \langle B' \rangle \right|.$$

Next we evaluate
$$\left| (\mathbb{P}V \cap \langle X \rangle^{\perp}) \setminus \langle B' \rangle \right| - q \left| (\mathbb{P}V \cap \langle v \rangle^{\perp} \cap \langle X \rangle^{\perp}) \setminus \langle B' \rangle \right|$$

for $B' \subset (\mathbb{P}V \cap \langle v \rangle^{\perp} \cap \langle X \rangle^{\perp})$ and $B' \in I_{m-1}$. Since $X \in I_k$, $\dim \langle X \rangle^{\perp} = n - k$. As $\langle B' \rangle$ is an $(m-1)$-dimensional subspace of $\langle X \rangle^{\perp}$, we have $\left| \langle X \rangle^{\perp} \setminus \langle B' \rangle \right| = q^{n-k} - q^{m-1}$. Hence
$$\left| (\mathbb{P}V \cap \langle X \rangle^{\perp}) \setminus \langle B' \rangle \right| = \left| \mathbb{P}V \cap (\langle X \rangle^{\perp} \setminus \langle B' \rangle) \right| = \frac{q^{n-k} - q^{m-1}}{q-1}.$$

In the case when $v \notin \langle X \rangle$, we have $\dim(\langle v \rangle^{\perp} \cap \langle X \rangle^{\perp}) = n - k - 1$. Since $\langle B' \rangle$ is an $(m-1)$-dimensional subspace of $\langle v \rangle^{\perp} \cap \langle X \rangle^{\perp}$, we have
$$\left| \langle v \rangle^{\perp} \cap \langle X \rangle^{\perp}) \setminus \langle B' \rangle \right| = q^{n-k-1} - q^{m-1},$$

which implies
$$\left| (\mathbb{P}V \cap \langle v \rangle^{\perp} \cap \langle X \rangle^{\perp}) \setminus \langle B' \rangle \right| = \frac{q^{n-k-1} - q^{m-1}}{q-1}.$$

Hence we obtain
$$\left| (\mathbb{P}V \cap \langle X \rangle^{\perp}) \setminus \langle B' \rangle \right| - q \left| (\mathbb{P}V \cap \langle v \rangle^{\perp} \cap \langle X \rangle^{\perp}) \setminus \langle B' \rangle \right|$$
$$= \frac{q^{n-k} - q^{m-1}}{q-1} - q \frac{q^{n-k-1} - q^{m-1}}{q-1}$$
$$= q^{m-1}.$$

In the case when $v \in \langle X \rangle$, $\langle v \rangle^{\perp} \cap \langle X \rangle^{\perp} = \langle X \rangle^{\perp}$ one has $\dim(\langle v \rangle^{\perp} \cap \langle X \rangle^{\perp}) = \dim(\langle X \rangle^{\perp}) = n-k$. Since $\langle B' \rangle$ is an $(m-1)$-dimensional subspace of $\langle v \rangle^{\perp} \cap \langle X \rangle^{\perp}$, we have

$$\left| (\langle v \rangle^\perp \cap \langle X \rangle^\perp) \setminus \langle B' \rangle \right| = q^{n-k} - q^{m-1},$$

which implies

$$\left| (\mathbb{P}V \cap \langle v \rangle^\perp \cap \langle X \rangle^\perp) \setminus \langle B' \rangle \right| = \frac{q^{n-k} - q^{m-1}}{q - 1}.$$

Hence we obtain

$$\left| (\mathbb{P}V \cap \langle X \rangle^\perp) \setminus \langle B' \rangle \right| - q \left| (\mathbb{P}V \cap \langle v \rangle^\perp \cap \langle X \rangle^\perp) \setminus \langle B' \rangle \right|$$

$$= \frac{q^{n-k} - q^{m-1}}{q - 1} - q \frac{q^{n-k} - q^{m-1}}{q - 1}$$

$$= -q^{n-k} + q^{m-1}. \qquad \qquad \qquad \square$$

By Lemma 1.84, we obtain $P_B F = F_{n-m}(\langle B \rangle^\perp)$ for $B \in \boldsymbol{B}$. Since $\dim \langle B \rangle^\perp = \dim \langle B' \rangle^\perp = n - m$ for $B, B' \in \boldsymbol{I}_m$, we have

$$\left\{ X \in \boldsymbol{I}_{n-m} \mid \langle X \rangle = \langle B \rangle^\perp \right\} \cap \left\{ X \in \boldsymbol{I}_{n-m} \mid \langle X \rangle = \langle B' \rangle^\perp \right\} = \emptyset$$

if $\langle B \rangle \neq \langle B' \rangle$. As $F_{n-m}(\langle B \rangle^\perp)$ and $F_{n-m}(\langle B' \rangle^\perp)$ do not have a common monomial, $\left\{ F_{n-m}(\langle B \rangle^\perp) \mid B \in \boldsymbol{B}_m \right\}$ is $\mathbb{Q}$-linearly independent. Hence $\{ P_B F \mid B \in \boldsymbol{B} \}$ is a $\mathbb{Q}$-linearly independent set consisting of $|\boldsymbol{B}|$ elements. Therefore $\dim_{\mathbb{Q}}((Q/\operatorname{Ann}_Q(F))_k) \geq |\boldsymbol{B}_k|$, which implies that $Q/\operatorname{Ann}_Q(F)$ and $R/(I_0 + I_1 + I_2)$ are isomorphic.

Remark 1.85. Since $Q/\operatorname{Ann}_Q(F)$ and $R/(I_0 + I_1 + I_2)$ are isomorphic, the ring $R/(I_0 + I_1 + I_2)$ is Gorenstein. This in particular proves that the sequence

$$\left(\begin{bmatrix} n \\ 0 \end{bmatrix}_q, \begin{bmatrix} n \\ 1 \end{bmatrix}_q, \begin{bmatrix} n \\ 2 \end{bmatrix}_q, \ldots, \begin{bmatrix} n \\ n \end{bmatrix}_q \right)$$

is a **Gorenstein vector**. (See Theorem 2.71. See also [84].)

The next task is to find a strong Lefschetz element for $R/(I_0 + I_1 + I_2)$. Consider an element $P \in Q/\operatorname{Ann}_Q(F)$. Obviously, PF is well-defined as an element of R. Let $\left\{ P_1^{(k)}, \ldots, P_{m_k}^{(k)} \right\}$ be a $\mathbb{Q}$-basis of the homogeneous component of $Q/\operatorname{Ann}_Q F$ of degree k. Then the determinant $\det((P_i^{(k)} P_j^{(k)} F)_{i,j})$ is called the k-th Hessian. It is shown in Sect. 3.6, the Hessians characterize the strong Lefschetz elements. In particular, for $l = \sum_i a_i \partial_i$ and $k > 0$, $\times l^{n-2k} : (Q/\operatorname{Ann}_Q(F))_k \to (Q/\operatorname{Ann}_Q(F))_{n-k}$ is bijective if and only if the value of the k-th Hessian $\det((P_i^{(k)} P_j^{(k)} F)_{i,j})$ at $x_i = a_i$ is not zero. Moreover, $l^n \neq 0$ if and only if the value of F at $x_i = a_i$ does not equal zero. Hence we have

$$l = \sum_{i=1}^{m_1} a_i \partial_i \text{ is a strong Lefschetz element}$$

$$\Longleftrightarrow \begin{cases} \det((P_i^{(k)} P_j^{(k)} F)_{i,j})(a_1, a_2, \ldots, a_{m_1}) \neq 0 & \text{for all } k, \text{ and} \\ F(a_1, a_2, \ldots, a_{m_1}) \neq 0. \end{cases}$$

Here we consider only the first Hessian. If F and the first Hessian does not vanish at $x_v = 1$, then $\times l^n : A_0 \to A_n$ and $\times l^{n-2} : A_1 \to A_{n-1}$ are bijective, which implies the Strong Lefschetz property of $A = Q/\operatorname{Ann}_Q(F)$ for the case when $n \leq 4$. The set $\{ P_v \mid v \in \mathbb{P}V \}$ is a $\mathbb{Q}$-basis for the homogeneous component of $Q/\operatorname{Ann}_Q(F)$ of degree one. It follows from direct calculation that

$$F_m(W)(1, 1, \ldots, 1) = \frac{q^{\frac{m(m-1)}{2}}}{m!} \frac{[k]_q!}{[k-m]_q!}$$

for a k-dimensional subspace W in V and $m \leq k$. By Lemma 1.84, we have

$$P_v P_w F = q^{n-1} q^{n-2} F_{n-2}(\langle v \rangle^\perp \cap \langle w \rangle^\perp),$$

$$P_v P_v F = q^{n-1} q^{n-2}(-q + 1) F_{n-2}(\langle v \rangle^\perp)$$

for $\{ v, w \} \in I$ with $v \neq w$. Hence we obtain that

$$(P_v P_u F)(1, 1, \ldots, 1) = \begin{cases} \frac{1}{(n-2)!} q^{\frac{n(n-1)}{2}} [n-2]_q! & (v \neq u), \\ \frac{1}{(n-2)!} q^{\frac{n(n-1)}{2}} (-q+1) [n-1]_q! & (v = u) \end{cases}$$

for $v, w \in \mathbb{P}V$. Let

$$f_{u,v} = \frac{(n-2)!}{q^{\frac{n(n-1)}{2}} [n-2]_q!} (P_v P_w F)(1, 1, \ldots, 1) = \begin{cases} 1 & (u \neq v), \\ 1 - q^{n-1} & (u = v). \end{cases}$$

Since the eigenvalues of the $N \times N$-matrix $A = (1)_{i,j=1,2,\ldots,N}$ are N and 0, the determinant $\det(A - tI)$ is zero only if $t = 0$ or N, where I stands for the identity matrix of size N. Hence, if $q^{n-1} - [n]_q \neq 0$, then the determinant $\det((f_{v,w})_{v,w \in \mathbb{P}V}) = \det(A - q^{n-1}I)$ of the $[n]_q \times [n]_q$-matrix does not vanish.

Hence, if $n > 1$, then

$$\det((P_u P_v F)_{u,v \in \mathbb{P}V})(1, 1, \dots, 1) \neq 0.$$

So the first Hessian does not vanish, and we obtain that $\times l^{n-1} : (Q / \operatorname{Ann}_Q F)_1 \to (Q / \operatorname{Ann}_Q F)_{n-1}$ is bijective. Hence $l = \sum_{v \in \mathbb{P}V} x_v$ is a strong Lefschetz element of $R/(I_0 + I_1 + I_2)$ if $n = 3$ or 4. On the other hand, l satisfies the condition of Lemma 1.50. Hence, in this case, the poset $\mathscr{V}(n, q)$ has the Sperner property.

Problem 1.86. Given integers n, q, does there exist an algebraic variety such that the even part of the cohomology ring is isomorphic to the ring $A = Q / \operatorname{Ann}_Q(F)$ constructed in the proof of Theorem 1.83?

Chapter 2
Basics on the Theory of Local Rings

2.1 Minimal Generating Set of an Ideal and Number of Generators

The reader is assumed to have basic knowledge of the theory of commutative rings.

Let R be a commutative ring with an identity element and let $f_1, f_2, \ldots, f_m$ be elements of R. The ideal generated by $f_1, f_2, \ldots, f_m$ is denoted

$$(f_1, f_2, \ldots, f_m) \quad \text{or} \quad Rf_1 + Rf_2 + \cdots + Rf_m.$$

It is the smallest ideal that contains the elements $f_1, f_2, \ldots, f_m$. Thus

$$(f_1, f_2, \ldots, f_m) = \{ r_1 f_1 + r_2 f_2 + \cdots + r_m f_m \mid r_i \in R \}.$$

We regard R itself as an ideal of R. R is called **Noetherian** if every ideal of R is generated by finitely many elements. The Hilbert basis theorem says that the polynomial ring over a field K

$$K[x_1, x_2, \ldots, x_n]$$

is a Noetherian ring. The formal power series ring $K[[x_1, x_2, \ldots, x_n]]$ over a field K is another typical example of a Noetherian ring.

A set

$$\{ f_1, f_2, \ldots, f_m \}$$

is a **minimal generating set** of an ideal if

$$f_i \notin (f_1, f_2, \ldots, \check{f_i}, \ldots, f_m)$$

for $i = 1, 2, \ldots, m$, where $\check{f_i}$ means that f_i is being omitted.

T. Harima et al., *The Lefschetz Properties*, Lecture Notes in Mathematics 2080, DOI 10.1007/978-3-642-38206-2_2, © Springer-Verlag Berlin Heidelberg 2013

An ideal I is **irreducible** if I cannot be an intersection of two properly larger ideals. Namely, I is irreducible if $I = I_1 \cap I_2$ forces either $I = I_1$ or $I = I_2$. An intersection of ideals

$$I = I_1 \cap I_2 \cap \cdots \cap I_t$$

is **irredundant** if

$$I \subsetneq I_1 \cap I_2 \cap \cdots \cap \check{I}_i \cap \cdots \cap I_t$$

for $i = 1, 2, \ldots, t$. In a Noetherian commutative ring, every ideal is the intersection of a finitely many irreducible ideals. This can be proved by Noetherian induction.

In a commutative ring R an ideal I is a **prime ideal** if R/I is an integral domain or equivalently the set $R \setminus I$ is closed under multiplication. An ideal I is a **maximal ideal** if

$$I \subsetneq J \implies J = R$$

for any ideal J. It is easy to see that I is a maximal ideal if and only if the quotient ring R/I is a field. The **Krull dimension** of a commutative ring is the maximum length n of a chain of prime ideals:

$$\mathfrak{p}_0 \subsetneq \mathfrak{p}_1 \subsetneq \cdots \subsetneq \mathfrak{p}_n.$$

For example the Krull dimension of the polynomial ring $K[x_1, x_2, \ldots, x_n]$ over a field has Krull dimension n, and

$$(0) \subset (x_1) \subset (x_1, x_2) \subset \cdots \subset (x_1, x_2, \ldots, x_n)$$

is a saturated chain of prime ideals of maximum length.

A local ring is a commutative ring with a unique maximal ideal. It is often denoted by a triple $(R, \mathfrak{m}, K)$ to mean that R is a local ring, $\mathfrak{m}$ is the maximal ideal and $K = R/\mathfrak{m}$ is the residue field.

Let $(R, \mathfrak{m}, K)$ be a local ring of Krull dimension n. Then there exist elements $f_1, f_2, \ldots, f_n \in \mathfrak{m}$ such that $\sqrt{(f_1, f_2, \ldots, f_n)} = \mathfrak{m}$. Here

$$\sqrt{I} = \left\{ x \in R \mid x^k \in I, \text{ some } k \right\}.$$

The set $\sqrt{I}$ is called the **radical** of an ideal I. Such a set of elements $f_1, f_2, \ldots, f_n$ is called a **system of parameters** of R.

Definition 2.1. Suppose that $(R, \mathfrak{m}, K)$ is a local ring of Krull dimension n. R is called a **regular** local ring if there exist n elements $f_1, f_2, \ldots, f_n$ such that

$$\mathfrak{m} = (f_1, f_2, \ldots, f_n).$$

The formal power series ring

$$K[[x_1, x_2, \ldots, x_n]]$$

over a field K is a typical example of a regular local ring as well as the polynomial ring

$$K[x_1, x_2, \ldots, x_n]$$

localized at a maximal ideal.

Definition 2.2. A Noetherian local ring $(R, \mathfrak{m}, K)$ is called a **complete intersection**, if there exist a regular local ring R' and a system of parameter $f_1, f_2, \ldots, f_n$ for R' such that $R \cong R'/(f_1, f_2, \ldots, f_m)$. In this case

$$\text{Krull dim } R = n - m.$$

Let R be a commutative ring. An element $u \in R$ is a **unit** if there exists v such that $uv = 1$. In a local ring $(R, \mathfrak{m}, K)$, an element $a \in R$ is a unit if and only if $a \notin \mathfrak{m}$. (To prove this suppose that a is not a unit. Then the ideal $(a) \neq R$. Hence $(a) \subset \mathfrak{m}$, since $\mathfrak{m}$ is the unique maximal ideal. Conversely if a is a unit, then obviously $a \notin \mathfrak{m}$.)

We are interested in the number of generators of an ideal in a local ring. A basic fact is the following

Proposition 2.3. *Let $(R, \mathfrak{m}, K)$ be a local ring and I an ideal of R. Then*

$$\{ f_1, f_2, \ldots, f_m \} \subset I$$

is a minimal generating set for I if and only if the images of these elements are a basis for the vector space $I/\mathfrak{m}I$ over K.

Proof. Assume that R is a local ring. Let $f_1, f_2, \ldots, f_m$ be a representative of a basis for the vector space $I/\mathfrak{m}I$. Put $J = (f_1, f_2, \ldots, f_m)$. We have to show that $I = J$. By assumption we have

$$I = J + \mathfrak{m}I.$$

Thus $I = J$ follows immediately by **Nakayama's lemma** which we prove below. □

Lemma 2.4 (Nakayama's lemma). *Let $(R, \mathfrak{m}, K)$ be a Noetherian local ring. Let I, J be ideals of R. If $I \subset J + \mathfrak{m}I$, then $I \subset J$.*

Proof. Let $a_1, a_2, \ldots, a_r$ be a generating set for I. Then we have

$$I = Ra_1 + Ra_2 + \cdots + Ra_r,$$

and

$$\mathfrak{m}I = \mathfrak{m}a_1 + \mathfrak{m}a_2 + \cdots + \mathfrak{m}a_r.$$

Since we assume that $I \subset J + \mathfrak{m}I$, there exist elements $b_{ij} \in \mathfrak{m}$ and $c_i \in J$ such that

$$a_i = c_i + b_{i1}a_1 + b_{i2}a_2 + \cdots + b_{ir}a_r,$$

for each a_i. In the matrix notation we have

$$\begin{pmatrix} a_1 \\ a_2 \\ \vdots \\ a_r \end{pmatrix} = \begin{pmatrix} c_1 \\ c_2 \\ \vdots \\ c_r \end{pmatrix} + M \begin{pmatrix} a_1 \\ a_2 \\ \vdots \\ a_r \end{pmatrix},$$

where $M = (b_{ij})$. Let E be the $r \times r$ identity matrix. Then we have

$$(E - M) \begin{pmatrix} a_1 \\ a_2 \\ \vdots \\ a_r \end{pmatrix} = \begin{pmatrix} c_1 \\ c_2 \\ \vdots \\ c_r \end{pmatrix}.$$

Since $\det(E - M) \equiv 1 \bmod \mathfrak{m}$, it follows that $\det(E - M) \notin \mathfrak{m}$. Thus $\det(E - M)$ is a unit element of A and the matrix $M - E$ is invertible, which shows that $a_i \in J$ and $I \subset J$ as desired. $\qquad\qquad\qquad\qquad\qquad\qquad\qquad\qquad\qquad\qquad\qquad$ □

A consequence of Proposition 2.3 is that the minimal number of generators of I is equal to $\dim_K I / \mathfrak{m}I$. We denote it by $\mu(I)$.

Definition 2.5. Let $(R, \mathfrak{m}, K)$ be a Noetherian local ring. Let I be an ideal of R. We define the integer $\mu(I)$ by

$$\mu(I) = \dim_K (I / \mathfrak{m}I).$$

The number $\mu(I)$ is called the **minimal number of generators** of the ideal I

Remark 2.6. Suppose that R is a local ring and I is an ideal of R. Proposition 2.3 says that any minimal generating set for I consists of the same number of elements. It should be noted that if R is not a local ring, or if I is not homogeneous in a graded ring, then the number of elements in a minimal generating set is not necessarily the minimal number of generators. For example let $R = K[x]$ be the polynomial ring in one variable over a field K, and consider the ideal

$$I = (x^2, x + x^3).$$

Then $\{ x^2, x + x^3 \}$ is a minimal generating set for the ideal they generate in the sense that neither of the elements can be omitted and still generate I. However the minimal number of generators of I is one since it is a principal ideal: $I = (x)$.

2.1.1 Graded Rings

A **graded ring** is a ring R with a decomposition

$$R = R_0 \oplus R_1 \oplus R_2 \oplus R_3 \oplus \cdots$$

as a direct sum of additive groups R_i, such that the multiplication is defined so that if $a \in R_i$ and $b \in R_j$, then $ab \in R_{i+j}$. Thus the multiplication induces a bilinear map:

$$R_i \times R_j \to R_{i+j}.$$

Note that R_0 is a subring of R and R_i are modules over R_0. Thus R is an R_0-algebra. Unless otherwise specified, we will be assuming that R_0 is a field and R is Noetherian. In this case R_i are finite dimensional vector spaces over R_0. R_i is called the **homogeneous part** of degree i and an element in it a homogeneous element of degree i.

A graded algebra $R = \bigoplus_{i \geq 0} R_i$ over $K := R_0$ is called **standard** if

$$R = K[R_1],$$

i.e., R is generated by homogeneous elements of degree one as an algebra.

An ideal I in a graded algebra $R = \bigoplus_{i=0}^{\infty} R_i$ is a **homogeneous ideal** if

$$I = \bigoplus_{i=0}^{\infty} I \cap R_i.$$

It is easy to see that an ideal $I \subset R$ is homogeneous if and only if I can be generated by homogeneous elements. A graded Noetherian algebra R is a quotient of a polynomial ring over R_0 by a homogeneous ideal. If $R = \bigoplus_{i \geq 0} R_i$ is a graded ring over a field $K = R_0$, then

$$\mathfrak{m} := \bigoplus_{i=1}^{\infty} R_i$$

is the unique homogeneous maximal ideal of R. It is often the case that a proposition that holds for a local ring can be translated for a graded ring and homogeneous ideals. The following is a variant of Nakayama's lemma.

Lemma 2.7 (homogeneous version of Nakayama's lemma). *Let* $R = \bigoplus_{i=0}^{\infty} R_i$ *be a Noetherian graded ring,* R_0 *a field, and* $\mathfrak{m} = \bigoplus_{i \geq 1} R_i$. *Let* $I, J \subset R$ *be graded ideals of* R. *If* $I \subset J + \mathfrak{m}I$, *then* $I \subset J$.

Proof. Same as Lemma 2.4. $\square$

Here is a translation of Proposition 2.3.

Proposition 2.8. *Let* $R = \bigoplus_{i=0}^{\infty} R_i$ *be a graded ring,* $K = R_0$ *a field, and* $\mathfrak{m} = \bigoplus_{i=1}^{\infty} R_i$ *the homogeneous maximal ideal. Let* I *be a homogeneous ideal of* R. *Then a set of homogeneous elements* $\{ f_1, f_2, \ldots, f_m \} \subset I$ *is a minimal generating set for* I *if and only if the images of these elements are a basis for the vector space* $I/\mathfrak{m}I$ *over* K.

Proof. Let $f_1, f_2, \ldots, f_m$ be a homogeneous representative of a basis for the vector space $I/\mathfrak{m}I$. Put $J = (f_1, f_2, \ldots, f_m)$. We have to show that $I = J$. By assumption we have

$$I = J + \mathfrak{m}I.$$

Thus $I = J$ follows immediately by (the homogeneous version of) Nakayama's lemma. $\square$

The minimal number $\mu(I)$ of an ideal I in a local ring has been defined. We use the same notation for a homogeneous ideal in a graded ring indicated next.

Definition 2.9. Let $R = \bigoplus_{i=0}^{\infty} R_i$ be a graded ring, $\mathfrak{m} := \bigoplus_{i=1}^{\infty} R_i$, and $K := R/\mathfrak{m} \cong R_0$. Let I be a homogeneous ideal of R. Define the number $\mu(I)$ by

$$\mu(I) = \dim_K(I/\mathfrak{m}I),$$

which is called the **minimal number of generators** of the homogeneous ideal $I \subset R$ as well as for an ideal in a local ring.

2.1.2   Artinian Local Rings

A Noetherian local ring $(A, \mathfrak{m}, K)$ is **Artinian** if it has finite length. This means that there exists a series of ideals

$$A = \mathfrak{a}_0 \supset \mathfrak{a}_1 \supset \cdots \supset \mathfrak{a}_l = 0$$

such that $\mathfrak{a}_{i-1}/\mathfrak{a}_i \cong K$ for all $i = 1, 2, \ldots, l$. Such a sequence of ideals is called a **composition series**. It is easy to see that l does not depend on the choice of the composition series. In this case l is the **length** of A and is denoted by length(A). Here are some basic properties of Artinian local rings.

Proposition 2.10. *Suppose that $(A, \mathfrak{m}, K)$ is a Noetherian local ring. Then the following conditions are equivalent.*

1. *A is Artinian.*
2. *Every descending chain of ideals in A stops.*
3. *$\mathfrak{m}^k = 0$ for some integer k.*
4. *$\mathfrak{m}$ is the only prime ideal of A.*
5. *Krull dim $A = 0$.*

Proof. (1) $\implies$ (2) Let $l = $ length A. Suppose that

$$A = \mathfrak{a}_0 \supset \mathfrak{a}_1 \supset \cdots$$

is a descending chain of ideals. Then since length $A \geq \sum_{j=1}^{i}$ length $\mathfrak{a}_{j-1}/\mathfrak{a}_j$, a proper inclusion $\mathfrak{a}_{j-1} \supsetneq \mathfrak{a}_j$ occurs at most l times. Thus there is i such that $\mathfrak{a}_{j-1} = \mathfrak{a}_j$ for all $j > i$.

(2) $\implies$ (3) Consider the descending chain of ideals:

$$A \supsetneq \mathfrak{m}^1 \supsetneq \mathfrak{m}^2 \supsetneq \cdots .$$

Since it should stop, there exists k such that $\mathfrak{m}^k = \mathfrak{m}^{k+1}$. By Nakayama's lemma, this implies that $\mathfrak{m}^k = 0$.

(3) $\implies$ (4) Recall that an ideal I is a prime ideal if and only if the set $A \setminus I$ is multiplicatively closed. We have seen that $A \setminus \mathfrak{m}$ is the set of units in A. Thus $\mathfrak{m}$ is a prime ideal. Let $\mathfrak{p}$ be any prime ideal of A. Then we have $\mathfrak{p} \supset 0 = \mathfrak{m}^k$ for some k. Hence $\mathfrak{p} \supset \mathfrak{m}$. Thus $\mathfrak{m}$ is the unique prime ideal of A.

(4) $\implies$ (1) First we claim that $\mathfrak{m}$ consists of nilpotent elements. By way of contradiction assume that there exists $a \in \mathfrak{m}$ such that $a^k \neq 0$ for all $k \geq 1$. Put $S = \{a, a^2, a^3, \ldots\}$. S is a multiplicatively closed set. Let F be the set of ideals $\mathfrak{a}$ such that $\mathfrak{a} \cap S = \emptyset$. One sees easily that F is an inductive set in the sense that it is a partially ordered set (with respect to inclusion) in which any totally ordered subset has an upper bound in F. Thus by Zorn's lemma, it has a maximal element, say $\mathfrak{p}$. We claim it is a prime ideal. We have to show that the complement of $\mathfrak{p}$ is multiplicatively closed. Let $x \notin \mathfrak{p}$ and $x' \notin \mathfrak{p}$. Then since $Ax + \mathfrak{p} \supsetneq \mathfrak{p}$ and $Ax' + \mathfrak{p} \supsetneq \mathfrak{p}$, $Ax + \mathfrak{p}$ and $Ax' + \mathfrak{p}$ meet S. Thus we have $bx + p = a^k$ and $b'x' + p' = a^{k'}$ for some integers k, k' and for some elements $b, b' \in A$ and $p, p' \in \mathfrak{p}$. Hence $a^{k+k'} = (bx + p)(b'x' + p') \notin \mathfrak{p}$. Hence $xx' \notin \mathfrak{p}$ so we have shown that $\mathfrak{p}$ is a prime ideal. Then, however, we have $\mathfrak{p} = \mathfrak{m}$ by assumption, which contradicts $a \in \mathfrak{m}$. This contraction shows that $\mathfrak{m}$ consists of nilpotent elements. Since $\mathfrak{m}$ is finitely generated, there is k such that $\mathfrak{m}^k = 0$. The sequence of ideals

$$A \supset \mathfrak{m}^1 \supset \mathfrak{m}^2 \supset \cdots \supset \mathfrak{m}^k = 0$$

shows that A is Artinian.

Finally, the equivalence (5) $\iff$ (1) is clear by the definition of Krull dimension. $\qquad\square$

Remark 2.11. 1. Let $(R, \mathfrak{m}, K)$ be a local ring, and I an ideal of R. Then R/I is an Artinian local ring if and only if I contains some power of the maximal ideal. Namely,

$$R/I \text{ is Artinian} \iff \exists k, I \supset \mathfrak{m}^k.$$

Such an ideal I is called an $\mathfrak{m}$-**primary ideal**. It is the same as saying that I is **of finite colength**. (The colength of I is the length of R/I.)

2. Let $(R, \mathfrak{m}, K)$ be a Noetherian local ring and let M be an R-module. As for ideals, we denote by length(M) the **length** of the composition series M. Namely if length$(M) = l$, then there exists a sequence of submodules

$$M = M_0 \supset M_1 \supset \cdots \supset M_l = 0$$

such that $M_{i-1}/M_i \cong K$ for $i = 1, 2, \ldots, l$.

3. Let $(A, \mathfrak{m}, K)$ be an Artinian local ring. The ideal $(0 : \mathfrak{m})$ is called the **socle** of A. The ideal $(0 : \mathfrak{m})$ is by definition

$$(0 : \mathfrak{m}) = \{ x \in A \mid \mathfrak{m}x = 0 \}.$$

It is regarded as a vector space over $K = A/\mathfrak{m}$.

4. A graded algebra $A = \bigoplus_{i \geq 0} A_i$ (with A_0 a field) is an Artinian local ring if $A = \bigoplus_{i=0}^{c} A_i$ for some integer $c \geq 0$. The maximal integer c such that $A_c \neq 0$ is called the **maximal socle degree** of A, since it is obvious that A_c is in the socle of A. Sometimes it is clear that $A_c = 0 : \mathfrak{m}$ from the context. (For example A is Artinian Gorenstein if and only if the socle of A is one dimensional, as we will see in Theorem 2.58.) In such a case we will simply call c the **socle degree** of A.

5. Let A be a graded algebra $A = \bigoplus_{i \geq 0} A_i$ (with A_0 a field). Then

$$\{ a_0 + a_1 + \cdots + a_c \mid a_0 \neq 0, a_j \in A_j \}$$

is the set of units.

6. The **embedding dimension** of a graded K-algebra A is the least integer n such that A is a homomorphic image of the polynomial ring $K[x_1, x_2, \ldots, x_n]$ as a graded algebra.

2.1.3 The Type of an $\mathfrak{m}$-Primary Ideal

For ideals I, J in a commutative ring, $(I : J)$ denotes the ideal

$$(I : J) = \{ x \in R \mid xJ \subset I \}.$$

If $J = (y)$ is a principal ideal, then we also write $I : y$ for $I : J$. Recall that we have defined $\mu(I)$ to be the dimension of the vector space $I/\mathfrak{m}I$. The dual notion to the vector space $I/\mathfrak{m}I$ is $(I : \mathfrak{m})/I$. Notice that we have the canonical isomorphisms:

$$I/\mathfrak{m}I \cong I \otimes_R R/\mathfrak{m},$$

and

$$(I : \mathfrak{m})/I \cong \mathrm{Hom}_R(R/\mathfrak{m}, R/I).$$

Definition 2.12. Let $(R, \mathfrak{m}, K)$ be a Noetherian local ring. Let I be an ideal of R. Define $\tau(I)$ to be

$$\tau(I) = \dim_K((I : \mathfrak{m})/I),$$

or equivalently,

$$\tau(I) = \dim_K(\mathrm{Hom}_R(K, R/I)).$$

If I is of finite colength, $\tau(I)$ is called the **type** of I. It is just the dimension of the socle of the Artinian algebra R/I.

Remark 2.13. Mostly we are concerned with ideals of finite colength. For such ideals we have $\tau(I) > 0$. In fact if I is of finite colength, then there exists an ideal $I' \supset I$ such that $I'/I \cong K$, so $(I : \mathfrak{m}) \supset I' \supsetneqq I$ and $(0 : \mathfrak{m})/I \neq 0$.

Proposition 2.14. *Suppose that $(R, \mathfrak{m})$ is a local ring and I an ideal of finite colength. Then I is irreducible if and only if $\tau(I) = 1$.*

Proof. Suppose that $\tau(I) = 1$. This means that $I : \mathfrak{m}$ is the smallest ideal which contains I properly. Thus I cannot be the intersection of properly larger ideals. Conversely assume that $\tau(I) > 1$. Choose $e_1, e_2 \in (I : \mathfrak{m}) \setminus I$ so that their images are linearly independent in the vector space $(I : \mathfrak{m})/I$. Then we have $I + Re_1 \neq I + Re_2$ and $\dim_K(I + Re_i)/I = 1, i = 1, 2$. Hence $I = (I + Re_1) \cap (I + Re_2)$, which is an irredundant intersection. $\qquad\square$

Proposition 2.15. *Suppose that $(R, \mathfrak{m}, K)$ is a local ring and $I \subset R$ an ideal of finite colength.*

1. *If $I = I_1 \cap I_2 \cap \cdots \cap I_t$ is an irredundant intersection of irreducible ideals, then $t = \tau(I)$.*
2. *Suppose that R is Artinian. Let $t = \tau(0)$, where 0 is the zero ideal of A. Then t is the maximum number s such that 0 can be expressed as an irredundant intersection of s irreducible ideals:*

$$0 = I_1 \cap I_2 \cap \cdots \cap I_s.$$

Proof. (1) We may consider $A := R/I$ for R. So it suffices to prove (2).

(2) By Proposition 2.14 the ideal $q \subset A$ is irreducible if and only if $q : m/q \cong K$. Now suppose that $0 = q_1 \cap q_2 \cap \cdots \cap q_t$ is an irredundant intersection of irreducible ideals. We want to show that the homomorphism

$$0 \rightarrow A \rightarrow \bigoplus_{i=1}^{t} A/q_i$$

defined by

$$x \mapsto (x + q_1, x + q_2, \ldots, x + q_t)$$

induces a bijection

$$0 \rightarrow \operatorname{Hom}_A(K, A) \rightarrow \bigoplus_{i=1}^{t} \operatorname{Hom}_A(K, A/q_i).$$

Since $\bigcap_{j \neq i} q_j$ is not contained in q_i, there exists $e'_i \in \left(\bigcap_{j \neq i} q_j \right) \setminus q_i$ for every i. Since length $\left(Ae'_i + q_i/q_i \right)$ is finite, there exists a minimal ideal which contains q_i and is contained in $Ae'_i + q_i$. Let e_i be a generator for such an ideal mod q_i. Then we have $e_i \in (q_i : m) \setminus q_i$ and $e_i \in q_j$ for any $j \neq i$. This shows that the map above is a bijection or $\dim \operatorname{Hom}_A(K, A) = t$ as we wanted. $\qquad\square$

2.2 Complete Local Rings and Matlis Duality

In this section we introduce the Matlis duality over complete local rings and show that the number of generators and the type are dual notions.

Let R be a Noetherian local ring and m the maximal ideal of R. The projective limit of the projective system

$$\{ R/m^i \mid i = 1, 2, \ldots \},$$

denoted by $\varprojlim R/m^i$, is by definition the set of infinite sequences

$$(f_1 + m^1, f_2 + m^2, \ldots) \in \prod_{i=1}^{\infty} R/m^i$$

such that

$$f_{i+1} + m^i = f_i + m^i, i = 1, 2, \ldots$$

It is a ring with component-wise addition and multiplication. The projective limit is called the **completion of** R with respect to $\mathfrak{m}$ and is denoted by $\hat{R}$. It is a local ring with the maximal ideal $\mathfrak{m}\hat{R}$. The completion of R with respect to a maximal ideal $\mathfrak{m}$ coincides with the completion of the local ring $R_\mathfrak{m}$ with respect to the maximal ideal $\mathfrak{m}R_\mathfrak{m}$. The completion of the polynomial ring

$$K[x_1, x_2, \ldots, x_n]$$

with respect to the maximal ideal $(x_1, x_2, \ldots, x_n)$ can be identified as the formal power series ring $K[[x_1, x_2, \ldots, x_n]]$. There exists the natural homomorphism $R \to \hat{R}$ defined by

$$f \to (f + \mathfrak{m}^i)_{i=1,2,\ldots}.$$

If $R = \hat{R}$, R is said to be a **complete local ring**. Suppose that $(A, \mathfrak{m}, K)$ is an Artinian local ring. Then high powers of $\mathfrak{m}$ are 0. Thus trivially we have

$$\varprojlim A/\mathfrak{m}^i = A.$$

Thus all Artinian local rings are complete. In the next subsection we review what the complete structure theorem says when it is applied to Artinian local rings.

2.2.1 Application of the Structure Theorem

Following is a statement of Cohen's structure theorem of complete local rings. For completeness, we give a definition of a discrete valuation ring. A local ring $(S, \mathfrak{n})$ is a **discrete valuation ring** if it is an integral domain and $\mathfrak{n}$ is generated by a single element. If $(S, \mathfrak{n})$ is a discrete valuation ring, $\mathfrak{n}$ is the only prime ideal of S apart from $\{0\}$.

Theorem 2.16. *Suppose that* $(R, \mathfrak{m}, K)$ *is a complete local ring.*

1. *If R contains a field, then R contains a field K which is mapped isomorphically to the residue field $R/\mathfrak{m}$ by the composition map*

$$K \to R \to K.$$

   Moreover R is a homomorphic image of the formal power series ring

$$K[[x_1, x_2, \ldots, x_n]]$$

   such that the images of $x_1, x_2, \ldots, x_n$ generate $\mathfrak{m}$. So in particular R is Noetherian.

2. *Even if R does not contain any field, R contains a subring which is the image of a discrete valuation ring $(S, pS, S/pS)$ such that S/pS is isomorphic to $K = R/\mathfrak{m}$. Moreover R is a homomorphic image of the formal power series ring $S[[x_1, x_2, \ldots, x_n]]$ such that $\mathfrak{m}$ is generated by the images of $p, x_1, x_2, \ldots, x_n$. (Here p is in fact a prime number considered as an element of S.)*

(We do not give a proof for this theorem. The interested reader may consult e.g. Matsumura [93].)

We would like to see what the theorem says for Artinian local rings. Any subfield K of R as in Theorem 2.16 (1) of the theorem is called a **coefficient field** of R. Let $(A, \mathfrak{m}, K)$ be an Artinian local ring. Suppose for a moment that Char $K = p \neq 0$. Then $\mathfrak{m}$ contains $p = \underbrace{1 + 1 + \cdots + 1}_{p}$, but since a high power of $\mathfrak{m}$ is 0, this means that $p^m = 0$ in R for some m. It is not difficult to see that if Char $K = p > 0$, then Char A is a power of p. (We omit the details.)

Next we apply the structure theorem to an Artinian local ring $(A, \mathfrak{m}, K)$. Let $R \to A \to 0$ be a surjection with R a regular local ring. There are three cases:

1. Char R = Char K = 0. In this case R contains $\mathbb{Q}$; and hence contains a coefficient field.
2. Char R = Char K = p. In this case R contains $\mathbb{Z}/p\mathbb{Z}$; and hence contains a coefficient field.
3. Char $R = p^m >$ Char $K = p$. In this case R does not contain a field.

In this book we are mostly interested in the case (1) but at times (2) appears. We do not treat the case (3).

Let S be either a field or a discrete valuation ring, $(A, \mathfrak{m}, K)$ an Artinian local ring and let

$$\phi: S[[x_1, x_2, \ldots, x_n]] \to A$$

be a surjective homomorphism such that $\phi(x_1), \phi(x_2), \ldots, \phi(x_n) \in \mathfrak{m}$. Then since a high power of $\mathfrak{m}$ is 0, ϕ factors through

$$\phi: S[[x_1, x_2, \ldots, x_n]]/(x_1, x_2, \ldots, x_n)^m \to A$$

for some m.

In particular Ker ϕ contains high power of the variables, and hence we may think of Ker ϕ as consisting of polynomials (rather than power series). Thus we can state this as a structure theorem for Artinian local rings.

Theorem 2.17. *Let $(A, \mathfrak{m}, K)$ be an Artinian local ring. Then there exists a field or a discrete valuation ring S with pS the maximal ideal, and a positive integer m such that A is a homomorphic image of*

$$S[x_1, x_2, \ldots, x_n]/(p^m, x_1^m, x_2^m, \ldots, x_n^m)$$

and the images of $p, x_1, x_2, \ldots, x_n$ *generate* $\mathfrak{m}$. *It should be understood if* $p = 0$ *then* S *is a field and otherwise* $p = \operatorname{Char} K > 0$. *Moreover* S *is a field if and only if* $\operatorname{Char} R = \operatorname{Char} K$. *In the case* $\operatorname{Char} R \neq \operatorname{Char} K$, *the generator of the maximal ideal of* S *is* $p = \operatorname{Char} K$.

As a corollary we have the following

Corollary 2.18. *1. Any Artinian local ring is a homomorphic image of a complete regular local ring.*
2. Any Artinian local ring is a homomorphic image of an Artinian complete intersection local ring.

2.2.2   Injective Modules over Commutative Noetherian Rings

Let R be a Noetherian ring. Any R-module M can be embedded in an injective module. The "smallest injective module" that contains M is called the **injective hull** (or also **injective envelope**) of M and is denoted by $E(M)$. To be precise, $E(M)$ is characterized by the following properties.

1. $E(M)$ is an injective R-module.
2. $M \subset E(M)$.
3. If $N \subset E(M)$ is any non-trivial R-module, then $N \cap M \neq 0$.

It is known that there is a one-to-one correspondence between the set of prime ideals of R and the set of indecomposable injective modules given by $\mathfrak{p} \leftrightarrow E(R/\mathfrak{p})$. For details see Matlis [92].

Example 2.19. Consider $\mathbb{Z}$ as a Noetherian ring. Then

$$E(\mathbb{Z}/(\mathfrak{p})) = \begin{cases} \mathbb{Q} & \text{if } \mathfrak{p} = (0), \\ \mathbb{Q}/\mathbb{Z}_{(p)} & \text{if } \mathfrak{p} = (p). \end{cases}$$

Here if $p \in \mathbb{Z}$ is a prime, $\mathbb{Z}_{(p)}$ denotes $\mathbb{Z}\left[\frac{a}{b} \,\middle|\, b \notin (p)\right]$.

Example 2.20. Let $R = K[[x]]$ be the formal power series ring in one variable over a field K. Then

$$E(R/(\mathfrak{p})) = \begin{cases} R[x^{-1}] & \text{if } \mathfrak{p} = (0), \\ R[x^{-1}]/R & \text{if } \mathfrak{p} = (x). \end{cases}$$

The injective hull $E(R/\mathfrak{m})$ of the residue field plays in important role in the theory of Gorenstein rings. For a description of $E(R/\mathfrak{m})$ for the local ring $R = K[x_1, x_2, \ldots, x_n]_{(x_1, x_2, \ldots, x_n)}$ see Sect. 2.4.1.

2.2.3 Gorenstein Local Rings and Cohen–Macaulay Rings

We review some basic facts about Cohen–Macaulay rings and Gorenstein local rings.

Definition 2.21. Noetherian local ring $(R, \mathfrak{m}, K)$ of dimension n is a **Cohen–Macaulay** ring if there exists a system of parameters $a_1, a_2, \ldots, a_n$ such that

$$(a_1, a_2, \ldots, a_{k-1}) : a_k = (a_1, a_2, \ldots, a_{k-1})$$

for $i = 1, 2, \ldots, n$.

The condition in the definition is the same as the claim that the induced map

$$\times a_i \colon R/(a_1, a_2, \ldots, a_{i-1}) \to R/(a_1, a_2, \ldots, a_{i-1})$$

defined by $x \mapsto a_i x$ is injective. Thus, the definition can be rephrased as follows: *A Noetherian local ring R is Cohen–Macaulay if there is a system of parameters which is a regular sequence.* It is remarkable that if R is Cohen–Macaulay then every system of parameters is a regular sequence. (The interested reader may consult Matsumura [93, p. 135, Theorem 17.4].)

Definition 2.22. A local ring R is **Gorenstein** if it has a finite injective resolution as a module over itself.

This is the simplest definition for Gorenstein local ring. However, many other equivalent definitions are possible and it seems important to understand them all as a whole.

Since we are interested mostly in Artinian Gorenstein rings, we do not go into details of the general theory of Gorenstein rings, and we confine ourselves to giving here some basic properties of Gorenstein local rings without proof. (Interested readers may consult Bass [5] and Matsumura [93].)

Theorem 2.23. *Let $(R, \mathfrak{m}, K)$ be a Noetherian commutative local ring. Then the following conditions are equivalent.*

1. *R is Gorenstein.*
2. *R is a Cohen–Macaulay ring in which some system of parameters generates an irreducible ideal.*
3. *R is a Cohen–Macaulay ring in which every system of parameters generates an irreducible ideal.*
4. *If*

$$0 \to R \to E_0 \to E_1 \to \cdots \to E_h \to \cdots$$

is a minimal injective resolution, then

$$E_h = \bigoplus_{\mathrm{ht}(\mathfrak{p})=h} E(R/\mathfrak{p}).$$

The sum is over all prime ideals $\mathfrak{p}$ of R of height h, where the height of $\mathfrak{p}$ is:

$$\mathrm{ht}(\mathfrak{p}) = (\text{Krull dim } R) - (\text{Krull dim } R/\mathfrak{p}).$$

Remark 2.24. Theorem 2.23 implies the following:

$$\text{Regular local} \implies \text{Complete intersection} \implies \text{Gorenstein.}$$

The following is very important for us to characterize Artinian Gorenstein rings as quotients of a regular local ring in terms of a minimal free resolution.

Theorem 2.25. *Let $(R, \mathfrak{m}, K)$ be a Noetherian local ring of Krull dimension n and let E be the injective hull of K. Then the following conditions are equivalent.*

1. *R is Gorenstein.*
2. *R is Cohen–Macaulay and there exists a canonical isomorphism $\mathrm{Ext}_R^n(M, R) \cong \mathrm{Hom}_R(M, E)$ for any R-module M of finite length.*

Proof. We outline a proof. Details are found in Bass [5] or Matsumura [93].
Assume that inj. dim $R < \infty$. Let

$$0 \to R \to E_0 \to E_1 \to \cdots \to E_{n'} \to 0$$

be the minimal injective resolution of A. Then it is can be shown that $n' = n$ and moreover

$$E_h = \bigoplus_{\mathrm{ht}(\mathfrak{p})=h} E(R/\mathfrak{p}).$$

In particular $E_n = E$ is the injective hull of K. (See Theorem 2.23. This is a striking fact in the theory of Gorenstein ring.) Since $\mathrm{Ext}_R^n(M, R)$ can be computed by applying the functor $\mathrm{Hom}_R(M, -)$ to an injective resolution of R, we have the natural isomorphism

$$\mathrm{Ext}_R^n(M, R) \cong \mathrm{Hom}_R(M, E).$$

Thus we have proved (1) $\implies$ (2). The converse is true if $n = 0$. Indeed the isomorphism $\mathrm{Ext}_R^n(M, R) \cong \mathrm{Hom}_R(M, E)$ is nothing but $\mathrm{Hom}_R(M, R) \cong \mathrm{Hom}_R(M, E)$. So if we let $M = R$, it says $R = E$. Thus inj. dim$_R R = 0 < \infty$.

Assume $n > 0$. Let x be a non-zero divisor of R contained in the annihilator of M. Then it is not difficult to prove that

$$\text{Ext}_R^n(M, R) \cong \text{Ext}_{R/(x)}^{n-1}(M, R/(x)).$$

On the other hand $\text{Hom}_R(M, E)$ is the same as $\text{Hom}_{R/(x)}(M, \text{Hom}_R(R/(x), E))$. This proves $R/(x)$ is Gorenstein with x a non-zero divisor. Thus R is Gorenstein. $\square$

Let $(R, \mathfrak{m}, K)$ be a complete local ring and let E be the injective hull of K. Put $T(-) = \text{Hom}_R(-, E)$, where $T(-)$ is regarded as a functor. Then for a finitely generated R-module M, we have a canonical isomorphism

$$M \cong T(T(M)).$$

This is known as **Matlis duality**, which is a consequence of the following:

Theorem 2.26 (Matlis [92]). *Suppose that $(R, \mathfrak{m}, K)$ is a complete local ring and let E be the injective hull of K. Then*

$$\text{Hom}_R(E, E) \cong R.$$

Note that any module of finite length over a regular local ring has finite projective dimension which is equal to the Krull dimension of R. (See for example Matsumura [93, Theorem 19.2] and Bruns–Herzog [12, Theorem 1.3.3].)

Proposition 2.27. *Suppose that $(R, \mathfrak{m}, K)$ is a regular local ring of Krull dimension n and $I \subset R$ is an ideal of finite colength. Let*

$$0 \to F_n \to F_{n-1} \to \cdots \to F_1 \to F_0 \to R/I \to 0$$

be a minimal free resolution of R/I. Then we have

1. *The rank of F_1 is equal to $\mu(I)$, and*
2. *The rank of F_n is equal to $\tau(I)$,*

where $\tau(I)$ is the type of I (see Definition 2.12) and $\mu(I)$ the minimal number of generators for I as an ideal.

Proof. If $n = 0$, R is a field and there is nothing to prove. If $n = 1$, R is a discrete valuation ring and any ideal is principal and irreducible. Thus minimal free resolution for $R/(x)$ is simply $0 \to R \to R \to R/(x) \to 0$. Hence the assertion is clear. We assume that $n \geq 2$.

(1) We have the exact sequence

$$F_2 \to F_1 \to I \to 0$$

as R-modules. We apply the tensor product $- \otimes_R R/\mathfrak{m}$ and get

$$F_2 \otimes_R R/\mathfrak{m} \to F_1 \otimes_R R/\mathfrak{m} \to I/\mathfrak{m}I \to 0.$$

Since the first map is a zero map, we have $\mu(I) = \dim_K I/\mathfrak{m}I$ is equal to rank F_1.

(2) Recall that $\mathrm{Ext}_R^n(R/I, R) \to 0$ can be obtained by applying $\mathrm{Hom}_R(-, R)$ to a free resolution of R/I. Let $\phi \colon F_n \to F_{n-1}$ be the last homomorphism in the minimal free resolution of R/I. Then

$$F_{n-1}^* \to F_n^* \to \mathrm{Ext}_R^n(R/I, R) \to 0 \qquad (2.1)$$

is exact, where the first map is the dual to ϕ. We apply the exact functor $\mathrm{Hom}_R(-, E)$ to the exact sequence (2.1). Then we have the exact sequence:

$$0 \to \mathrm{Hom}_R(\mathrm{Ext}_R^n(R/I, R), E) \to E^{\beta_n} \to E^{\beta_{n-1}}, \qquad (2.2)$$

where $\beta_i = \mathrm{rank}\ F_i$. By Theorem 2.25, we have

$$\mathrm{Ext}_R^n(R/I, R) \cong \mathrm{Hom}_R(R/I, E),$$

since R is Gorenstein. Thus (2.2) is the same as

$$0 \to R/I \to E^{\beta_n} \to E^{\beta_{n-1}}. \qquad (2.3)$$

Hence applying $\mathrm{Hom}_R(K, -)$ to (2.2) or (2.3), we get

$$\mathrm{Hom}_R(K, R/I) \cong K^{\beta_n},$$

since $\mathrm{Hom}_R(K, E) \cong K$. $\square$

This is one of the most striking results of homological algebra and was established in the early 1970s. It explains in what sense the type of an ideal of finite colength is dual to the minimal number of generators.

2.3   Ideals of Finite Colength and Artinian Local Rings

Suppose that

$$0 \to I \to R \to A \to 0$$

is an exact sequence, where $(R, \mathfrak{m}, K)$ a regular local ring and I an ideal of R. Generally speaking I is easier to deal with than the quotient $A = R/I$ and to get information about A it is necessary to look at I, but some properties of I can be understood by studying the quotient ring A. The number of generators $\mu(I)$ becomes obscure in A, but in fact $\mu(I)$ is equal to the first **deviation** $\epsilon_1(A)$, provided that $I \subset \mathfrak{m}^2$. The condition that $I \subset \mathfrak{m}^2$ has to do with minimality of the number of generators of I or the presentation of A as R/I. See Bruns–Herzog [12, §2.3 pp. 72–73]. (We do not use this number $\epsilon_1(A)$, but interested readers may consult [12], 2.3.2(b).) Information about an ideal basis of I is not readily available in R/I, but it is visible in the quotient ring $R/\mathfrak{m}I$. Roughly speaking we may regard the ring $R/\mathfrak{m}I$ as $R/\mathfrak{m}I = R/I \sqcup I/\mathfrak{m}I$ in the sense that the ring $R/\mathfrak{m}I$ is R/I with additional information concerning a generating set for I. (See e.g., Corollary 2.31.) Thus by studying Artinian rings in general we are studying $\mathfrak{m}$-primary ideals in regular local rings also.

2.3.1 Dilworth Number and Rees Number of Artinian Local Rings

In 1983, David Rees asked the following question:

Question 2.28. For which ideals I is it true that

$$J \supset I \implies \mu(J) \le \mu(I).$$

A related question is:

Question 2.29. For which ideals I is it true that

$$J \supsetneq I \implies \mu(J) < \mu(I).$$

The following was first proved by G. Trung [143] as early as in 1984. It was proved in Watanabe [147] independently.

Proposition 2.30. *Let $(A, \mathfrak{m}, K)$ be an Artinian local ring. For any ideal $I \subset \mathfrak{m}$ and for any element $y \in \mathfrak{m}$ we have*

$$\mu(I) \le \text{length}(A/yA) \ (= \dim_K(A/yA) \ \text{in the graded case}).$$

Proof (Ikeda [66]). One looks at the exact sequence

$$0 \to (0 :_I y) \to I \xrightarrow{\cdot y} I \to I/yI \to 0.$$

Taking Euler characteristics shows

$$\dim_K(0 :_I y) = \dim_K(I/yI), \tag{2.4}$$

so one gets

$$\mu(I) = \dim I/\mathfrak{m}I \leq \text{length } I/yI = \text{length}(0 :_I y)$$
$$\leq \text{length}(0 :_A y) = \text{length } A/yA.$$

(The first "$\leq$" because $y \in \mathfrak{m}$, and the second "$\leq$" because $I \subset A$.) The last equality following from (2.4) after setting $I = A$.   □

Corollary 2.31. *Let* $(R, \mathfrak{m}, K)$ *be a local ring. For any* $\mathfrak{m}$*-primary ideal* $I \subset \mathfrak{m}$ *and for any element* $y \in \mathfrak{m}$ *we have*

$$\mu(I) \leq \text{length}(R/\mathfrak{m}I + yR).$$

Proof. Put $A = R/\mathfrak{m}I$. Then since $\mu(I) = \dim I/\mathfrak{m}I$, we have

$$\mu(I) = \mu(IA).$$

Thus by Proposition 2.30, we have

$$\mu(I) = \mu(IA) \leq \text{length}(A/yA) = \text{length}(R/\mathfrak{m}I + yR).$$ □

The inequality in Proposition 2.30 leads us to consider the two numbers: $\max\{\mu(I) \mid I \subset A\}$ and $\min\{\text{length}(A/yA) \mid y \in \mathfrak{m}\}$.

Definition 2.32. Suppose that $(A, \mathfrak{m}, K)$ is an Artinian local ring.

1. We call

$$\max\{\mu(I) \mid \text{ideal } I \subset A\}$$

the **Dilworth number** of A and denoted it by $d(A)$.
2. We call

$$\min\{\text{length}(A/lA) \mid l \in \mathfrak{m}\}$$

the **Rees number** of A and denote it by $r(A)$.

Proposition 2.33. *Let* $(A, \mathfrak{m})$ *be an Artinian local ring. Then*

$$d(A) \leq r(A).$$

If there are an ideal $\mathfrak{a} \subset A$ *and an element* $y \in \mathfrak{m}$ *such that*

$$\mu(\mathfrak{a}) = \text{length } A/yA,$$

then $\mu(\mathfrak{a}) = d(A)$ *and* $\text{length}(A/yA) = r(A)$.

Proof. The first part follows immediately from Proposition 2.30. The second part
follows from the first part.   □

Before we proceed, we need to make a remark on the terminology "general
element" for an Artinian local ring.

Remark 2.34. In many situations in the theory of commutative rings it is necessary
to make use of an element which is sufficiently general in some sense depending
on the situation involved. In other words we need to define a "general element"
of an Artinian local ring. For an Artinian local ring A, it seems natural to define
a general element $g \in A$ to be an element which makes length(A/gA) least
possible. However, this contains a subtle problem and will be discussed in Chap. 5.
(See Definitions 5.4 and 5.7 for a weak Rees element and a strong Rees element
of A.) Throughout this book we need to use the term "general element" only for
Artinian local K-algebras, where K is a field. In this case a general element can be
explained as follows. Suppose that $(A, \mathfrak{m})$ is an Artinian local K-algebra such that
the composition of maps $K \to A \to A/\mathfrak{m}$ induces an isomorphism $K \cong A/\mathfrak{m}$.
Let $n = \mu(\mathfrak{m})$ and introduce new variables

$$X_1, X_2, \ldots, X_n.$$

Let $K' = K(X_1, \ldots, X_n)$ be the rational function field and $A' = A \otimes_K K'$. Choose
a set of generators $\mathfrak{m} = (m_1, \ldots, m_n)$ and let $x_1, x_2, \ldots, x_n$ be any elements in K
or in A, and $Y = X_1 m_1 + X_2 m_2 + \cdots + X_n m_n$ and $y = x_1 m_1 + x_2 m_2 + \cdots + x_n m_n$.
Note $Y \in A'$ and $y \in A$. Then it can be proved (Sect. 5.2, Theorem 5.3) that

$$\text{length}_{A'}(A'/YA') \leq \text{length}_A(A/yA). \tag{2.5}$$

In [147] $y \in \mathfrak{m}$ was defined to be a general element of A if equality holds in (2.5).
This is equivalent to saying that y is a "weak Rees element" as defined in Sect. 5.2.
For most purposes of this book the term "general element" is used in this sense.
For example if the equality (2.5) holds then the Rees number $r(A)$ of A is equal
to length(A/yA). However depending on the situation the element just satisfying
the equality (2.5) may not be "general enough". In such a case we are to choose
coefficients $(x_1, \ldots, x_n)$ of y suitably so that the element y meets the requirement
of the purpose. Most likely the situation is not affected if A is replaced by A'. Then
definitely $Y \in A'$ should be a general element for A', and if $y \in A$ plays the same
role as Y for A', then y should be a general enough element. From the viewpoint of
the theory of commutative algebra, probably it is best to define a general element as
a "strong Rees element" as defined in Sect. 5.2, Definition 5.7. However we prefer
not to give a precise definition to "general element" but to use it in preference to
"strong or weak Rees element" in an appeal to intuition. The same remark applies
to "general linear form." Here are some examples.

1. Let K be any field and $A = K[x_1, x_2, \ldots, x_n]/I$, where I is an ideal (of finite
   colength) generated by monomials in $x_1, x_2, \ldots, x_n$. Then it can be proved that

$g := x_1 + x_2 + \cdots + x_n$ is a strong Rees element. So, g may be called a general element of A. In fact it is "general enough" for most purposes.

2. Let $A = \mathbb{Q}[x, y]/(x^2, y^2)$ and set $g = x$. Then it is easy to see that

$$\text{length}(A/gA) = \min\{\,\text{length}(A/zA) \mid z : \text{non-unit element}\,\}.$$

Thus $r(A) = \text{length } A/gA = 2$ and g is a weak Rees element, but it is not a strong Rees element. In this sense g may not be general enough.

3. Let $K = \mathbb{Z}/2\mathbb{Z}$ and $A = K[x, y]/((x^2y + xy^2)A + (x, y)^4)$. Then for any element $g \in \mathfrak{m} \setminus \mathfrak{m}^2$, we have $\text{length}(A/gA) = r(A) = 4$, but no weak/strong Rees elements exist in A. The point is that if the residue field is finite, then no "strong Rees elements" may exist in the ring, but they exist only in a faithfully flat extension of the ring. Replace K by the function field $K' := K(t)$, and let $A' := K' \otimes_K A$. Then $g' = x + ty$ is a strong Rees element of A'. We have $r(A') = 3$. We should say that general enough elements do not exist in A.

Proposition 2.35. *Suppose that A is a graded Artinian ring. Then $d(A)$ coincides with*

$$\max\{\,\mu(I) \mid \text{homogeneous ideal } I \subset A\,\}.$$

Proof. Suppose that $d(A) = m$ and $I = (f_1, f_2, \ldots, f_m)$. We have to show that there exists an ideal generated by m homogeneous elements. Write f° for the homogeneous part of f of the highest degree. If $\{\,f_1^\circ, f_2^\circ, \ldots, f_m^\circ\,\}$ is a minimal generating set we are done. Otherwise suppose that $f_1^\circ \in (f_2^\circ, f_3^\circ, \ldots, f_m^\circ)$ for example. Then we may write

$$f_1^\circ = \sum_{j=2}^{m} h_j f_j^\circ,$$

with homogeneous elements h_j. Put

$$f_1' = f_1 - \sum_{j=2}^{m} h_j f_j.$$

Then we have

$$(f_1, f_2, f_3, \ldots, f_m) = (f_1', f_2, f_3, \ldots, f_m),$$

and $\deg f_1' < \deg f_1$ Hence we may induct on $\sum_{i=1}^{m} \deg f_i$ to obtain an ideal generator $\{\,f_i\,\}$ for I such that

$$\mu(f_1^\circ, f_2^\circ, \ldots, f_m^\circ) = m. \qquad \square$$

2.3.2 Monomial Artinian Rings and the Dilworth Number

Let K be an arbitrary field and let $R = K[x_1, x_2, \ldots, x_n]$ be the polynomial ring in the variables $x_1, x_2, \ldots, x_n$. We let $\deg x_i = 1$ for all $i = 1, 2, \ldots, n$. The degree makes the polynomial ring a graded ring:

$$R = \bigoplus_{i=0}^{\infty} R_i.$$

An ideal $I \subset R$ is a **homogeneous ideal** if I is generated by homogeneous elements. I is a **monomial ideal** if I is generated by monomials.

Put $\mathfrak{m} = (x_1, x_2, \ldots, x_n)$, which is the unique homogeneous maximal ideal of R. We denote by $R_\mathfrak{m}$ the polynomial ring R localized at the maximal ideal $\mathfrak{m}$. Suppose that I is a homogeneous ideal with homogeneous generators:

$$f_1, f_2, \ldots, f_m.$$

Then it is a minimal generating set for I if and only if it is a minimal generating set for $IR_\mathfrak{m}$. Also one notices that I is irreducible if and only if $IR_\mathfrak{m}$ is irreducible. For homogeneous ideals in R, we use the same notation $\mu(I)$ and $\tau(I)$ as set out in the Definitions 2.5 and 2.12. Namely,

$$\mu(I) = \dim I/\mathfrak{m}I = \mu(IR_\mathfrak{m}),$$

and

$$\tau(I) = \dim(I : \mathfrak{m})/I = \tau((I : \mathfrak{m})R_\mathfrak{m}/IR_\mathfrak{m}).$$

This means that a minimal generating set for I consisting of homogeneous elements contains a minimal generating set for the ideal and the number of elements in it is equal to $\mu(I)$, and similarly if $I = \mathfrak{q}_1 \cap \mathfrak{q}_2 \cap \cdots \cap \mathfrak{q}_t$ is an intersection of homogeneous irreducible ideals, we may delete superfluous components so that it is an irredundant intersection with $t = \tau(I)$ components.

Suppose that I is a monomial ideal in R. One sees easily that

1. I is $\mathfrak{m}$-primary (i.e., of finite colength) if and only if I contains a power of x_i for all $i = 1, 2, \ldots, n$.
2. An ideal I is irreducible if and only if it is generated by powers of the variables. (Proof is left to the reader.)

Notice that the set of all monomials in R forms a poset with divisibility as an order, and if I is a monomial ideal and f, g are monomials in R, then

$$f \in I, f \text{ divides } g \implies g \in I.$$

Hence the set of "minimal" monomials in I is a minimal generating set for the ideal I. Furthermore the monomials not in I form a poset with divisibility as the order. This motivates the following definition.

Definition 2.36. Let $R = K[x_1, x_2, \ldots, x_n]$ be the polynomial ring over a field K and $I \subset R$ a monomial ideal. We denote by $P(R/I)$ the poset consisting of the monomials not in I. Namely

$$P(R/I) = \{ \text{monomial } f \in R \mid f \notin I \}.$$

A monomial in $P(R/I)$ is called a **standard monomial** for I.

Put $A = R/I$. The set $P(A)$ is a K-basis for the ring A. If A is Artinian, $P(A)$ is a finite poset. The following shows that the Dilworth number $d(A)$ of the Artinian ring A coincides with the Dilworth number of $P(A)$ in the original sense.

Theorem 2.37. *As above let R be a polynomial ring and let $I \subset R$ be an ideal of finite colength generated by monomials. Let $A = R/I$ and $P = P(A)$. Then we have $d(A) = d(P(A))$.*

Proof. An antichain in P is a minimal generating set of an ideal of A. Hence $d(A) \geq d(P)$. Suppose that $I = (f_1, f_2, \ldots, f_m)$ with $m = \mu(I)$. By Proposition 2.35 we may assume that $f_1, f_2, \ldots, f_m$ are homogeneous. For $f \in A$ we may write f uniquely as a sum of monomials in P. Let f° denote the initial monomial that occurs in f with respect to the lexicographic order (or any term order) and let $I^\circ = (f^\circ \mid f \in I)$. Then one sees easily that $\mu(I) \leq \mu(I^\circ)$. This show that $d(P) \geq d(A)$. $\square$

Next we show what posets are possible for $P(A)$.

2.3.3   Poset of Standard Monomials as a Basis for Monomial Artinian Rings

As before let I be an $\mathfrak{m}$-primary ideal generated by monomials in a polynomial ring. We know that I is irreducible if and only if

$$I = (x_1^{e_1+1}, x_2^{e_2+1}, \ldots, x_n^{e_n+1})$$

for some non-negative integers $(e_1, e_2, \ldots, e_n) \in \mathbb{Z}_+^n$. Note that (with I as above)

$$(I : \mathfrak{m}) = I + (x_1^{e_1} x_2^{e_2} \cdots x_n^{e_n})$$

and

$$\mu(I : \mathfrak{m}) = n + 1.$$

Let $f \in R$ be a monomial. Then one notices easily that

$$f \notin I \iff f \text{ divides } x_1^{e_1} x_2^{e_2} \cdots x_n^{e_n}.$$

Choose n distinct prime numbers $p_1, p_2, \ldots, p_n$ in $\mathbb{Z}_+$ and put

$$\nu = p_1^{e_1} p_2^{e_2} \cdots p_n^{e_n}.$$

Then we have

$$P(R/I) \cong \mathscr{L}(\nu),$$

where $\mathscr{L}(\nu)$ is the divisor lattice as defined in Sect. 1.4.2. This can be generalized to an arbitrary $\mathfrak{m}$-primary monomial ideal as follows.

Theorem 2.38. *Let R be a polynomial ring and $I \subset R$ a monomial ideal of finite colength. Let $t = \tau(I)$. Then there are t positive integers $\nu_1, \nu_2, \ldots, \nu_t$ such that*

$$P(A) \cong \mathscr{L}(\nu_1) \cup \mathscr{L}(\nu_2) \cup \cdots \cup \mathscr{L}(\nu_t).$$

Proof. Write I as the intersection of t monomial irreducible ideals $\mathfrak{q}_1, \mathfrak{q}_2, \ldots, \mathfrak{q}_t$. Then

$$P(A) := \{ \text{monomial } f \notin I \} = \bigcup_{i=1}^{t} P(R/\mathfrak{q}_i).$$

Fix n distinct prime numbers $p_1, p_2, \ldots, p_n$ and consider the map ϕ

$$\phi : \{ \text{monomials of } R \} \to \mathbb{Z}_+$$

defined by

$$x_1^{i_1} x_2^{i_2} \cdots x_n^{i_n} \mapsto p_1^{i_1} \cdots p_n^{i_n}.$$

ϕ is an injective homomorphism of posets. With this map each $P(R/\mathfrak{q}_i)$ is mapped onto $\mathscr{L}(\nu)$ for some integer ν. Thus $P(R/I)$ is mapped onto the union of such $\mathscr{L}(\nu)$. $\qquad\qquad\square$

2.3.4 The Sperner Property of Artinian Local Rings

Definition 2.39. Let $(A, \mathfrak{m})$ be an Artinian local ring. The **Sperner number** of A is defined by

$$\operatorname{Sperner}(A) := \max_k \left\{ \mu(\mathfrak{m}^k) \right\}.$$

We say that A has the **Sperner property** if $d(A) = \text{Sperner}(A)$.

So A has the Sperner property if among the ideals $I \subset A$ with $\mu(I)$ maximal one finds a power of the maximal ideal $\mathfrak{m}$.

Proposition 2.40. *Suppose that A is an Artinian local ring. If there exists an element $y \in \mathfrak{m}$ such that $\text{length}(A/yA) = \mu(\mathfrak{m}^r)$ for some r, then A has the Sperner property.*

Proof. By Proposition 2.30 we have

$$\mu(\mathfrak{m}^r) \leq d(A) \leq \text{length}(A/yA) = \mu(\mathfrak{m}^r). \qquad \square$$

Proposition 2.41. *Let R be the polynomial ring and I an $\mathfrak{m}$-primary monomial ideal of R. Put $A = R/I$. If $P(A)$ has the Sperner property as a poset, then A has the Sperner property.*

Proof. By Theorem 2.37 we have $d(A) = d(P(A))$. By Theorem 2.38 $P(A)$ is a ranked poset and if $P(A)_k$ is the k-th rank set, then $|P(A)_k| = \mu(\mathfrak{m}^k)$. Hence the assertion follows. $\qquad \square$

Let $R = K[x_1, x_2, \ldots, x_n]$ and I an $\mathfrak{m}$-primary monomial ideal of R. Let $A = R/I$ and let $P(A) = \bigsqcup_{i=0}^{c} P_i$ be a rank decomposition for A. For a subset $S \subset P_k$, we denote

$$A_1 S = \bigcup_{i=1}^{n} \{ x_i f \mid f \in S \} \subset A_{k+1}$$

and

$$\partial S = \bigcup_{i=1}^{n} \{ x_i^{-1} f \mid f \in S \cap (x_i) \} \subset A_{k-1}.$$

Proposition 2.42. *Let R be the polynomial ring and $I \subset R$ an $\mathfrak{m}$-primary monomial ideal. Put $A = R/I$ and $P = P(A)$. Let $P = \bigsqcup_{i=0}^{c} P_i$ is the rank decomposition of P. Suppose that the rank function of P is unimodal and assume that*

1. if $|P_k| \leq |P_{k+1}|$, then $|S| \leq |A_1 S|$ for any subset $S \subset P_k$.
2. if $|P_k| \geq |P_{k+1}|$, then $|\partial S| \geq |S|$ for any subset $S \subset P_{k+1}$.

Then P has the Sperner property.

Proof. Immediate by Corollary 1.31. $\qquad \square$

Proposition 2.43. *Let R be the polynomial ring and I an $\mathfrak{m}$-primary monomial ideal. Let $A = R/I$ and $P = P(A)$. Let M be the incidence matrix for P. Then A has the Sperner property if and only if*

$$\max_k |P_k| = |P| - \beta(M).$$

Proof. By Theorem 2.37 we have $d(A) = d(P)$ and by Theorem 1.57 $d(P) = |P| - \beta(M)$. $\qquad\square$

Remark 2.44. 1. Suppose that $(R, \mathfrak{m}, K)$ is a local ring and I an ideal of R. Then

$$d(R/\mathfrak{m}I) \le d(R/I) + \mu(I).$$

To see this, let $\mathfrak{a}$ be an ideal of $R/\mathfrak{m}I$ such that $\mu(\mathfrak{a}) = d(R/\mathfrak{m}I)$. Let $\mathfrak{a}'$ be the natural image of $\mathfrak{a}$ in R/I. Let $r = \mu(\mathfrak{a}'), s = \mu(\mathfrak{a})$. Then one may choose generators $(a_1, a_2, \ldots, a_r, b_{r+1}, \ldots, b_s)$ of $\mathfrak{a}$ so that the natural images of $a_1, \ldots, a_r$ in R/I generate $\mathfrak{a}'$ and $b_{r+1}, \ldots, b_s \in I/\mathfrak{m}I$. The set of preimages of $b_{r+1}, \ldots, b_s$ in R can be a part of a minimal generating set for I. So $s - r \le \mu(I)$. Hence the assertion follows.

2. Let $R = K[x, y, z]$ be the polynomial ring, $\mathfrak{m} = (x, y, z)$, where K is a field of arbitrary characteristic. Let $I = (x^3, y^3, z^3, xyz)$. Then we have

$$\max_i \{\mu(\mathfrak{m}^i)\} = d(R/I) = 6 < r(R/I) = 7,$$

and

$$\max_i \{\mu(\mathfrak{m}^i)\} = d(R/\mathfrak{m}I) = 10 = r(R/\mathfrak{m}I).$$

Both R/I and $R/\mathfrak{m}I$ have the Sperner property. To see this, let $P := P(R/I)$ be the poset of the standard monomials for R/I. (See Definition 2.36.) Let $P = \bigsqcup_{k=0}^s P_k$ be the rank decomposition. Then it is easy to see that $s = 4$, and the rank numbers are $|P_0| = 1, |P_1| = 3, |P_2| = 6, |P_3| = 6, |P_4| = 3$. Let $M^{(k)}$ be the incidence matrices for $B_k(P)$. (See Definition 1.48.) Then it is easy to see that all $M^{(k)}$ contain full matchings. Hence P has the Sperner property by Theorem 1.31 and we have $d(R/I) = 6$. On the other hand it turns out that the 6×6 matrix $M^{(3)}$ has rank 5, while $M^{(1)}, M^{(2)}$, and $M^{(4)}$ have full rank. Let $L = x + y + z$. The linear form L induces a homomorphism $\times L : R/I \to R/I$ by multiplication: $f \mapsto fL$. The matrices for the graded homogeneous components

$$\times L : (R/I)_{k-1} \to (R/I)_k$$

may be regarded as the same as $M^{(k)}$ if we choose P as a basis. Hence we have that $\dim R/(I + LR) = |P| - \mathrm{rank}(\times L) = 19 - 12 = 7$. It is not difficult to see that L is sufficiently general for the monomial ideal. Thus we have $r(R/I) = 7$. For the poset $P := P(R/\mathfrak{m}I)$, the rank numbers $|P_k|$, for $k = 0, 1, \ldots, 4$, are $(1, 3, 6, 10, 3)$, and one sees that the incidence matrices $M^{(k)}$ have full rank (hence contain full matchings) for $k = 1, \ldots, 4$. (Cf. Theorem 1.31 and Lemma 1.51.) Hence $d(P/\mathfrak{m}I) = r(P/\mathfrak{m}I) = 10$. Generally speaking $r(A)$ depends on the characteristic of K. In this example, it happens that $r(R/P)$ does not depend on the characteristic. In characteristic zero, we have not found any counter examples for the equality $d(A) = r(A)$ if A is an Artinian local ring of the form $A = R/\mathfrak{m}I$ for some ideal I in the polynomial ring R.

3. Let $R = K[u, v, x, y, z]$ be the polynomial ring, K a field of any characteristic. Put

$$I = (u, v)^3 + (x, y, z)^2 + (vx, uz, uy - vz, ux - vy)$$

Then we have $\max_i \{ \mu(\mathfrak{m}^i) \} = 5 < d(R/I) = 6 = r(R/I)$, so R/I does not have the Sperner property. To see this, put $A = R/I$. It is easy to check that the ideal $\mathfrak{a}$ of A generated by the images of u^2, uv, v^2, x, y, z is minimally generated by these six elements. Hence $d(A) \geq 6$. Hence A does not have the Sperner property. Let $L = u + v + x + y + z$. Then is not difficult to compute

$$\dim_K A/LA = \dim_K(R/I + LR) = 6.$$

Thus we have $6 \leq d(A) \leq r(A) \leq 6$. Hence we conclude $d(A) = r(A) = 6$. In this example also the integer $r(A)$ does not depend on the characteristic.

2.3.5 The Dilworth Lattice of Ideals

Definition 2.45. Let $(A, \mathfrak{m}, K)$ be an Artinian local ring. Define the family $\mathscr{F}(A)$ of ideals by

$$\mathscr{F}(A) = \{ \text{ideal } I \subset A \mid \mu(I) = d(A) \} .$$

If A is graded, we make an additional definition:

$$\mathscr{F}_H(A) = \{ \text{homogeneous ideal } I \subset A \mid \mu(I) = d(A) \} .$$

The following example shows that in most cases $\mathscr{F}(A)$ and even $\mathscr{F}_H(A)$ are not finite.

Example 2.46. Let K be a field and let $A = K[x, y, z]/(x^5, y^3, z^2)$. Then $d(A) = 6$. Put

$$I_{(a,b)} = (axy^2 + bxyz, y^2z) + (x, y, z)^4,$$

for $a, b \in K$. It is easy to see that $\mu(I_{(a,b)}) = 6$, but for different $b \in K$ the ideals $I_{(1,b)}$ are all different. Thus $\mathscr{F}(A)$ is infinite provided that K is infinite.

The following was proved in [67], but it was preceded by Gunston [45].

Theorem 2.47. *Let A be an Artinian local ring. Then $\mathscr{F}(A)$ is a lattice with sum and intersection as its join and meet. (In the usual definition of a lattice operations $\wedge$ and $\vee$ are called the join and the meet. See e.g., Aigner [2].)*

Proof. Let $I, J \in \mathscr{F}(A)$. We have to show that $I \cap J \in \mathscr{F}(A)$ and $I + J \in \mathscr{F}(A)$. Consider the exact sequence:

$$0 \to I \cap J \to I \oplus J \to I + J \to 0,$$

where the third map is defined by $(i, j) \mapsto i + j$. This gives rise to the exact sequence

$$(I \cap J) \otimes_A A/\mathfrak{m} \to (I \oplus J) \otimes_A A/\mathfrak{m} \to (I + J) \otimes_A A/\mathfrak{m} \to 0.$$

So we have, by taking the Euler characteristic,

$$2d(A) = \mu(I) + \mu(J) \leq \mu(I \cap J) + \mu(I + J) \leq 2d(A).$$

Hence

$$\mu(I \cap J) + \mu(I + J) = 2d(A).$$

In addition $\mu(I \cap J) \leq d(A)$ and $\mu(I + J) \leq d(A)$, so we must have

$$\mu(I \cap J) = \mu(I + J) = d(A),$$

as desired. □

Corollary 2.48. *Let A be an Artinian local ring. Then $\mathscr{F}(A)$ has maximum and minimum members and they are unique.*

Proof. Since A is Artinian, the sum of all ideals in $\mathscr{F}(A)$ is in fact a finite sum, so by Theorem 2.47 it is in $\mathscr{F}(A)$. Hence $\mathscr{F}(A)$ has a unique maximum member. Similarly, an infinite intersection of ideals is in fact finite, namely the intersection of all members of $\mathscr{F}(A)$. Thus $\mathscr{F}(A)$ has a unique minimum member. □

Remark 2.49. If $A = \bigoplus_{i=0}^{c} A_i$ is a graded ring, A_0 a field, then $\mathscr{F}(A)$ contains a homogeneous ideal. (See Proposition 2.35.) Thus $\mathscr{F}_H(A)$ is a (non-empty) sublattice of $\mathscr{F}(A)$.

Theorem 2.50. *Suppose that $(A, \mathfrak{m}, K)$ is an Artinian local ring. Then $d(A)$ coincides with*

$$\max \{ \tau(I) \mid I \subset \mathfrak{m} \}.$$

Proof. This was proved in [146, Theorem 2.6]. Here we give another proof. Let $\mathfrak{a} \subset A$ be any ideal. We note that $(\mathfrak{m}\mathfrak{a} : \mathfrak{m}) \supset \mathfrak{a}$, so $(\mathfrak{m}\mathfrak{a} : \mathfrak{m})/\mathfrak{m}\mathfrak{a} \supset \mathfrak{a}/\mathfrak{m}\mathfrak{a}$.

This shows that $\tau(\mathfrak{m}\mathfrak{a}) \geq \mu(\mathfrak{a})$, and $\max\{\tau(I) \mid I \subset \mathfrak{m}\} \geq d(A)$. Similarly we have $\mathfrak{m}(\mathfrak{a} : \mathfrak{m}) \subset \mathfrak{a}$ for any ideal $\mathfrak{a}$. Hence there is a surjective map $\mathfrak{a} : \mathfrak{m}/(\mathfrak{m}(\mathfrak{a} : \mathfrak{m})) \to (\mathfrak{a} : \mathfrak{m})/\mathfrak{a}$. This shows that $\mu(\mathfrak{a} : \mathfrak{m}) \geq \tau(\mathfrak{a})$ and $\max\{\tau(I) \mid I \subset \mathfrak{m}\} \leq d(A)$. $\qquad\square$

Define the family $\mathscr{G}(A)$ of ideals by

$$\mathscr{G}(A) = \{\text{ideal } I \subset A \mid \tau(I) = d(A)\}.$$

Proposition 2.51. *Let $(A, \mathfrak{m}, K)$ be an Artinian local ring. Then $\mathscr{G}(A)$ is a lattice with sum and intersection as its join and meet.*

Proof. Let $I, J \in \mathscr{G}(A)$. We show that $I \cap J \in \mathscr{G}(A)$ and $I + J \in \mathscr{G}(A)$. Consider the exact sequence:

$$0 \to A/(I \cap J) \to A/I \oplus A/J \to A/(I + J) \to 0,$$

where the third map is defined by

$$(x \bmod I, y \bmod J) \mapsto x + y \bmod (I + J).$$

This gives rise to the exact sequence

$$0 \to \mathrm{Hom}_A(K, A/(I \cap J)) \to \mathrm{Hom}_A(K, A/I) \oplus \mathrm{Hom}_A(K, A/J)$$
$$\to \mathrm{Hom}_A(K, A/(I + J)).$$

So we have

$$\tau(I) + \tau(J) \leq \tau(I \cap J) + \tau(I + J),$$

but the last sum does not exceed $2d(A)$. Hence as for Theorem 2.47 we get

$$\mu(I \cap J) = \mu(I + J) = d(A),$$

as desired. $\qquad\square$

Theorem 2.52. *Let $(A, \mathfrak{m}, K)$ be an Artinian local ring. Then the correspondence*

$$\begin{cases} \mathscr{F}(A) \ni I \mapsto \mathfrak{m}I \in \mathscr{G}(A) \\ \mathscr{G}(A) \ni J \mapsto J : \mathfrak{m} \in \mathscr{F}(A) \end{cases}$$

is a lattice isomorphism between $\mathscr{F}(A)$ and $\mathscr{G}(A)$.

Proof. We divide the proof into four steps.

Step 1

$I \in \mathscr{F}(A) \implies \mathfrak{m}I \in \mathscr{G}(A)$. Suppose that $I \in \mathscr{F}(A)$. Then we have

$$\tau(\mathfrak{m}I) = \text{length}((\mathfrak{m}I : \mathfrak{m})/\mathfrak{m}I) \geq \text{length}(I/\mathfrak{m}I) = d(A).$$

Step 2

$J \in \mathscr{G}(A) \implies J : \mathfrak{m} \in \mathscr{F}(A)$. Suppose that $J \in \mathscr{G}(A)$. Then

$$\mu(J : \mathfrak{m}) = \text{length}((J : \mathfrak{m})/\mathfrak{m}(J : \mathfrak{m})) \geq \text{length}((J : \mathfrak{m})/J) = d(A).$$

Step 3

$\mathfrak{m}I : \mathfrak{m} = I$ for $I \in \mathscr{F}(A)$ and $\mathfrak{m}(J : \mathfrak{m}) = J$ for $J \in \mathscr{G}(A)$. These are immediate from the proof of Steps 1 and 2.

Step 4

For $I_1, I_2 \in \mathscr{F}(A)$, we have $\mathfrak{m}(I_1 + I_2) = \mathfrak{m}I_1 + \mathfrak{m}I_2$ and $\mathfrak{m}(I_1 \cap I_2) = \mathfrak{m}I_1 \cap \mathfrak{m}I_2$. The first assertion is obvious. The second follows from Step 3 which shows that the correspondence is one to one. □

2.3.6 $\mathfrak{m}$-*Full Ideals*

Proposition 2.53. *Let $(R, \mathfrak{m}, K)$ be a local ring, not necessarily Artinian. Suppose that I is an $\mathfrak{m}$-primary ideal. Then*

$$\max \{ \mu(J) \mid J \supset I \} = d(R/\mathfrak{m}I).$$

Proof. Let $\bar{\ } : R \to R/\mathfrak{m}I$ denote the natural surjection. If $J \supset I$, then

$$\mu(J) = \dim_K J/\mathfrak{m}J = \dim_K (J + \mathfrak{m}I)/(\mathfrak{m}J + \mathfrak{m}I) = \mu(\bar{J})$$

Thus we have

$$\max \{ \mu(J) \mid J \supset I \} \leq d(R/\mathfrak{m}I).$$

To show the converse let $\bar{J} = (\bar{b}_1, \bar{b}_2, \ldots, \bar{b}_m)$ with $m = d(R/\mathfrak{m}I)$. We claim that $\bar{J}$ contains $\bar{I}$. By way of contradiction assume that there is an element

$$\bar{a} \in \bar{I} \text{ and } \bar{a} \notin \bar{J}.$$

Then since

$$\mu((\bar{a}, \bar{b}_1, \bar{b}_2, \ldots, \bar{b}_m)) \leq m,$$

one of the $\bar{b}_i$, say, $\bar{b}_1$, is superfluous as a generator. Thus we can write

$$\bar{b}_1 = \alpha\bar{a} + \beta_2\bar{b}_2 + \beta_3\bar{b}_3 + \cdots + \beta_m\bar{b}_m,$$

where $\alpha, \beta_i \in R$. If α is a unit, then we have $\bar{a} \in \bar{J}$, a contradiction. Thus $\alpha \in \mathfrak{m}$, which says $\alpha\bar{a} = 0$ in $\bar{R}$, and this contradicts to the fact $\{\bar{b}_1, \bar{b}_2, \ldots, \bar{b}_m\}$ is a minimal generating set for $\bar{J}$. Put $J = (b_1, b_2, \ldots, b_m)$. Then

$$J + \mathfrak{m}I \supset I.$$

Hence by Nakayama's lemma $J \supset I$ with $\mu(J) = m$.   □

Definition 2.54. Let $(R, \mathfrak{m}, K)$ be a local ring.

1. An ideal I is $\mathfrak{m}$-**full** if there exists an element $y \in \mathfrak{m}$ such that $(\mathfrak{m}I : y) = I$.
2. An ideal I has the **Rees property** if

$$J \supset I \implies \mu(J) \leq \mu(I).$$

Proposition 2.55. *Let $(R, \mathfrak{m}, K)$ is a local ring. If I is $\mathfrak{m}$-primary and $\mathfrak{m}$-full, then I has the Rees property.*

Proof. We have the exact sequence

$$0 \to R/(\mathfrak{m}I : y) \to R/\mathfrak{m}I \to R/(\mathfrak{m}I + yR) \to 0.$$

Since $(\mathfrak{m}I : y) = I$, this shows that

$$\mu(I) = \dim(I/\mathfrak{m}I) = \text{length}(R/(\mathfrak{m}I + yR)).$$

Let $J \supset I$. Then by Corollary 2.31 we have

$$\mu(I) = \text{length}(R/\mathfrak{m}I + yR) \geq \text{length}(R/\mathfrak{m}J + yR) \geq \mu(J).$$

Thus I has the Rees property.   □

Question 2.56. 1. Characterize the ideals which have the Rees property. In regular local rings, are they precisely the m-full ideals? If Krull dim $R = 2$, this is known to be true [147, Theorem 4].
2. Suppose that $(R, \mathfrak{m}, K)$ is a regular local ring. For which ideals I is it true that

$$\max \{ \, \mu(J) \mid J \supset I \, \} = \text{length } R/(\mathfrak{m}I + yR)$$

for some $y \in \mathfrak{m}$?

Concerning m-full ideals, the following is an amusing observation about a local ring.

Proposition 2.57. *Suppose that $(R, \mathfrak{m})$ is a local ring of Krull dimension n. Suppose that I is a parameter ideal, i.e., I is an $\mathfrak{m}$-primary ideal generated by n elements. If I is $\mathfrak{m}x$-full, then R is a regular local ring and there exists a regular system of parameters $x_1, x_2, \ldots, x_n$ such that $I = (x_1, x_2, \ldots, x_{n-1}, x_n^k)$ for some integer k.*

Proof. Since I is m-full, it has the Rees property by Proposition 2.55. Hence m is generated by n elements. This shows that R is regular. As in the proof of Proposition 2.55, we have $n = \mu(I) = \text{length}(R/\mathfrak{m}I + yR)$ for some element $y \in \mathfrak{m}$. Since $\mathfrak{m}I \subset \mathfrak{m}^2$,

$$n = \mu(I) = \text{length}(R/(\mathfrak{m}I + yR)) \geq \text{length}(R/(\mathfrak{m}^2 + yR)).$$

If $y \in \mathfrak{m}^2$, then $\text{length}(R/(\mathfrak{m}^2 + yR)) = n + 1$. Hence the above inequality forces in fact that $y \in \mathfrak{m} \setminus \mathfrak{m}^2$ and $\mathfrak{m}I + yR = \mathfrak{m}^2 + yR$. Since elements of $\mathfrak{m}I + yR$ are superfluous as generators of $I + yR$, this shows that $I + yR$ is generated by a regular system of parameters. If $I + yR = (x_1, x_2, \ldots, x_{n-1}, y)$, then this implies that $I = (x_1, \ldots, x_{n-1}, y^k)$ for some k. $\square$

2.4 Artinian Gorenstein Rings

In the famous paper of Bass [5], a Gorenstein local ring is defined to be a local ring which has finite self-injective dimension. In Theorem 2.25 we showed that a Gorenstein local ring is characterized by the isomorphism

$$\text{Ext}_R^n(M, R) \cong \text{Hom}_R(M, E),$$

where E is the injective hull of K. Another characterization (Theorem 2.23) is: A local ring $(A, \mathfrak{m}, K)$ is Gorenstein if there exists a system of parameters $f_1, f_2, \ldots, f_n$ which forms a regular sequence and generates an irreducible ideal.

In this book we are primarily interested in Artinian Gorenstein algebras. Here are some characterization of Artinian Gorenstein local rings.

Theorem 2.58. *Suppose that* $(A, \mathfrak{m}, K)$ *is an Artinian local ring. Then the following conditions are equivalent.*

1. *A is Gorenstein.*
2. *A is injective as an A-module.*
3. *$\mathrm{Hom}_A(-, A)$ is an exact functor.*
4. *$(0 : \mathfrak{m}) \cong K$ as A-modules.*
5. *The ideal 0 is irreducible.*
6. *A has a unique minimal non-zero ideal.*

Proof. By definition we have (1) $\Longleftrightarrow$ (2) and (2) $\Longleftrightarrow$ (3). The equivalence of (4) and (5) is proved in Proposition 2.15. The equivalence (5) $\Longleftrightarrow$ (6) is easy. As for (3) $\Longrightarrow$ (4). Since $\mathrm{Hom}_A(-, A)$ is exact, we have

$$\mathrm{length}(M) = \mathrm{length}(\mathrm{Hom}_A(M, A))$$

for any finite module M. Thus $0 : \mathfrak{m} \cong \mathrm{Hom}_A(K, A) \cong K$. Finally we prove the essential part (3) $\Longleftarrow$ (4). $\mathrm{Hom}_A(K, A) \cong K$ implies

$$\mathrm{Hom}_A(\mathfrak{m}^i/\mathfrak{m}^{i+1}, A) \cong \mathfrak{m}^i/\mathfrak{m}^{i+1}$$

in the sense that they are the vector spaces of the same dimension. The exact sequence

$$0 \to \mathfrak{m}^{i+1} \to \mathfrak{m}^i \to \mathfrak{m}^i/\mathfrak{m}^{i+1} \to 0 \tag{2.6}$$

gives rise to the exact sequence:

$$0 \to \mathrm{Hom}_A(\mathfrak{m}^i/\mathfrak{m}^{i+1}, A) \to \mathrm{Hom}_A(\mathfrak{m}^{i+1}, A) \to \mathrm{Hom}_A(\mathfrak{m}^i, A)$$
$$\xrightarrow{\phi_i} \mathrm{Ext}_A^1(\mathfrak{m}^i/\mathfrak{m}^{i+1}, A) \to \mathrm{Ext}_A^1(\mathfrak{m}^i, A). \tag{2.7}$$

Put

$$l_i = \mathrm{length}\,\mathrm{Hom}_A(\mathfrak{m}^i, A), \epsilon_i = \mathrm{length}\,\mathrm{Im}(\phi_i).$$

Then, by the exact sequence (2.7),

$$\mathrm{length}(\mathfrak{m}^i/\mathfrak{m}^{i+1}) = \mathrm{length}\,\mathrm{Hom}_A(\mathfrak{m}^i/\mathfrak{m}^{i+1}, A) = l_{i+1} - l_i + \epsilon_i.$$

Sum up these terms for $i = 0, 1, 2, \ldots, c$ where $\mathfrak{m}^{c+1} = 0$. Then we have

$$\mathrm{length}\,A = l_0 - l_{c+1} + \sum_{i=0}^{c} \epsilon_i.$$

But l_0 = length A and $l_{c+1} = 0$. Hence we have $\sum_{i=0}^{c+1} \epsilon_i = 0$. This shows that $\mathrm{Ext}_A^1(\mathfrak{m}^i/\mathfrak{m}^{i+1}, A) = 0$ for all i. In particular $\mathrm{Ext}_A^1(K, A) = 0$. This shows that all ϕ_i are injective. For $i = 0$ we have

$$0 \xrightarrow{\phi_0} \mathrm{Ext}_A^1(A/\mathfrak{m}, A) \longrightarrow \mathrm{Ext}_A^1(A, A) = 0.$$

This shows that $\mathrm{Ext}_A^1(K, A) = 0$. By induction on the length, this furthermore shows that $\mathrm{Ext}_A^1(M, A) = 0$ for any finite M. Thus we have proved that A is self-injective. $\qquad\square$

We may write an Artinian local ring A as a homomorphic image of a regular local ring. The previous proposition can be converted into the following statement.

Corollary 2.59. *Suppose that $(R, \mathfrak{m})$ is a regular local ring of Krull dimension n and I is an $\mathfrak{m}$-primary ideal. Then the following conditions are equivalent.*

1. *I is irreducible*
2. *R/I is a Gorenstein Artinian local ring.*
3. *$\mathrm{Ext}_R^n(R/I, R) \cong R/I$.*
4. *$\mathrm{Ext}_R^n(R/I, R)$ is a cyclic R-module, i.e, generated by a single element.*

Proof. By Theorem 2.58 (1) is equivalent to (2). For any R-module of finite length the annihilator of $\mathrm{Ext}_R^n(M, R)$ is the same as that of M. So if there is a surjective map $R \to \mathrm{Ext}_R^n(R/I, R)$, its kernel is I. Thus we have the equivalence (3) $\iff$ (4). By Proposition 2.27 I is irreducible if and only if the rank of the last module in a minimal free resolution of R/I is one. This proves the equivalence (1) $\iff$ (4). $\qquad\square$

2.4.1 The Inverse System of Macaulay

To study Artinian Gorenstein algebras, the theory of the inverse system is a very powerful tool. First we introduce the inverse system of Macaulay [84] in the original form, but shortly we adopt a variant appropriate to characteristic zero only. Throughout the book we use this variation of the inverse system. There are still other variations of Macaulay's inverse system, using Hopf algebra duality and local cohomology. See Meyer and Smith [99]. Other contemporary treatments can be found in [38, 65].

Let K be a field. Let

$$Q = K[X_1, X_2, \dots, X_n]$$

be the polynomial ring in the variables

$$X_1, X_2, \dots, X_n.$$

Let $R = K[X_1^{-1}, X_2^{-1}, \ldots, X_n^{-1}]$ be the polynomial ring in the variables

$$X_1^{-1}, X_2^{-1}, \ldots, X_n^{-1}.$$

We make R into a Q-module by defining the operation of $\phi \in Q$ on R as follows: Let $\phi = X^E \in Q$ be a monomial. X^E operates on an inverse monomial $X^{-F} \in R$ by the contraction

$$X^E \cdot X^{-F} = \begin{cases} X^{-F+E} & \text{if all the components of } -F + E \text{ are non-positive,} \\ 0 & \text{otherwise.} \end{cases}$$

Here $E = (e_1, e_2, \ldots, e_n)$, likewise $F = (f_1, f_2, \ldots, f_n)$ and

$$X^E = X_1^{e_1} X_2^{e_2} \cdots X_n^{e_n}, X^{-F} \qquad = X_1^{-f_1} X_2^{-f_2} \cdots X_n^{-f_n}.$$

For a polynomial $\phi \in Q$ and an inverse polynomial $f \in R$, the operation $\phi \cdot f$ is defined to be the element of R obtained by formally expanding ϕf as a Laurent polynomial, and monomial-wise applying the rule for $X^E \cdot X^{-F} = X^{-F+E}$ or 0.

The Q-module R is an injective module, which is called the **inverse system**. In fact it is the injective envelope of the residue field $K = Q/(X_1, \ldots, X_n)$ as a Q-module.

2.4.2 A Variation of the Inverse System

In Sect. 2.4.1 we set $R = K[X_1^{-1}, X_2^{-1}, \ldots, X_n^{-1}]$, and let $x_i = X_i^{-1}$. Then we have

$$R = K[x_1, x_2, \ldots, x_n].$$

A monomial in R is a power product:

$$x_1^{i_1} x_2^{i_2} \cdots x_n^{i_n}$$

with non-negative exponents $i_1, i_2, \ldots, i_n$. The set of all monomials form a K-basis of the polynomial ring R. Sometimes a constant multiple of a power product is also called a **monomial**, and as a basis element for R, a nonzero constant multiple can be regarded as the same basis element. Suppose that K is a field of characteristic zero. Put

$$\partial_i = \frac{\partial}{\partial x_i},$$

and let

$$Q = K[\partial_1, \partial_2, \ldots, \partial_n],$$

while we keep the notation

$$R = K[x_1, x_2, \ldots, x_n].$$

Obviously R is a Q-module. With the following correspondence R becomes the inverse system for Q:

$$\partial_i \leftrightarrow X_i,$$

$$x_i \leftrightarrow X_i^{-1},$$

$$\frac{1}{a_1! a_2! \cdots a_n!} x_1^{a_1} x_2^{a_2} \cdots x_n^{a_n} \leftrightarrow X_1^{-a_1} X_2^{-a_2} \cdots X_n^{-a_n}.$$

This correspondence gives us an isomorphism of rings

$$K[X_1, X_2, \ldots, X_n] \to K[\partial_1, \partial_2, \ldots, \partial_n]$$

as well as an isomorphism

$$K[X_1^{-1}, X_2^{-1}, \ldots, X_n^{-1}] \to K[x_1, x_2, \ldots, x_n]$$

as Q-modules. Thus R is an injective Q-module. (We give a detailed proof below. See Theorem 2.60 and remarks that follow.) We understand that the variables $\partial_1, \partial_2, \ldots, \partial_n$ have degree one, and when we consider R as a Q-module, we give $x_1, x_2, \ldots, x_n$ degree -1.

Theorem 2.60. *With the same notation as above R is the injective hull of the residue field in the category of graded Q-modules.*

Proof. Let I be any homogeneous ideal of Q. We have to show that the diagram

$$
\begin{array}{ccccc}
0 & \longrightarrow & I & \longrightarrow & Q \\
& & & & \downarrow \\
& & & & R
\end{array}
\tag{2.8}
$$

can be embedded in the diagram

$$
\begin{array}{ccccc}
0 & \longrightarrow & I & \longrightarrow & Q \\
& & & & \downarrow \swarrow \\
& & & & R
\end{array}
$$

where the horizontal map is the natural injection.

Suppose that $I = (\phi_1, \phi_2, \ldots, \phi_m)$, where ϕ_i are homogeneous elements of Q, and suppose that

$$Q^{m'} \xrightarrow{\;\Psi\;} Q^m \xrightarrow{\;\Phi\;} Q \longrightarrow Q/I \qquad (2.9)$$

is exact, where Φ is the column vector

$$\Phi = \begin{pmatrix} \phi_1 \\ \phi_2 \\ \vdots \\ \phi_m \end{pmatrix},$$

and $\Psi = (\psi_{ij})$ is an $m' \times m$ matrix with homogeneous entries ψ_{ij}. Elements of $Q^{m'}$ and Q^m are written as row vectors, and the matrices Φ and Ψ take their argument on the left. Let Q_d be the homogeneous part of Q of degree d, and $Q_d = 0$ if $d < 0$. (We let $\deg \partial_i = 1$ and $\deg x_i = -1$.) The exact sequence (2.9) consists of graded pieces of finite dimensional vector spaces:

$$\bigoplus_{j=i}^{m'} Q_{d_j'} \xrightarrow{\;\Psi\;} \bigoplus_{j=i}^{m} Q_{d_j} \xrightarrow{\;\Phi\;} Q_d \longrightarrow Q_d/I_d, \qquad (2.10)$$

where $d - d_i = \deg \phi_i$ and $d_i = d_j' = \deg \psi_{ij}$. Since ϕ_{iv} and ϕ_i act on R in the opposite direction, we may reverse the arrows to obtain an exact sequence:

$$\bigoplus_{j=i}^{m'} R_{-d_j'} \xleftarrow{\;\Psi\;} \bigoplus_{j=i}^{m} R_{-d_j} \xleftarrow{\;\Phi\;} R_{-d}. \qquad (2.11)$$

These pieces altogether gives us the exact sequence

$$R^{m'} \xleftarrow{\;\Psi\;} R^m \xleftarrow{\;\Phi\;} R. \qquad (2.12)$$

Now the elements of $R^{m'}$ are written as column vectors and the matrices Ψ and Φ await their argument on the right. Since the sequence (2.11) is the dual of (2.10), the sequence (2.12) is exact. Now let f_i be the image of ϕ_i, under the vertical map $I \to R$ in the diagram (2.8), for $i = 1, 2, \ldots, m$, and consider the column vector $\begin{pmatrix} f_1 \\ f_2 \\ \vdots \\ f_m \end{pmatrix} \in R^m$. Notice that $\Psi \begin{pmatrix} \phi_1 \\ \phi_2 \\ \vdots \\ \phi_m \end{pmatrix} = 0$ implies $\Psi \begin{pmatrix} f_1 \\ f_2 \\ \vdots \\ f_m \end{pmatrix} = 0$. By the exactness of (2.12), there exists $F \in R$ such that $\phi_i F = f_i$ for $i = 1, 2, \ldots, m$. Thus we may define the map $Q \to R$ by $1 \to F$ to obtain the map we wanted. This shows that R is an injective Q-module in the category of graded Q-modules. R contains the smallest Q-module isomorphic to $K := Q/\mathfrak{m}$. Obviously every element of R is annihilated by a power of m. Thus R is the injective envelope of K in the category of graded Q-modules.   □

Example 2.61. (The purpose of this example is to show that if elements of Q^m are written as row vectors and elements of R^m as column vectors then the same matrix Φ is used to represent a homomorphism $Q^{m'} \xrightarrow{\Phi} Q^m$ and its dual $R^{m'} \xleftarrow{\Phi} R^m$.)

Let $Q = K[\partial_1, \partial_2]$, where $\partial_1 = \frac{\partial}{\partial x_1}, \partial_2 = \frac{\partial}{\partial x_2}$. The elements of Q^2 and Q^3 are written as row vectors and they are to be written on the left of the matrices. Consider the maps

$$
0 \longrightarrow Q^2 \xrightarrow{\begin{pmatrix} -\partial_2 & \partial_1 & 0 \\ 0 & -\partial_2 & \partial_1 \end{pmatrix}} Q^3 \xrightarrow{\begin{pmatrix} \partial_1^2 \\ \partial_1\partial_2 \\ \partial_1^2 \end{pmatrix}} Q .
$$

The cokernel of the second map is $Q/(\partial_1, \partial_2)^2$.

Let $R = K[x_1, x_2]$. The elements of R^2 and R^3 are written as column vectors and they are to be written on the right of the matrices.

$$
0 \longleftarrow R^2 \xleftarrow{\begin{pmatrix} -\partial_2 & \partial_1 & 0 \\ 0 & -\partial_2 & \partial_1 \end{pmatrix}} R^3 \xleftarrow{\begin{pmatrix} \partial_1^2 \\ \partial_1\partial_2 \\ \partial_2^2 \end{pmatrix}} R .
$$

The kernel of the second map from left is the Q-module $K + Kx_1 + Kx_2$.

Remark 2.62. 1. Suppose that $(f_1, f_2, \ldots, f_m) \in R^m$ and $(\phi_1, \phi_2, \ldots, \phi_m) \in Q^m$ are homogeneous elements. Then we have actually shown that there exists a form $F \in R$ such that

$$
\phi_1 F = f_1, \phi_2 F = f_2, \ldots, \phi_m F = f_m,
$$

if and only if

$$
\Psi \begin{pmatrix} f_1 \\ f_2 \\ \vdots \\ f_m \end{pmatrix} = 0.
$$

This is equivalent to claiming that R is an injective Q-module in the category of graded Q-modules.

2. The sequence (2.12) should not be confused with the dual complex:

$$
\operatorname{Hom}_Q(Q^{m'}, Q) \xleftarrow{{}^t\Psi} \operatorname{Hom}_Q(Q^m, Q) \xleftarrow{{}^t\Phi} Q. \qquad (2.13)
$$

Remark 2.63. Let $Q = K[\partial_1, \partial_2, \ldots, \partial_n]$ and $R = K[x_1, x_2, \ldots, x_n]$. In the proof above we assumed that I is a homogeneous ideal of Q. Thus it only proves that R is an injective module in the category of graded Q-modules. Nonetheless the assumption the objects to be graded can be dropped. The interested reader may consult Bruns–Herzog [12, Corollary 3.6.7]. So in fact R is an injective module in the category of Q-modules. Hence the exactness of (2.9) for arbitrary matrices Φ, Ψ implies the exactness of (2.12) and any free resolution of a finite module over Q implies the exactness of its dual sequence over R. Moreover R is also an injective module over the localization of Q at the maximal ideal $\mathfrak{m}$ as well as over the formal power series ring $\hat{Q} := K[[\partial_1, \partial_2, \ldots, \partial_n]]$.

Definition 2.64. Let Q and R be the same as before.

1. A vector subspace $I \subset R$ is an R^*-**ideal** if I is closed under the operation

$$\partial_i : I \to I$$

   for all $i = 1, 2, \ldots, n$. In other words, $I \subset R$ is an R^*-ideal if and only if it is a Q-module. (Note that an R^*-ideal is *not* an ideal of the ring R. In many cases they are finite dimensional vector spaces.)
2. An R^*-ideal I is a **finite** R^*-**ideal** if I is a finite dimensional vector space.
3. For a finite dimensional vector space $I \subset R$, the **annihilator** of I in Q is

$$\mathrm{Ann}_Q(I) := \{ f(\partial_1, \partial_2, \ldots, \partial_n) \in Q \mid f(\partial_1, \partial_2, \ldots, \partial_n)I = 0 \}.$$

   If I is the span of the single element $F \in R$, we may write $\mathrm{Ann}_Q F$ for $\mathrm{Ann}_Q I$.
4. For an ideal $\mathfrak{a}$ of Q, the **annihilator** of $\mathfrak{a}$ in R is

$$\mathrm{Ann}_R(\mathfrak{a}) := \{ f \in R \mid \phi f = 0 \text{ for all } \phi \in \mathfrak{a} \}.$$

As one will see, an R^*-ideal is nothing but the injective hull of the residue field of $Q/\mathrm{Ann}_Q(I)$. For Artinian rings the canonical module and the injective hull of the residue field coincide. Thus if I is finite, it is the canonical module of the Artinian ring $Q/\mathrm{Ann}_Q(I)$. The correspondence $\mathfrak{a} \mapsto \mathrm{Ann}_R(\mathfrak{a})$ and $I \mapsto \mathrm{Ann}_Q(I)$ establishes a bijective correspondence between the ideals of Q and the R^*-ideals of R.

Proposition 2.65. *Let* $Q = K[\partial_1, \partial_2, \ldots, \partial_n]$ *be the polynomial ring and* $R = K[x_1, x_2, \ldots, x_n]$. *Let* $I \subset R$ *be an* R^*-*ideal, and let* $\mathfrak{a} = \mathrm{Ann}_Q(I)$. *Then* I *may be viewed as the injective hull of the residue field of* $Q/\mathfrak{a}$.

Proof. Note that I may be viewed as a module over $Q/\mathfrak{a}$. To prove that it is an injective module over $Q/\mathfrak{a}$, we have to embed every diagram

$$0 \longrightarrow \mathfrak{b}/\mathfrak{a} \longrightarrow Q/\mathfrak{a}$$
$$\downarrow$$
$$I$$

in a diagram

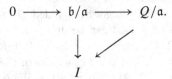

The composition $\mathfrak{b}/\mathfrak{a} \to I \to R$ may be extended to a homomorphism $Q \to R$. Since the image is annihilated by $\mathfrak{a}$, it is contained in I. Thus we get the desired homomorphism $Q/\mathfrak{a} \to I$. Since I is annihilated by a power of the maximal ideal of $Q/\mathfrak{a}$ and since I contains the smallest Q-module, it follows that I is the injective hull of the residue field of $Q/\mathfrak{a}$.   □

Proposition 2.66. *Let $Q = K[\partial_1, \partial_2, \ldots, \partial_n]$ be the polynomial ring and $\mathfrak{m} = (\partial_1, \partial_2, \ldots, \partial_n)$. Suppose that $\mathfrak{a} \subset Q$ is an ideal. Then $\sqrt{\mathfrak{a}} = \mathfrak{m}$ if and only if there is a finite R^*-ideal $I \subset R$ such that $\mathfrak{a} = \mathrm{Ann}_Q(I)$.*

Proof. Notice that $\mathrm{Ann}_R \, \mathfrak{a} = \mathrm{Hom}_Q(Q/\mathfrak{a}, R) = \mathrm{Ext}_Q^n(Q/\mathfrak{a}, Q)$. Thus the assertion is essentially Matlis duality. We omit the details.   □

Proposition 2.67. *With the same notation as before let $I \subset R$ be an R^*-ideal. Then, as a Q-module, I has finite length if and only if I is finitely generated. In this case*

$$\mu(I) = \tau(\mathrm{Ann}_Q I).$$

Proof. It is trivial that if I has finite length, then it is finitely generated. We prove the converse. Write $I = Qf_1 + Qf_2 + \cdots + Qf_t$ for some elements $f_1, f_2, \ldots, f_t \in I$. The vector space $Q f_i$ is nothing but the span of all derivatives of f_i. Thus I has finite length.

Let $t = \mu(I)$. If $t = 1$, then I is an image of Q and $I \cong Qf_1 \cong Q/\mathrm{Ann}_Q(I)$ has a smallest non-zero submodule. Thus $Q/\mathrm{Ann}_Q(I)$ is an Artinian Gorenstein ring (by Proposition 2.58) and $\mathrm{Ann}_Q(I)$ is an irreducible ideal of finite colength, and moreover $\mu(I) = 1 = \tau(\mathrm{Ann}_Q I)$. For $t > 1$, let $I_i = Qf_i$. Then we have

$$\mathrm{Ann}_Q(I) = \bigcap_{i=1}^t \mathrm{Ann}_Q(I_i).$$

This is an irredundant intersection because the correspondence $I \mapsto \mathrm{Ann}_Q(I)$ is one-to-one. Thus we have $t = \tau(\mathrm{Ann}_Q(I))$.   □

Theorem 2.68. *Suppose that A is an Artinian Gorenstein algebra and $\mathfrak{a} \subset A$ an ideal. Then*

$$(0 : (0 : \mathfrak{a})) = \mathfrak{a}.$$

Moreover $\mu(\mathfrak{a}) = \tau(0 : \mathfrak{a})$ and $\tau(\mathfrak{a}) = \mu(0 : \mathfrak{a})$.

Proof. We may identify $0 : \mathfrak{a}$ with $\operatorname{Hom}_A(A/\mathfrak{a}, A)$ by the correspondence

$$(0 : \mathfrak{a}) \ni f \mapsto (x \bmod \mathfrak{a} \mapsto f) \in \operatorname{Hom}_A(A/\mathfrak{a}, A).$$

Since A is self-injective, $\operatorname{Hom}_A(-, A)$ is a contravariant exact functor. Hence the assertion follows. □

Remark 2.69 (Comments by L. Smith). Theorem 2.68 was proved in slightly more generality by E. Noether; but she never published her proof. W. Gröbner in [44] quotes her as saying (translated into English) "the proof is too ugly to publish." W. Krull [76] on p. 32 "Idealtheory", also attributes this to E. Noether. Macaulay [83] had done this for homogeneous ideals in polynomial rings earlier. The proof in the graded case is very simple [99, p. 15].

Example 2.70. Let $f = x_1^{d_1} x_2^{d_2} \cdots x_n^{d_n}$. Let $I = Qf$. Then

$$\operatorname{Ann}_Q I = (\partial_1^{d_1+1}, \partial_2^{d_2+1}, \dots, \partial_n^{d_n+1}).$$

In particular if $f = x_1 x_2 \cdots x_n$, and $I = Qf$, then

$$\operatorname{Ann}_Q I = (\partial_1^2, \partial_2^2, \dots, \partial_n^2).$$

Theorem 2.71. *Let $R = K[x_1, x_2, \dots, x_n]$ be the polynomial ring, $\mathfrak{a}$ an ideal, and $A = R/\mathfrak{a}$. Suppose that A is an Artinian Gorenstein ring. Then there exists an element $F \in R$ such that*

$$\mathfrak{a} = \left\{ f \in R \,\middle|\, f\left(\frac{\partial}{\partial x_1}, \frac{\partial}{\partial x_2}, \dots, \frac{\partial}{\partial x_n}\right) F = 0 \right\}.$$

Proof. As before, let $Q = K[\partial_1, \partial_2, \dots, \partial_n]$, where $\partial_i = \frac{\partial}{\partial x_i}$, and let

$$\mathfrak{a}' = \{ f(\partial_1, \partial_2, \dots, \partial_n) \mid f(x_1, x_2, \dots, x_n) \in \mathfrak{a} \}.$$

Let $I = \operatorname{Ann}_R(\mathfrak{a}')$. Since $\mathfrak{a}'$ is irreducible, I is generated by a single element. In other words, $I = QF$ for some $F \in R$. We have

$$\operatorname{Ann}_Q F = \operatorname{Ann}_Q I = \operatorname{Ann}_Q(\operatorname{Ann}_R(\mathfrak{a}')) = \mathfrak{a}'.$$

Hence the assertion follows. □

2.4.3 The Ring of Invariants of Binary Octavics and Height Three Gorenstein Ideals

The purpose of this subsection is to show the importance of Gorenstein algebras in the theory of commutative rings. By a series of theorems above we have established a one-to-one correspondence

$$Q \supset \mathfrak{a} \mapsto \mathrm{Ann}_R(\mathfrak{a}) \subset R$$

and

$$R \supset I \mapsto \mathrm{Ann}_Q(I) \subset Q.$$

between the family of ideals of finite colength in $Q = K[\partial_1, \partial_2, \ldots, \partial_n]$ and the family of finite Q-modules in $R = K[x_1, x_2, \ldots, x_n]$.

In particular we have a one-to-one correspondence between the set of graded Artinian Gorenstein algebras of embedding dimension at most n and the set of homogeneous forms in n variables up to linear change of variables. If $F_1, F_2 \in K[x_1, x_2, \ldots, x_n]$ are homogeneous forms of degree d such that $F_2 = F_1^\sigma$ with $\sigma \in GL(n, K)$, then $Q/\mathrm{Ann}_Q(F_1)$ and $Q/\mathrm{Ann}_Q(F_2)$ are isomorphic by a linear change of variables.

Put $G = SL(n, K)$ and let $V = K[x_1, x_2, \ldots, x_n]_d$ be the space of symmetric tensors of degree d in n variables. Note $\dim_K V = \binom{n+d-1}{d}$. Then G acts on V, and it induces the automorphisms of the symmetric K-algebra $\mathrm{Sym}(V)$. ($\mathrm{Sym}(V)$ is the quotient of the tensor algebra $T(V) := \sum_{k=0}^\infty V^{\otimes k}$ by the ideal generated by $\{f - f^\sigma\}_{f \in T(V)}$, where σ runs over permutations of the components. We simply regard $\mathrm{Sym}(V)$ as the polynomial ring in $\binom{n+d-1}{d}$ variables, with $\mathrm{Sym}(V)_1 = V$ acted on by G.) We may think that the standard graded Gorenstein algebras of embedding dimension at most n with socle degree d are parameterized by the closed points of

$$\mathrm{Proj}(\mathrm{Sym}(V)^G),$$

where $\mathrm{Sym}(V)^G$ is the ring of invariants of the polynomial ring $\mathrm{Sym}(V)$. Thus

$$\mathrm{Proj}(\mathrm{Sym}(V)^G)$$

is the moduli space for such algebras. It is remarkable that $\mathrm{Sym}(V)^G$ itself is a (non-Artinian) Gorenstein algebra. This follows from three very general theorems:

1. The ring of invariants $\mathrm{Sym}(V)^G$ is Cohen–Macaulay for any reductive linear algebraic group G. (See e.g. [12].)
2. The ring of invariants $\mathrm{Sym}(V)^G$ is a UFD if the reductive linear algebraic group G is connected and semi-simple. (See e.g. [116, p. 376].)
3. A Cohen–Macaulay UFD is a Gorenstein ring. (See e.g. [106].)

Let $G = SL(2, \mathbb{C})$ and let $V = \mathbb{C}[x, y]_8$. V is the space of binary octavics. $\mathrm{Sym}(V)$ is the polynomial ring in nine variables $x^i y^{8-i}$ for $i = 0, 1, \ldots, 8$.

T. Shioda [125] obtained an integrity basis for the ring of invariants $\mathrm{Sym}(V)^G$. It turned out to consist of nine invariants $J_2, J_3, \ldots, J_{10}$ (subscripts indicating the degrees).

Thus $\text{Sym}(V)^G$ is an image of the polynomial ring $S = \mathbb{C}[z_1, z_2, \ldots, z_9]$. He also obtained a minimal free resolution of the ring $\text{Sym}(V)^G$ as a quotient ring of S. The homological dimension turned out to be three and moreover a minimal free resolution has the form

$$0 \to S \to S^5 \to S^5 \to S \to \text{Sym}(V)^G \to 0.$$

The second syzygy (the matrix at the middle) is a 5×5 skew-symmetric matrix and the third syzygy (the last one-row matrix) is the dual to the first syzygy!

Much later height three Gorenstein ideals were studied by Buchsbaum and Eisenbud [13] and Watanabe [144], and the result of Buchsbaum and Eisenbud says that height three Gorenstein ideals are generated by the Pfaffians of skew symmetric matrices, that necessarily appear at the middle of finite free resolutions of length three. It is amazing that such an example had preceded their result by many years.

2.4.4 The Principle of Idealization and Level Algebras

We keep the notation $Q = K[\partial_1, \partial_2, \ldots, \partial_n]$ and $R = K[x_1, x_2, \ldots, x_n]$ as above.

Definition 2.72. A graded Artinian algebra $A = \bigoplus_{i=0}^{c} A_i$ with $A_c \neq 0$ is called a **level algebra** if

$$0 : \mathfrak{m} = A_c,$$

where $\mathfrak{m} = \bigoplus_{i>0} A_i$. (So the socle of A is concentrated in a single homogeneous degree.)

Definition 2.73. A local ring $(A, \mathfrak{m}, K)$ with $\mathfrak{m}^c \neq 0, \mathfrak{m}^{c+1} = 0$ is called a **level ring** if

$$0 : \mathfrak{m} = \mathfrak{m}^c.$$

Proposition 2.74. *Let $A = \bigoplus_{i=0}^{c} A_i$ with $A_c \neq 0$ be a graded Artinian ring. Then the following conditions are equivalent:*

1. A is a level ring.
2. $A \simeq Q/\text{Ann}_Q(f_1, f_2, \ldots, f_m)$, where $f_1, f_2, \ldots, f_m \in R$ are homogeneous forms of the same degree, namely c.

Proof. Write $A \cong Q/\mathfrak{a}$ for some ideal $\mathfrak{a} \subset Q$. The assertion follows from the Matlis duality: $\text{Ext}_Q^n(A, Q) \cong \text{Hom}_R(A, R)$. (See Theorem 2.25.) □

Definition 2.75. Let S be a ring and M an S-module. The **idealization** of M, denoted by $S \ltimes M$, is the product set $S \times M$ in which addition and multiplication are defined as follows:

$$(a, x) + (b, y) = (a + b, x + y),$$

$$(a, x)(b, y) = (ab, ay + bx).$$

The principle of idealization was systematically used in Nagata's famous book [107]. For more information concerning the principle of idealization see [4].

Theorem 2.76. *Suppose that* $(A, \mathfrak{m}, K)$ *is an Artinian local ring and* E *the indecomposable injective* A*-module, i.e., the injective hull of* K*. Then* $A \ltimes E$ *is an Artinian Gorenstein local ring with the maximal ideal* $\mathfrak{m} \times E$.

Proof (Bruns–Herzog [12], pp. 110–111). Let M be the socle of $A \ltimes E$. We have to show that the dimension of M is one. Let $(b, y) \in M$. Then $(a, x) \cdot (b, y) = 0$ for any $a \in \mathfrak{m}$ and $x \in E$. Thus $b \in 0 : \mathfrak{m}$. We claim that $b = 0$. Indeed, if we assume that $b \neq 0$, then there exists $x \in E$ such that $bx \neq 0$. Hence we get $(0, x) \cdot (b, y) = (0, bx) \neq 0$, a contradiction. So $(0, y) \in M$ implies $y \in (0 : E)$. Hence M is one-dimensional.   □

Theorem 2.77. *As before put* $Q = K[\partial_1, \partial_2, \ldots, \partial_n]$, $R = K[x_1, x_2, \ldots, x_n]$, *where*

$$\partial_i = \frac{\partial}{\partial x_i}, i = 1, 2, \ldots, n.$$

Introduce new variables $y_1, y_2, \ldots, y_m$ *and put*

$$R' = K[x_1, x_2, \ldots, x_n, y_1, y_2, \ldots, y_m],$$

and

$$Q' = K[\partial_1, \partial_2, \ldots, \partial_n, \delta_1, \delta_2, \ldots, \delta_m].$$

where

$$\delta_j = \frac{\partial}{\partial y_j}, j = 1, 2, \ldots, m.$$

Let $I = Qf_1 + Qf_2 + \cdots + Qf_m \subset R$ *be the* Q*-module generated by homogeneous forms* $f_1, f_2, \ldots, f_m \in R$ *of a same degree. Let* $A = Q/\operatorname{Ann}_Q(I)$ *and let* $A' = A \ltimes I$. *Furthermore put*

$$F = f_1 y_1 + f_2 y_2 + \cdots + f_m y_m \in R'.$$

F *is considered as a form in* R'. *Then we have*

$$A \ltimes I \cong Q'/\operatorname{Ann}_{Q'}(F).$$

Proof. For $p \in Q'$, we may write

$$p = \alpha + \beta,$$

where $\alpha \in Q$, and $\beta \in (\delta_1, \delta_2, \ldots, \delta_m)Q'$. This decomposition is unique, since if $p = p(\partial_1, \partial_2, \ldots, \partial_n, \delta_1, \delta_2, \ldots, \delta_m)$, then $\alpha = p(\partial_1, \partial_2, \ldots, \partial_n, 0, 0, \ldots, 0)$, and $\beta = p - \alpha$. Define the map

$$\Phi: Q' \to A \ltimes I$$

by

$$\Phi(p(\partial, \delta)) = (\overline{\alpha}, \beta F),$$

where $\overline{\alpha} \in A$ is the canonical image of $\alpha \in Q$ in A. Notice that $\beta F \in R$, since F is linear in $y_1, y_2, \ldots, y_m$ and that the product defined by

$$(\overline{\alpha}_1, \beta_1 F)(\overline{\alpha}_2, \beta_2 F) = (\overline{\alpha}_1 \overline{\alpha}_2, \alpha_1 \beta_2 F + \alpha_2 \beta_1 F)$$

is indeed well defined. We claim that Φ is a ring homomorphism. Let $p_1, p_2 \in Q'$. We have to show that $\Phi(p_1 p_2) = \Phi(p_1)\Phi(p_2)$. Write $p_i = \alpha_i + \beta_i$, for $i = 1, 2$ as in the definition of Φ. In the polynomial $p_1 p_2$, if $\delta_1, \delta_2, \ldots, \delta_m$ are substituted by 0, $p_1 p_2$ reduces to $\alpha_1 \alpha_2$. Thus the first component of $\Phi(p_1 p_2)$ coincides with the first component of $\Phi(p_1)\Phi(p_2)$. The second component of $\Phi(p_1 p_2)$ is, by definition,

$$(p_1 p_2 - \alpha_1 \alpha_2)F = (\alpha_1 \beta_2 + \beta_1 \alpha_2 + \beta_1 \beta_2)F,$$

but the last summand $\beta_1 \beta_2 F = 0$, since $\delta_i \delta_j F = 0$. Thus it coincides with the second component of the product $(\overline{\alpha}_1, \beta_1 F)(\overline{\alpha}_2, \beta_2 F)$. The map Φ is surjective, since we have $\Phi(\partial_i) = \partial_i$ and $\Phi(\delta_i) = f_i$. Next we show that $\mathrm{Ker}\,\Phi = \mathrm{Ann}_{Q'} F$. Let $p = p(\partial_1, \partial_2, \ldots, \partial_n, \delta_1, \delta_2, \ldots, \delta_m) \in \mathrm{Ker}\,\Phi$. This implies

$$p(\partial_1, \partial_2, \ldots, \partial_n, 0, \ldots, 0) \in \mathrm{Ann}_Q F$$

and

$$(p(\partial_1, \partial_2, \ldots, \partial_n, \delta_1, \delta_2, \ldots, \delta_m) - p(\partial_1, \partial_2, \ldots, \partial_n, 0, 0, \ldots, 0))F = 0.$$

Hence it follows that

$$p(\partial_1, \partial_2, \ldots, \partial_n, \delta_1, \delta_2, \ldots, \delta_m) \in \mathrm{Ann}_{Q'} F.$$

We show the converse. Write

$$p = p(\partial_1, \partial_2, \ldots, \partial_n, \delta_1, \delta_2, \ldots, \delta_m).$$

and assume that $p \in \mathrm{Ann}_{Q'} F$. Let

$$\alpha = p(\partial_1, \partial_2, \ldots, \partial_n, 0, 0, \ldots, 0),$$

and

$$\beta = p - \alpha.$$

We have $\alpha F = -\beta F$ from $p \in \mathrm{Ann}_{Q'} F$. Since βF does not involve $y_1, y_2, \ldots, y_m$ and since αF is a linear form in $y_1, y_2, \ldots, y_m$, this means that $\alpha F = \beta F = 0$. This shows that $\alpha \in \mathrm{Ann}_Q(I)$ and $\beta F = 0$. Hence $p \in \mathrm{Ker}\, \Phi$.   □

Definition 2.78 (See [127]). Suppose that $A = \bigoplus_{i=0}^{c} A_i$ is an Artinian algebra. A is called a **Poincaré duality algebra** if the map

$$A_i \times A_{c-i} \rightarrow A_c$$

defined by the multiplication gives a perfect pairing for every $i = 0, 1, 2, \ldots, c$.

Theorem 2.79. *A graded Artinian algebra $A = \bigoplus_{i=0}^{c} A_i$ is Gorenstein if and only if A is a Poincaré duality algebra.*

Proof. Assume that A is Gorenstein. Put $\mathfrak{m} = \bigoplus_{i=1}^{c} A_i$. A basis of $\mathfrak{m}^i / \mathfrak{m}^{i+1}$ and the dual basis for $\mathrm{Hom}_K(\mathfrak{m}^i / \mathfrak{m}^{i+1}, K)$ give us a perfect pairing

$$\mathfrak{m}^i / \mathfrak{m}^{i+1} \times \mathrm{Hom}_K(\mathfrak{m}^i / \mathfrak{m}^{i+1}, K) \rightarrow K.$$

Thus it suffices to prove that $\mathrm{Hom}_K(\mathfrak{m}^i / \mathfrak{m}^{i+1}, K) \cong \mathfrak{m}^{c-i} / \mathfrak{m}^{c-i+1}$ canonically. Using the fact that $\mathrm{Hom}_A(-, A)$ is an exact functor, one sees easily that

$$0 : \mathfrak{m}^i \cong \mathrm{Hom}_A(A/\mathfrak{m}^i, A) \cong \mathfrak{m}^{c+1-i}.$$

Then it immediately follows that

$$\mathfrak{m}^{c-i} / \mathfrak{m}^{c-i+1} \cong \mathrm{Hom}_A(\mathfrak{m}^i / \mathfrak{m}^{i+1}, A) = \mathrm{Hom}_K(\mathfrak{m}^i / \mathfrak{m}^{i+1}, K),$$

as desired. Conversely assume that A is a Poincaré duality algebra. Put $\mathfrak{m} = \bigoplus_{i=1}^{c} A_i$. A is Gorenstein if $0 : \mathfrak{m} = \mathfrak{m}^c$. Let $x \in 0 : \mathfrak{m}$. We have to show that $x \in \mathfrak{m}^c$. By way of contradiction assume that $x \in \mathfrak{m}^j \setminus \mathfrak{m}^{j+1}$, $j < c$. Then by the perfectness of the pairing $A_j \times A_{c-j} \rightarrow K$, there is $y \in A_{c-j}$ such that $xy \neq 0$, and get a contradiction.   □

2.5 Complete Intersections

Suppose that R is a regular local ring of Krull dimension n, and suppose that $I = (f_1, f_2, \ldots, f_n)$ is generated by a system of parameters. Then R/I is an Artinian Gorenstein algebra. If $R = K[x_1, x_2, \ldots, x_n]$ is the polynomial ring and if I can be generated by monomials, then R/I is Artinian Gorenstein if and only if

$$I = (x_1^{d_1+1}, x_2^{d_2+1}, \ldots, x_n^{d_n+1}).$$

In particular it is necessarily a complete intersection. By Theorem 2.37 and the fact that a finite chain product has the Sperner property, we have that R/I has the Sperner property. (See Sect. 2.3.3.) We conjecture that this should be true for all standard graded Artinian complete intersections in any characteristic.

A finite poset P with rank function ρ is said to have the **LYM property** if

$$\sum_{x \in X} \frac{1}{|P_{\rho(x)}|} \leq 1,$$

for any antichain $X \subset P$. The LYM property was used to prove Sperner's theorem independently by Lubell, Yamamoto, and Meshalkin. (See [42].) From the LYM property it follows that the maximum and the minimum member of the Dilworth lattice are rank sets. (Recall Definition 1.24; If $P = \sqcup P_i$, then P_i is a rank set.) In particular it implies that P has the Sperner property. To see this, let $P = \sqcup P_i$ be a poset with rank function and let $s = \max_i |P_i|$, the Sperner number of P. Then $\frac{1}{s} \leq \frac{1}{|P_k|}$ for any k and

$$\frac{|A|}{s} \leq \sum_{x \in A} \frac{1}{|P_{\rho(x)}|}.$$

If we assume the LYM property on P, the RHS does not exceed one, hence

$$|A| \leq s.$$

This proves that the LYM property implies the Sperner property. If $A \subset P$ is an antichain which contains an element $x \in P_i$ with $|P_i| < s$, then obviously the LYM property fails to hold. Thus the LYM property implies that the minimal member of $\mathscr{F}(A)$ is a rank set. It is known that the divisor lattice has the LYM property. (Cf. [3, 42].) Thus a natural question arises:

Question 2.80. For which complete intersection A is it true that the maximum and the minimum members of the Dilworth lattice $\mathscr{F}(A)$ are powers of the maximal ideal? Intuitively this should be the case for all complete intersections. In particular we conjecture that all complete intersections have the Sperner property. (Some results are found in [67].)

One of the properties that are common to Artinian complete intersections is that one may construct a minimal free resolution of R/I as an exterior algebra, known as the Koszul complex. This can be explained as follows:

Let $\Omega = Ru_1 \oplus Ru_2 \oplus \cdots \oplus Ru_n$ be a free R-module with a free basis $u_1, u_2, \ldots, u_n$. Put $F_k = \bigwedge^k \Omega$. F_k is a free R-module with a free basis

$$\{ u_{i_1} \wedge u_{i_2} \wedge \cdots \wedge u_{i_k} \mid 1 \leq i_1 < i_2 < \cdots < i_k \leq n \}.$$

Fix elements $f_1, f_2, \ldots f_n \in R$ and define the map

$$d_k \colon F_k \to F_{k-1}$$

by

$$u_{i_1} \wedge u_{i_2} \wedge \cdots \wedge u_{i_k} \mapsto \sum_j (-1)^{j-1} f_{i_j} (u_{i_1} \wedge u_{i_2} \wedge \cdots \wedge \check{u}_{i_j} \wedge \cdots \wedge u_{i_k}).$$

Thus, for a given $(f_1, f_2, \ldots, f_n)$, we have constructed a complex

$$0 \to F_n \to F_{n-1} \to \cdots \to F_1 \to F_0 \tag{2.14}$$

with an augmentation map

$$F_0 = R \to R/(f_1, f_2, \ldots, f_n)$$

that is the cokernel of the map $F_1 \to F_0$. If $(f_1, f_2, \ldots, f_n)$ is a regular sequence, then the complex

$$0 \to F_n \to F_{n-1} \to \cdots \to F_1 \to F_0 \to R/I \to 0 \tag{2.15}$$

is exact and it is a minimal free resolution of $R/(f_1, f_2, \ldots, f_n)$. In Sect. 9.4, the sequence (2.15) will be written as

$$\to \overset{3}{\bigwedge} \Omega \to \overset{2}{\bigwedge} \Omega \to \overset{1}{\bigwedge} \Omega \to \overset{0}{\bigwedge} \Omega \to R/I \to 0. \tag{2.16}$$

One sees that the coefficients in the map $F_n \to F_{n-1}$ in (2.14) is, as a matrix,

$$(f_1, -f_2, f_3, -f_4, \ldots, (-1)^{n-1} f_n).$$

This implies that the cokernel of the dual map to $\Omega^n \to \Omega^{n-1}$, which is $\mathrm{Ext}_R^n(R/I, R)$, is isomorphic to R/I. Recall that the isomorphisms

$$\mathrm{Ext}_R^n(R/I, R) \cong R/I$$

characterizes the irreducibility of an ideal I of finite colength (Corollary 2.59). Thus, the exterior algebra as a minimal free resolution of a complete intersection also proves that a complete intersection is a Gorenstein ring. Suppose that I, J are irreducible ideals of finite colength such that $J \supset I$. Then the surjection

$$R/I \to R/J \to 0$$

induces

$$0 \longrightarrow \operatorname{Ext}_R^n(R/I, R) \longrightarrow \operatorname{Ext}_R^n(R/I, R)$$

$$\cong \uparrow \qquad\qquad \cong \uparrow$$

$$0 \longrightarrow \qquad R/J \qquad \longrightarrow \qquad R/I$$

where the vertical maps are isomorphisms. The inclusion shows that there exists $f \in R$ such that

$$J = (I : f).$$

In particular if I is an irreducible ideal of finite colength, then for any system of parameters $(x_1, x_2, \ldots, x_n)$ of R, there exists $f \in R$ such that

$$I = ((x_1^d, \ldots, x_n^d) : f)$$

for some integer d, since I contains a high power of any system of parameters.

Remark 2.81. The basic idea for the transition element goes back to the paper of Smith [2]. In the paper K. Kuhnigk [77] and in the manuscript D.M. Meyer and L. Smith [99], an element f is called a transition element for J over I, if $J = (I : f)$.

Let K be a field of characteristic zero and let $R = K[x_1, x_2, \ldots, x_n]$ be the polynomial ring in the variables $\{ x_1, x_2, \ldots, x_n \}$. In the previous section we have introduced the polynomial ring Q and the notation $\operatorname{Ann}_Q(F)$ for a homogeneous polynomial $F \in R$. To deal with the Artinian Gorenstein rings $Q/\operatorname{Ann}_Q(F)$ without referring to Q, we introduce the notation:

$$\operatorname{Ann}_R(F) = \{ p(x_1, x_2, \ldots, x_n) \in R \mid p(\partial_1, \partial_2, \ldots, \partial_n)F = 0 \}.$$

If F is a monomial it is easy to see that $\operatorname{Ann}_R(F)$ is a complete intersection generated by monomials. If $n = 2$, then $\operatorname{Ann}_R(F)$ is a complete intersection since any Gorenstein ideal in $R = K[x_1, x_2]$ is a complete intersection. In fact suppose that R/I is an Artinian Gorenstein ring and let

$$0 \to F_2 \to F_1 \to R \to R/I \to 0$$

be a minimal free resolution of R/I. Then by Proposition 2.59, we have rank $F_2 = 1$. Thus rank $F_1 = 2$. This shows that I is a complete intersection. Except for these two cases, however, it seems difficult to find a reasonable sufficient condition on $F \in R$ such that

$$I = \operatorname{Ann}_R(F)$$

is a complete intersection.

Here are some known examples of F such that $\mathrm{Ann}_R(F)$ is a complete intersection. In these examples

$$e_i = e_i(x_1, x_2, \ldots, x_n)$$

denotes the elementary symmetric polynomial of degree i.

Example 2.82 (Type A). Let

$$F = \prod_{1 \le i < j \le n} (x_i - x_j).$$

Then

$$\mathrm{Ann}_R(F) = (e_1, e_2, \ldots, e_n).$$

In particular I is a complete intersection. This was known to Stong [141].

Proof. As before we put $\partial_i = \frac{\partial}{\partial x_i}$. Let $I = (e_1, e_2, \ldots, e_n)$. Since $e_i(\partial_1, \partial_2, \ldots, \partial_n)$ F is an alternating form of degree less than $(n - 1)n/2$, it forces that $e_i(\partial_1, \partial_2, \ldots, \partial_n)F = 0$. Thus $I \subset \mathrm{Ann}_R(F)$. The algebra R/I is a complete intersection with the same socle degree as $R/\mathrm{Ann}_R(F)$. Thus $I = \mathrm{Ann}_R(F)$. $\quad\square$

Example 2.83 (Type B). Let

$$F = x_1 x_2 \cdots x_n \prod_{1 \le i < j \le n} (x_i^2 - x_j^2).$$

Then $I = \mathrm{Ann}_R(F)$ is generated by $f_1, f_2, \ldots, f_n$, where

$$f_i = e_i(x_1^2, x_2^2, \ldots, x_n^2).$$

Proof. Notice that

$$F = \det(x_j^{2i-1}).$$

Generally speaking, suppose that $G = \det(x_j^{a_i})$, where a_i are integers. Then one notices that

$$(\partial_1^k + \partial_2^k + \cdots + \partial_n^k)G = d_1 + d_2 + \cdots + d_n,$$

where d_i is the determinant of the matrix $(x_j^{a_i})$ with the i-th row replaced by

$$a_i(a_i - 1) \cdots (a_i - k + 1)(x_1^{a_i-k}, x_2^{a_i-k}, \ldots, x_n^{a_i-k}).$$

Since elementary symmetric polynomials are expressed as polynomials of power sums $p_i = x_1^i + x_2^i + \cdots + x_n^i$ (Remark 1.72), this shows that

$$f_i \in \mathrm{Ann}_R(F), \text{ for } i = 1, 2, \ldots, n.$$

Thus we have that $(f_1, f_2, \ldots, f_n) \subset \mathrm{Ann}_R(F)$. Since deg F coincides with the degree of the Jacobian of $(f_1, f_2, \ldots, f_n)$, we get the assertion.

In these examples we have been assuming that characteristic of K is zero. In characteristic 2, this becomes Frobenius Reciprocity case. See [99], Chap. 2. $\quad\square$

Example 2.84 (Type D). Let

$$F = \prod_{1 \le i < j \le n} (x_i^2 - x_j^2).$$

Then $I = \mathrm{Ann}_R(F)$ is generated by $f_1, f_2, \ldots, f_{n-1}, f_n$ where

$$f_i = \begin{cases} e_i(x_1^2, x_2^2, \ldots, x_n^2) & \text{for } i < n, \\ x_1 x_2 \cdots x_n & \text{for } i = n. \end{cases}$$

Proof. Notice that $F = \det(x_j^{2i})$. Thus it is easy to see that

$$f_i \in \mathrm{Ann}_R F, \text{ for } i = 1, 2, \ldots, n - 1.$$

As to f_n, notice that no monomials in the expansion of F involve all the variables $x_1, x_2, \ldots, x_n$. Thus $f_n \in \mathrm{Ann}_R(F)$ and we have shown that

$$(f_1, f_2, \ldots, f_{n-1}) \subset \mathrm{Ann}_R(F).$$

As in the preceding examples, we have the equality for degree reasons. $\quad\square$

Example 2.85. Let $a > 0$ be an integer. Let

$$F = \prod_{1 \le i < j \le n} (x_i^a - x_j^a).$$

Then $I = \mathrm{Ann}_R F$ is generated by $f_1, f_2, \ldots, f_{n-1}, f_n$ where

$$f_i = \begin{cases} e_i(x_1^a, x_2^a, \ldots, x_n^a) & \text{for } i < n, \\ x_1 x_2 \cdots x_n & \text{for } i = n. \end{cases}$$

Proof. Notice that $F = \det(x_j^{a(i-1)})$. Thus the same proof works as in the case $a = 2$. $\quad\square$

The following is a generalization of the examples above.

Proposition 2.86. *Consider*

$$F = (x_1 x_2 \cdots x_n)^b \prod_{1 \le i < j \le n} (x_i^a - x_j^a).$$

Then $\operatorname{Ann}_R F$ is a complete intersection for all integers $a \ge 0$ and $b \ge 0$.

Proof. For $b = 1$ and $a > 0$, the same proof as Example 2.83 works. For $b = 0$ and $a > 0$, the same proof as Example 2.84 works. Proof for other cases are left to the reader.   □

Example 2.87. Let $a > 0$ be an integer. Let $F = \det(x_j^{a+i-1})$. Then $I = \operatorname{Ann}_R(F)$ is generated by $h_{a+1}, h_{a+2}, \ldots, h_{a+n}$ where h_i is the **complete symmetric polynomial** of degree i, that is, h_i is the sum of all the monomials of degree i.

Proof. Notice that F is the alternating sum of all monomials with the exponents

$$(a, a+1, \ldots, a+n-1).$$

Thus we have $h_k F = 0$ for $k > a$. Hence $(h_{a+1}, h_{a+2}, \ldots, h_{a+n}) \subset \operatorname{Ann}_R F$. The equality follows for the degree reasons.   □

Remark 2.88 (Comments by L. Smith). These computation of an apolar element for the Hilbert ideal of reflection groups also follows from the theorem of R. Steinberg ([140] Theorem 4.1). The product $\prod \ell_H^{m_H - 1}$ of the linear forms defining the set of reflecting hyperplanes of the group counted with multiplicity gives the Macaulay dual of the ring.

2.6 Hilbert Functions

Let $R = \bigoplus_{i \ge 0} R_i$ be a graded K-algebra with $R_0 = K$, a field, and let

$$M = \bigoplus_{i=-\infty}^{\infty} M_i$$

be a graded module over R. Put

$$h(M, i) := \dim_K(M_i).$$

The numerical function $i \mapsto h(M, i)$ is called the **Hilbert function** of M.

The **Hilbert series** $\operatorname{Hilb}(M, t)$ of M is the formal power series in t defined by

$$\operatorname{Hilb}(M, t) = \sum_{i=-\infty}^{\infty} h(M, i) t^i.$$

When $A = \bigoplus_{i=0}^{c} A_i$ is a graded Artinian ring, it is a polynomial

$$\text{Hilb}(A, t) = h_0 + h_1 t + \cdots + h_c t^c$$

where $h_i = h(A, i)$ and c is the maximal degree in which A has a nonzero component. We call this the maximal socle degree of A. For an Artinian graded ring A, the coefficients of $\text{Hilb}(A, t) = h_0 + h_1 t + \cdots + h_c t^c$ may be expressed as a vector (or a sequence)

$$(h_0, h_1, \ldots, h_c),$$

which is referred to as the Hilbert function of A. The Hilbert function $(h_0, h_1, \ldots, h_c)$ for an Artinian algebra A is said to be **unimodal** if the sequence $(h_0, h_1, \ldots, h_c)$ is unimodal. We say that the sequence $(h_0, h_1, \ldots, h_c)$ is **symmetric** if

$$h_i = h_{c-i}, \ i = 0, 1, \ldots, [c/2].$$

Proposition 2.89. *Suppose that*

$$R = K[x_1, x_2, \ldots, x_n]$$

is the polynomial ring with $\deg x_i = w_i$. *Then the Hilbert series of R is given by*

$$\text{Hilb}(R, t) = \frac{1}{(1 - t^{w_1})(1 - t^{w_2}) \cdots (1 - t^{w_n})}.$$

In particular if the grading is standard,

$$\text{Hilb}(R, t) = \frac{1}{(1 - t)^n}.$$

Proof. If $n = 1$, the assertion is obvious as

$$\text{Hilb}(R, t) = 1 + t^{w_1} + t^{2w_1} + \cdots = \frac{1}{1 - t^{w_1}}.$$

The general case follows from the identification of R as a tensor space:

$$R \cong \bigotimes_{j=1}^{n} K[x_j]. \qquad \square$$

Mostly in this book we work with the standard grading of R. In the case Char $K = 0$, let $R^* = K[\partial_1, \partial_2, \ldots, \partial_n]$, where $\partial_i = \frac{\partial}{\partial x_i}$. The algebra R^* is isomorphic to R and R can be viewed as an R^*-module. To consider R as an R^*-module means that the action of the polynomial ring R on itself is defined by

$$f\phi = f(\partial_1, \partial_2, \ldots, \partial_n)\phi(x_1, x_2, \ldots, x_n) \in R,$$

for $f \in R$ and $\phi \in R$. When R is viewed as an R^*-module, R is the injective hull of the residue field K of R at the homogeneous maximal ideal. This was explained in detail earlier in Sects. 2.4.1 and 2.4.2. We may give the variables ∂_i of R^* any degrees $w_i > 0$. However, if we look at R as an R^*-module, we assume $\deg x_i = -w_i$, so that the Hilbert series of the module R^* is given by

$$\mathrm{Hilb}(R, t) = \mathrm{Hilb}(R^*, t^{-1}).$$

Here $\mathrm{Hilb}(R^*, t^{-1})$ simply means the replacement of t by t^{-1} as a Laurent series in t. The action of an element of R^* of degree d increases the degrees of elements of R by d, but the elements of R have nonpositive degrees.

Suppose that $A = R/I$, where I is a homogeneous m-primary ideal of R. Sometimes we wish to identity the injective hull of A as a subspace of R itself. To do so, let

$$\phi : R \to R^*$$

be the isomorphism of algebras defined by $x_i \mapsto \partial_i$. Then $\phi(A) := \phi(R)/\phi(I)$ is a copy of A. The injective hull E_A of the residue field of A is isomorphic to

$$\mathrm{Hom}_{R^*}(\phi(A), R).$$

This may be identified as the subspace of R given by

$$E_A := \{\, \phi \in R \mid f(\partial_1, \partial_2, \ldots, \partial_n)\phi(x_1, x_2, \ldots, x_n) = 0, \ \ \forall f \in I \,\}.$$

The action of A/I on E_A is given by

$$\overline{f} \cdot \phi = f(\partial_1, \partial_2, \ldots, \partial_n)\phi(x_1, x_2, \ldots, x_n)$$

for $\overline{f} \in A/I$ and $\phi \in E_A$, where $f \in A$ is a representative of $\overline{f}$.

The Hilbert series of E_A is given by

$$\mathrm{Hilb}(E_A, t) = \mathrm{Hilb}(A, t^{-1}),$$

where c is the maximal socle degree of A. If A is Gorenstein, E_A is a free A-module of rank one and E_A is generated by an element of degree $-c$. Thus

$$E_A \cong A(c).$$

Generally speaking, if M is a graded R-module and i is an integer, then $M(i)$ denotes the graded R-module with grading given by $M(i)_k = M_{i+k}$. The above isomorphism $E_A \cong A(c)$ implies in particular that the Hilbert series of A is symmetric. We isolate this as a proposition.

Proposition 2.90. *Suppose that A is a standard graded Artinian Gorenstein ring. Then the Hilbert function of A is symmetric, i.e.,*

$$\mathrm{Hilb}(A, t) = t^c \, \mathrm{Hilb}(A, t^{-1}),$$

where c is the socle degree of A.

When the situation is clear from context, we do not have to introduce the new notation R^* and we simply continue to use R.

Example 2.91. Let $F := x^1 y^2 \in K[x, y]$. Then

$$\mathrm{Ann}_R F = \left\{ f(x, y) \in R \mid f(\partial_x, \partial_y)F = 0 \right\}.$$

Thus $\mathrm{Ann}_R F = (x^2, y^3)$. Let $A := R/\mathrm{Ann}_R F$. Then the canonical module E_A of A is a free module or rank one generated by an element of degree -3. Thus $E_A \cong A(3)$.

1. $\mathrm{Hilb}(A, t) = 1 + 2t + 2t^2 + t^3$.
2. $\mathrm{Hilb}(E_A, t) = \mathrm{Hilb}(A, t^{-1}) = 1 + 2t^{-1} + 2t^{-2} + t^{-3}$.
3. $\mathrm{Hilb}(A, t) = t^3 \, \mathrm{Hilb}(A, t^{-1})$.

Proposition 2.92. *Suppose that*

$$A = K[x_1, x_2, \ldots, x_n]/(f_1, f_2, \ldots, f_n)$$

is an Artinian complete intersection. Suppose that $\deg x_i = w_i$, and $\deg f_i = d_i$. Then the Hilbert series of A is given by

$$\mathrm{Hilb}(A, t) = \frac{(1 - t^{d_1})(1 - t^{d_2}) \cdots (1 - t^{d_n})}{(1 - t^{w_1})(1 - t^{w_2}) \cdots (1 - t^{w_n})}.$$

In particular if the grading is standard, this may be written as

$$\mathrm{Hilb}(A, t) = \left(\frac{1 - t^{d_1}}{1 - t} \right) \cdots \left(\frac{1 - t^{d_n}}{1 - t} \right)$$

$$= \prod_{i=1}^{n} \left(1 + t + t^2 + \cdots + t^{d_i - 1} \right).$$

Proof. Let $R = K[x_1, x_2, \ldots, x_n]$ and $S = K[f_1, f_2, \ldots, f_n]$. Then the inclusion $S \to R$ is flat with $R/(f_1, f_2, \ldots, f_n)$ as the fiber at the maximal ideal of S. As S-modules, $R \cong S \otimes_K R/(f_1, f_2, \ldots, f_n)$. Hence $\mathrm{Hilb}(R, t) = \mathrm{Hilb}(S, t) \cdot \mathrm{Hilb}(A, t)$. It is easy to see that

$$\text{Hilb}(R,t) = \prod_{i=1}^{n} \frac{1}{1 - t^{w_i}},$$

and

$$\text{Hilb}(S,t) = \prod_{i=1}^{n} \frac{1}{1 - t^{d_i}}.$$

Hence the assertion is proved. $\square$

Proposition 2.93. *Suppose that B is an Artinian level ring and let $A = B \ltimes E_B$ be the idealization of the canonical module of B. Then the Hilbert series of A is given by*

$$\text{Hilb}(A,t) = \text{Hilb}(B,t) + t^{c+1} \text{Hilb}(B,t^{-1}),$$

where c is the maximal socle degree of B.

Proof. From the construction of $B \ltimes E_B$, the assertion becomes obvious. See [134].
                                                                             $\square$

Example 2.94. This example shows that by the principle of idealization we may construct a graded Artinian ring with non-unimodal Hilbert function. Let $A = K[x_1, x_2, \ldots, x_n]/\mathfrak{m}^{d+1}$, where $\mathfrak{m}$ is the homogeneous maximal ideal. We assume $K[x_1, x_2, \ldots, x_n]/\mathfrak{m}^{d+1}$ has the standard grading. Then A is a level algebra with socle degree d.

The Hilbert function is $h(A, i) = \binom{i+n-1}{n-1}$, $0 \le i \le d$. Let $f(x)$ be the polynomial

$$f(x) = \frac{1}{(n-1)!}(x + n - 1)(x + n - 2) \cdots (x + 1)$$

in x which we regard as a real valued function. Noticing that $|f(x)|$ is symmetric about $x = -n/2$, the Hilbert function of $E_A(d + 1)$ is obtained by shifting of degrees:

$$h(E_A(d + 1), i) = (-1)^{n-1} f(i - n - d), \quad i = 0, 1, 2, \ldots, d.$$

This shows that

$$h(A \ltimes E_A, i) = \begin{cases} 1 & \text{for } i = 0, \\ f(i) + (-1)^{n-1} f(i - n - d - 1) & \text{for } 0 < i < d + 1, \\ 1 & \text{for } i = d + 1. \end{cases}$$

The real valued function $f(x) + (-1)^{n-1} f(x - n - d - 1)$ is symmetric about $x = (d + 1)/2$. Assume for the moment $n > 2$. Then the Hilbert function $h_i :=$ $h(A \ltimes E_A, i)$, has a unique minimum at $i = (d + 1)/2$ if d is odd and two minima at $i = d/2, d/2 + 1$ if d is even. Let $j = (d + 1)/2$ if d is odd and $j = d/2$ if d is even. Then we have

$$1 = h_0 < h_1 > h_2 > \cdots > h_j < h_{j+1} < \cdots < h_d > h_{d+1} = 1$$

if d is odd, and

$$1 = h_0 < h_1 > h_2 > \cdots > h_j = h_{j+1} < \cdots < h_d > h_{d+1} = 1$$

if d is even. If $n = 2$, then the Hilbert function is:

$$1, d + 3, d + 3, \ldots, d + 3, 1.$$

If $n = 3$ and $d = 3$, then $A \ltimes E$ has the Hilbert function

$$(1, 13, 12, 13, 1) = (1, 3, 6, 0) + (0, 6, 3, 1).$$

This is the first example of a graded Gorenstein algebra with a nonunimodal Hilbert function, discovered by Stanley [133]. Another approach to nonunimodal Gorenstein vectors was made by Boij and Laksov [7].

Let J be the ideal of $A \ltimes E_A$ generated by the set

$$\{ (a, 0) \mid a \in A_d \}.$$

Then $\mu(J) = f(d)$. Let $I = \{ (0, b) \mid b \in E_A \}$. Then I is an ideal of $A \ltimes E_A$ such that $\mu(I) = f(d)$. Thus $\mu(I + J) = 2f(c)$. Let $\mathscr{F} = \mathscr{F}(A \ltimes E_A)$. It is not difficult to see that

$$\mathscr{F}(A \ltimes E_A) = \{ I + J \}.$$

These are also examples of Gorenstein algebras which do not have the Sperner property.

Chapter 3
Lefschetz Properties

3.1 Weak Lefschetz Property

Definition 3.1. Let $A = \bigoplus_{i=0}^{c} A_i$, $A_c \neq 0$, be a graded Artinian algebra. We say that A has the **weak Lefschetz property** (**WLP**) if there exists an element $L \in A_1$ such that the multiplication map

$$\times L: A_i \to A_{i+1}$$

has full rank for all $0 \leq i \leq c - 1$. We call $L \in A_1$ with this property a **weak Lefschetz element**.

Proposition 3.2. *Suppose that A is a standard graded Artinian algebra over a field K. If A has the weak Lefschetz property then A has a unimodal Hilbert function.*

Proof. Let $\mathfrak{m}$ be the maximal ideal of A. Since A has the standard grading, $\mathfrak{m}^i$ is generated by the graded piece A_i. Let $j \geq 0$ be the smallest integer such that $\dim_K A_j > \dim_K A_{j+1}$. Since A has the WLP, there is an element $L \in A_1$ such that $\mathfrak{m}^{j+1} = L\mathfrak{m}^j$. Furthermore this implies that $\mathfrak{m}^{i+1} = L\mathfrak{m}^i$ for all $i \geq j$. Hence the map $A_i \to A_{i+1}$ is surjective for $i \geq j$. Hence

$$1 \leq \dim_K A_1 \leq \dim_K A_2 \leq \cdots \leq \dim_K A_j \geq \dim_K A_{j+1} \geq \cdots \geq \dim_K A_c.$$

$\square$

Remark 3.3. Consider $K[x, y]/(x^2, y^2)$, where we give x, y degrees $1, 3$ respectively. Then A is a graded algebra with non-unimodal Hilbert function $(1, 1, 0, 1, 1)$. It is easy to see that x is a weak Lefschetz element nonetheless. Consider the same algebra A with $\deg x = 1, \deg y = 2$. Then A has a unimodal Hilbert function $(1, 1, 1, 1)$ but does not have the WLP. If $\deg x = \deg y = 1$, then A has a unimodal Hilbert function $(1, 2, 1)$ and has the WLP (provided that Char $K = 0$).

T. Harima et al., *The Lefschetz Properties*, Lecture Notes in Mathematics 2080,
DOI 10.1007/978-3-642-38206-2_3, © Springer-Verlag Berlin Heidelberg 2013

Remark 3.4. An easy example shows that the weak Lefschetz property is very characteristic dependent. (See e.g., Example 3.17.) Remark 3.3 shows that it also depends on the grading. In this book we assume that the grading is standard, whenever we discuss weak and strong Lefschetz properties unless otherwise specified.

Proposition 3.5. *Let $A = \bigoplus_{i=0}^{c} A_i$ be a graded Artinian K-algebra with a unimodal Hilbert function, and let $L \in A_1$ be a linear element. Then the following are equivalent.*

1. *L is a weak Lefschetz element for A.*
2. *$\dim_K A/LA = \mathrm{Sperner}(A) := \max_i \{ \dim_K A_i \}$. (See Remark 3.7.)*
3. *$\dim_K LA = \mathrm{CoSperner}(A) := \sum_{i=0}^{c-1} \min \{ \dim_K A_i, \dim_K A_{i+1} \}$.*

Proof. (1) $\Longleftrightarrow$ (2): Let $\{ h_i \}$ be the Hilbert function of A, and let u be the smallest integer such that $h_u > h_{u+1}$. Then, noticing that $\mathrm{Sperner}(A)$ occurs in degree h_u we have

$$\mathrm{length}(A/LA) = \mathrm{length}(A_0)$$
$$+ \mathrm{length}(A_1/LA_0) + \mathrm{length}(A_2/LA_1) + \cdots + \mathrm{length}(A_c/LA_{c-1})$$
$$\geq h_0 + (h_1 - h_0) + (h_2 - h_1) + \cdots + (h_u - h_{u-1})$$
$$= \mathrm{Sperner}(A).$$

Furthermore we have the following equivalences:

L is a weak Lefschetz element.

$$\Longleftrightarrow \begin{cases} \mathrm{length}(A_{i+1}/LA_i) = h_{i+1} - h_i & i = 1, 2, \ldots, u-1, \\ \mathrm{length}(A_{i+1}/LA_i) = 0 & i = u, u+1, \ldots, s-1. \end{cases}$$

$$\Longleftrightarrow \dim_K A/LA = \mathrm{Sperner}(A).$$

(2) $\Longleftrightarrow$ (3): Easy.   □

Proposition 3.6. *Suppose that A is a standard graded Artinian algebra. Then A has the Sperner property if A has the weak Lefschetz property.*

Proof. Recall that $\mu(I) \leq \mathrm{length}\, A/(y)$ for any ideal $I \subset A$ and for any element $y \in \mathfrak{m}$ (Theorem 2.30). Choose y to be a weak Lefschetz element. Then $\mathrm{length}\, A/(y) = \max_k \{ \dim_K A_k \}$ by Proposition 3.5. Hence $\mu(I) \leq \max_k \{ \mu(\mathfrak{m}^k) \}$ because $\mathfrak{m}^k$ has A_k as a generating set, so $\max_k \{ \mu(\mathfrak{m}^k) \} = \mathrm{length}\, A/(y)$. According to Definition 2.39, we have to show that $\max_k \{ \mu(\mathfrak{m}^k) \} = d(A)$. Indeed, we have

$$d(A) =: \max_{I \subset A} \{ \mu(I) \} \leq \mathrm{length}(A/yA) = \max_k \{ \mu(\mathfrak{m}^k) \} \leq \max_{I \subset A} \{ \mu(I) \},$$

completing the proof.   □

Remark 3.7. In Definition 2.39, we defined Sperner(A) to be $\max_i \{ \mu(\mathfrak{m}^i) \}$. As long as $A = \bigoplus_{i=0}^{c} A_i$ has a standard grading, we have $\mu(\mathfrak{m}^i) = \dim A_i$. This is not necessarily true if the grading of A is not standard. Whenever we speak of the Sperner number for a graded ring A, we use the definition Sperner(A) = $\max_i \{ \dim_K A_i \}$. Mostly we deal with standard graded algebras, so there should be no danger of confusion. (Cf. Okon [111, 112])

3.2 Strong Lefschetz Property

Definition 3.8. Let $A = \bigoplus_{i=0}^{c} A_i$, $A_c \neq 0$, be a graded Artinian algebra. We say that A has the **strong Lefschetz property** (**SLP**) if there exists an element $L \in A_1$ such that the multiplication map

$$\times L^d : A_i \to A_{i+d}$$

has full rank for all $0 \leq i \leq c - 1$ and $1 \leq d \leq c - i$. We call $L \in A_1$ with this property a **strong Lefschetz element**. When A has the SLP and $L \in A_1$ is a strong Lefschetz element, we will say that the pair (A, L) has the SLP.

Proposition 3.9. *Let A be a (not necessarily standard) graded Artinian K-algebra. If A has the SLP, then the Hilbert function of A is unimodal.*

Proof. Suppose that the Hilbert function of A is not unimodal. Then there are integers $k < l < m$ such that

$$\dim_K A_k > \dim_K A_l < \dim_K A_m.$$

Hence the multiplication map $\times L^{m-k} : A_k \to A_m$ cannot have full rank for any linear element $L \in A$. Thus A cannot have the SLP. $\square$

Proposition 3.10. *Let $h = (1, h_1, h_2, \ldots, h_c)$ be a finite sequence of positive integers. The following conditions are equivalent.*

1. *h is the Hilbert function of a standard graded Artinian K-algebra with the WLP.*
2. *h is the Hilbert function of a standard graded Artinian K-algebra with the SLP.*
3. *h is a unimodal O-sequence such that the positive part of the first difference is an O-sequence. (See Definition 6.20 for the definition of O-sequence.)*

Proof. (2) $\implies$ (1) is obvious. (1) $\implies$ (3): Let (A, L) be a standard graded Artinian K-algebra with the WLP and the Hilbert function h. By Proposition 3.2 the Hilbert function of A is unimodal. Furthermore it is easy to show that the positive part of the first difference of h coincides with the Hilbert function of A/LA.

(3) $\implies$ (2): We give a sketch of the proof. For details see Proposition 4.5 in [55]. Since h is unimodal, there exist integers $u_1, u_2, \ldots, u_l$ such that

$$h_0 < h_1 < \cdots < h_{u_1} = h_{u_2} = \cdots = h_{u_2-1} > h_{u_2} = h_{u_2+1} = \cdots = h_{u_3-1}$$
$$> h_{u_3} > \cdots > h_{u_l} = h_{u_l+1} = \cdots = h_s > 0.$$

Define

$$\Delta h_j = \max\{h_j - h_{j-1}, 0\}, \qquad (3.1)$$

and

$$h^{(i)}(t) = \begin{cases} \min\{h_t, h_{u_i}\} & \text{if } t < u_i, \\ h_{u_i} & \text{otherwise,} \end{cases}$$

for $i = 2, 3, \ldots, l$. Let $n = h_1$ and $R = K[x_1, x_2, \ldots, x_n]$. Choose ideals of finite colength.

$$\bar{J}_1 \subset \bar{J}_2 \subset \cdots \subset \bar{J}_l \subset \bar{R} = K[x_2, \ldots, x_n]$$

such that $h_{\bar{R}/\bar{J}_1} = \Delta h$ and $h_{\bar{R}/\bar{J}_i}(t) = \Delta h^{(i)}(t)$ for all $i = 2, 3, \ldots, l$. (Such ideals certainly exist. For example, we can choose them as lex-segment ideals.) Now set $J_i = \bigoplus_{j \geq 0}(J_i)_j := \bar{J}_i R$ for all $i = 1, 2, \ldots, l$ and

$$I := J_1 + \sum_{i=2}^{l} \left(\bigoplus_{j \geq u_i}(J_i)_j \right) + \mathfrak{m}^{s+1},$$

where $\mathfrak{m} = (x_1, x_2, \ldots, x_n)$. Then $A = R/I$ has the SLP and the Hilbert function h. □

Proposition 3.11. *Let A be a graded algebra with the SLP and let L be an SL element of A. Then $A/(0 : L^k)$ has the SLP and $\bar{L}$ is an SL element. Here $\bar{L}$ denotes the image of L in the quotient ring.*

Proof. Since $0 : L^m = (0 : L^{m-1}) : L$, it suffices to prove the case $m = 1$. Let

$$(h_0, h_1, \ldots, h_c)$$

be the Hilbert function of A and let j be the integer such that

$$h_0 \leq h_1 \leq \cdots \leq h_j > h_{j+1} \geq \cdots \geq h_c.$$

Since the map $\times L: A_k \to A_{k+1}$ is injective for $k < j$, and surjective for $k \geq j$, we have

$$(A/(0 : L))_k \cong \begin{cases} A_k & \text{for } k < j, \\ LA_k = A_{k+1} & \text{for } j \leq k. \end{cases}$$

We consider the diagram

$$
\begin{array}{ccc}
A_i & \xrightarrow{\ \times L^k\ } & A_{i+k} \\[2pt]
\downarrow & & \downarrow \\[6pt]
(A/(0:L))_i & \xrightarrow{\ \times L^k\ } & (A/(0:L))_{i+k}
\end{array}
$$

We treat the following three cases separately: (1) $i + k < j$, (2) $i < j \leq i + k$, and (3) $j \leq i$. In case (1) the vertical maps are both bijective and the upper horizontal map is injective. Thus $(A/(0:L))_i \to (A/(0:L))_{i+k}$ is injective. In case (3) the vertical maps are surjective and the first horizontal map is surjective. Thus $(A/(0:L))_i \to (A/(0:L))_{i+k}$ is surjective. In case (2) the first vertical map is bijective and the second vertical map is surjective. Hence the second horizontal map is identified with the composition:

$$
A_i \xrightarrow{\ \times L^k\ } A_{i+k} \xrightarrow{\ \times L\ } A_{i+k+1}.
$$

Thus it is either injective or surjective. $\square$

Remark 3.12. If A is a Poincaré duality algebra, then the passage from A to $A/(0:L)$ is known among algebraic topologists as "dualizing a line bundle." It is an analogy of the following: Let M be a smooth closed manifold and $\lambda \downarrow M$ a line bundle. A submanifold $N \subset M$ with a normal bundle λ in M is called a dual for λ. The map $H^*(N) \to H^*(M)$ is then the analog of $A \to A/(0:L)$. The dualizing construction appeared in work of R.E. Stong [140] in connection with characteristic number computations.

Proposition 3.13. *Let A be a graded ring with a strong Lefschetz element L. Suppose that $e, e' \in A$ are homogeneous elements such that $e, e' \in (0:L)$ and $e \notin LA$ and $e' \notin LA$. Then $\deg e = \deg e'$.*

Proof. Suppose that $\deg e < \deg e'$ and consider the map

$$
\times L^{\deg e' - \deg e} : A_{(\deg e)} \to A_{(\deg e')}.
$$

This is neither injective nor surjective. Hence A cannot have the SLP. $\square$

Suppose that A is a graded Artinian algebra over K and L is a linear form. Let $S = K[z]$ be the polynomial ring in one variable z and consider A as an S-module by letting $z \mapsto L$. We assume S has the standard grading. If α is an integer, then $S(-\alpha)$ denotes the graded S-module with graded components $S(-\alpha)_i = S_{i-\alpha}$. Let

$$
\alpha_1, \alpha_2, \ldots, \alpha_n
$$

be the degrees of minimal homogeneous generators of A as an S-module, and let

$$0 \longrightarrow \bigoplus_{i=1}^{n} S(-\beta_i - \alpha_i) \xrightarrow{\ \Phi\ } \bigoplus_{i=1}^{n} S(-\alpha_i) \longrightarrow A \longrightarrow 0 \tag{3.2}$$

be the minimal free resolution of A as an S-module. The map Φ is defined by the diagonal matrix with diagonal entries $(z^{\beta_1}, z^{\beta_2}, \ldots, z^{\beta_n})$. Notice that in this case the Hilbert series of A is given by

$$\mathrm{Hilb}(A, t) = \sum_{i=1}^{n} t^{\alpha_i} \frac{1 - t^{\beta_i}}{1 - t}.$$

The linear map $\times L \colon A \to A$ can be represented by a matrix in Jordan canonical form with Jordan blocks of sizes $\{\beta_j\}_{j=1,2,\ldots,n}$ having 0 as diagonal entries. (Multiplication with L on A is of course nilpotent since A is Artinian, so zero is the only eigenvalue.)

Proposition 3.14. *With the above notation we have:*

1. *The Hilbert function of A is unimodal and L is a weak Lefschetz element of A if and only if*

$$\max_{i} \{\alpha_i\} \leq \min_{j} \{\alpha_j + \beta_j - 1\}.$$

2. *L is a strong Lefschetz element of A if and only if*

$$\alpha_i < \alpha_j \implies \alpha_i + \beta_i \geq \alpha_j + \beta_j.$$

Proof. (1) From the minimal free resolution (3.2) of A over S, it follows that the vector space $(0 : L)$ is spanned by homogeneous elements of degrees

$$\alpha_i + \beta_i - 1, \quad i = 1, 2, \ldots, n,$$

and $A/(L)$ has a basis consisting of homogeneous elements of degrees

$$\alpha_i, \quad i = 1, 2, \ldots, n.$$

Put $\mathrm{Hilb}(A, t) = \sum_{i=0}^{c} h_i t^i$. (For $i > c$, put $h_i = 0$.) Assume that the sequence $(h_0, h_1, \ldots, h_c)$ is unimodal and let j be the integer which satisfies

$$h_0 \leq h_1 \leq \cdots \leq h_j > h_{j+1} \geq h_{j+2} \geq \cdots \geq h_c.$$

Furthermore assume that L is a weak Lefschetz element. Then the homogeneous elements in $(0 : L)$ occur only in degrees $\geq j$. Dually, the nonzero homogeneous elements in $A/(L)$ occur in degrees $\leq j$, since A_i are contained in the ideal (L) if $i > j$. Thus we have

$$\alpha_i \leq j \leq \alpha_{i'} + \beta_{i'} - 1$$

for any pair (i, i').

The converse easily follows from the expression

$$\mathrm{Hilb}(A, t) = \sum_{i=1}^{n} t^{\alpha_i} (1 + t + t^2 + \cdots + t^{\beta_i - 1}).$$

(2) First assume that A has the SLP. Recall that in this case the Hilbert function is unimodal. Hence we may use (1) to get that if $\beta_i = 1$, then $\alpha_i = \alpha_i + \beta_i - 1$ is the maximum among $\{\alpha_k\}_{0 \leq k \leq n}$ and the least among $\{\alpha_k + \beta_k - 1\}_{0 \leq k \leq n}$. Moreover if $\beta_{i'} = 1$, then $\alpha_i = \alpha_{i'}$ by Proposition 3.13. We may assume that α_i, β_i are numbered so that

$$\beta_1 > 1, \beta_2 > 1, \ldots, \beta_{n'} > 1, \quad \beta_{n'+1} = \beta_{n'+2} = \cdots = \beta_n = 1.$$

Put $A' = A/(0 : L)$. Then a minimal free resolution of A' over S is given by

$$0 \longrightarrow \bigoplus_{i=1}^{n'} S(-\beta_i - \alpha_i - 1) \xrightarrow{\Phi'} \bigoplus_{i=1}^{n'} S(-\alpha_i) \longrightarrow A' \longrightarrow 0,$$

where the map Φ' is defined by a diagonal matrix with diagonal entries

$$z^{\beta_i - 1}, \quad i = 1, 2, \ldots, n'.$$

Since A' has the SLP by Proposition 3.11 we may induct on the length of A to prove the desired inequalities. The converse follows as for (1) from the expression

$$\mathrm{Hilb}(A, t) = \sum_{i=1}^{n} t^{\alpha_i} (1 + t + t^2 + \cdots + t^{\beta_i - 1}). \qquad \square$$

Proposition 3.15. *Let $K[x, y]$ be the polynomial ring in two variables over a field K of characteristic zero. Every Artinian K-algebra $K[x, y]/I$ with the standard grading has the SLP.*

Proof. First suppose that I is a Borel-fixed ideal in $R = K[x, y]$. (For definition of "Borel-fixed ideal," see Definition 6.4.) Since Char $K = 0$ and I is Borel-fixed, we may assume that I is generated by monomials and

$$f \in I \Rightarrow x \frac{\partial}{\partial y} f \in I.$$

In other words, if we fix the linear order for each degree d,

$$x^d > x^{d-1}y > \cdots > y^d,$$

then the set of monomials in I_d, for each d, consists of consecutive monomials from the first. (Say x^d is the first monomial and y^d the last.) So the vector space R/I_d is spanned by the consecutive monomials from the last.

Let $(h_0, h_1, \ldots, h_s)$ be the Hilbert function of $A = R/I$. Then it is well known (and easy to see) that it is unimodal. Assume first that $h_i \leq h_{i+d}$. Then $\times y^d : (R/I)_i \to (R/I)_{i+d}$ is injective, because if a monomial M is in $(R/I)_i$ then $y^d M$ is in $(R/I)_{i+d}$. (The point here is that if M is the t-th monomial of $(R/I)_i$ from the last then $y^d M$ is also the t-th monomial of $(R/I)_{i+d}$ from the last.)

Now assume that $h_i \geq h_{i+d}$. Suppose that a monomial M is in $(R/I)_{i+d}$. Say M is the t-th monomial from the last. Then the t-th monomial of $(R/I)_i$ from the last exists since $h_i > h_{i+d}$. Let it be N. Then we have $y^d N = M$. Thus the map $y^d : (R/I)_i \to (R/I)_{i+d}$ is surjective. Hence we have proved that if I is Borel-fixed in characteristic 0, then R/I has the strong Lefschetz property.

In the general case we have the fact that $\mathrm{gin}(I)$ is Borel-fixed, where $\mathrm{gin}(I)$ denotes the generic initial ideal of I. It is easy to see and well known (or see Eisenbud [31, Proposition 15.12]) that $\mathrm{in}(I : y^d) = \mathrm{in}(I) : y^d$ for $d = 1, 2, \ldots,$ where y is the last variable with respect to the reverse lexicographic order. Since the Hilbert function does not change by passing to $\mathrm{gin}(I)$ and since the strong Lefschetz property is characterized by the Hilbert functions of $A/(y^d)$, $d = 1, 2, \ldots$ (See Remark 6.11), the general case reduces to the case of Borel-fixed ideals. $\quad\square$

Remark 3.16. Proposition 3.15 is shown for Artinian Gorenstein algebras (not necessarily graded) in A. Iarrobino [64], Theorem 2.9.

Example 3.17. Theorem 3.15 does not hold in positive characteristic. Let $A = \mathbb{F}[x, y]/(x^3, y^3)$, where characteristic $\mathbb{F} = 2$ or 3. Then A fails to have the strong Lefschetz property. It does not even have the weak Lefschetz property.

Definition 3.18. Let $A = \bigoplus_{i=0}^{c} A_i$, $A_c \neq 0$, be a graded Artinian algebra. We say that A has the **strong Lefschetz property in the narrow sense** if there exits an element $L \in A_1$ such that the multiplication map

$$\times L^{c-2i} : A_i \to A_{c-i}$$

is bijective for $i = 0, 1, 2, \ldots, [c/2]$.

If a graded Artinian K-algebra A has the strong Lefschetz property in the narrow sense, then the Hilbert function of A is unimodal and symmetric. When a graded Artinian K-algebra A has a symmetric Hilbert function, the notion of the strong Lefschetz property on A coincides with the one in the narrow sense. Thus we have the equivalence:

$$\text{SLP} + \text{symmetric Hilbert function} \iff \text{SLP in the narrow sense.}$$

Hence the SLP for an Artinian Gorenstein algebra automatically means the SLP in the narrow sense. Algebras with the SLP in the narrow sense can be looked at as $\mathfrak{sl}_2$-modules whose weight space decomposition coincides with the grading decomposition. Here $\mathfrak{sl}_2$ is the special linear Lie algebra of degree 2. For those readers unfamiliar with it, we digress and discuss $\mathfrak{sl}_2$ independently of the theory of algebras with the SLP in the next section.

The following is an example of an Artinian ring with the SLP that does not have the SLP in the narrow sense.

Example 3.19. Let K be a field of any characteristic. Let $A = K[x, y]/(x^2, xy, y^a)$ with $a > 3$. The Hilbert function of A is

$$1, 2, \underbrace{1, 1, \ldots, 1}_{a-2}.$$

A has the SLP, but does not have the SLP in the narrow sense. For any $b \in K$, $y + bx$ is a strong Lefschetz element.

3.3 The Lie Algebra $\mathfrak{sl}_2$ and Its Representations

3.3.1 The Lie Algebra $\mathfrak{sl}_2$

The set of 2×2 matrices with trace zero forms a Lie algebra, which we denote by $\mathfrak{sl}_2$. The Lie algebra $\mathfrak{sl}_2$ has been a strong driving force in the general theory of Lie algebras, both structure theory and representation theory. (See e.g., [60, 68].) Artinian algebras with the SLP in the narrow sense may best be treated as modules over $\mathfrak{sl}_2$. We start with definition of Lie algebras in general. Throughout this section K denotes a field of characteristic zero.

Definition 3.20. A Lie algebra is a vector space $\mathfrak{g}$ equipped with a bilinear operator $[\cdot, \cdot] : \mathfrak{g} \times \mathfrak{g} \to \mathfrak{g}$ satisfying the following two conditions : $[x, y] = -[y, x]$ $(x, y \in \mathfrak{g})$, $[[x, y], z] + [[y, z], x] + [[z, x], y] = 0$ $(x, y, z \in \mathfrak{g})$.

The bilinear operator $[\cdot, \cdot]$ is called the **bracket product**, or simply the **bracket**. The second identity in the definition is called the **Jacobi identity**.

Any associative algebra $\mathfrak{A}$ has a Lie algebra structure with the bracket product defined by

$$[x, y] = xy - yx$$

for $x, y \in \mathfrak{A}$. The associativity implies the Jacobi identity. In this sense, the set of $n \times n$ matrices $M_n(K)$ over K forms a Lie algebra. This Lie algebra is denoted by $\mathfrak{gl}_n = \mathfrak{gl}_n(K)$. (As sets $\mathfrak{gl}_n(K) = M_n(K)$.) It is clear that a subspace of $\mathfrak{gl}_n$ closed under the bracket product forms a Lie algebra. Among those Lie algebras the most fundamental is the subspace consisting of all the matrices A having trace tr(A) equal to zero. One sees easily that tr$([A, B]) = 0$ for $A, B \in M_n(K)$. We denote this Lie algebra by $\mathfrak{sl}_n = \mathfrak{sl}_n(K)$. Namely,

$$\mathfrak{sl}_n = \{ A \in M_n(K) \mid \operatorname{tr} A = 0 \}.$$

In the case where $n = 2$, the dimension of $\mathfrak{sl}_2$ is three, and the matrices

$$e = \begin{pmatrix} 0 & 1 \\ 0 & 0 \end{pmatrix}, \qquad h = \begin{pmatrix} 1 & 0 \\ 0 & -1 \end{pmatrix}, \qquad f = \begin{pmatrix} 0 & 0 \\ 1 & 0 \end{pmatrix}$$

form a basis of $\mathfrak{sl}_2$. These elements satisfy the following three relations, which we call the **fundamental relations**.

$$[h, e] = 2e, \qquad [e, f] = h, \qquad [h, f] = -2f.$$

The algebra $\mathfrak{sl}_2$ is completely determined by these relations. The three elements $\{ e, h, f \}$ are called the $\mathfrak{sl}_2$-**triple**.

Let V be a vector space and End(V) the endomorphism algebra of V. Note that, since the algebra End(V) is associative, it admits a Lie algebra structure with the bracket product $[f, g] = fg - gf$ ($f, g \in \text{End}(V)$), where fg denotes the composition of f and g. This Lie algebra is denoted by $\mathfrak{gl}(V)$. Let $\mathfrak{g}$ and $\mathfrak{g}'$ be Lie algebras. A linear map $\rho : \mathfrak{g} \to \mathfrak{g}'$ is called a **homomorphism** of Lie algebras if $\rho([x, y]) = [\rho(x), \rho(y)]$ for $x, y \in \mathfrak{g}$. Two Lie algebras are **isomorphic** if there exists a bijective homomorphism between them. Fixing a basis for an n-dimensional vector space V, one sees that the Lie algebras $\mathfrak{gl}(V)$ and $\mathfrak{gl}_n(K)$ are isomorphic. A **representation** of a Lie algebra $\mathfrak{g}$ on a vector space V is a Lie algebra homomorphism $\rho : \mathfrak{g} \to \mathfrak{gl}(V)$, and the vector space V is called a $\mathfrak{g}$-**module**. If there exists a representation of $\mathfrak{g}$ on V, one say that $\mathfrak{g}$ **acts** on V. Let V be a $\mathfrak{g}$-module. A subspace W of V is called a $\mathfrak{g}$-**submodule** if $\rho(x)(W) \subset W$ for any $x \in \mathfrak{g}$. Any $\mathfrak{g}$-submodule is a $\mathfrak{g}$-module. It is clear that $\{ 0 \}$ and V themselves are $\mathfrak{g}$-submodules of V. The $\mathfrak{g}$-module V is called **irreducible** if V has no submodules except $\{ 0 \}$ and V. Two $\mathfrak{g}$-modules U, V are **isomorphic** if there exists a bijective linear map $\varphi : U \to V$ satisfying $\varphi(x \cdot u) = x \cdot \varphi(u)$ for any $x \in \mathfrak{g}$ and $u \in U$. The symbol "$x\cdot$" used here indicates the action of $x \in \mathfrak{g}$ on an element of a $\mathfrak{g}$-module V through the homomorphism ρ. Let

$$\rho : \mathfrak{sl}_2 \to \text{End}(V)$$

be a representation of $\mathfrak{sl}_2$. We call the set of elements

$$\{ \rho(e), \rho(h), \rho(f) \}$$

an $\mathfrak{sl}_2$-**triple**. In other words, any three elements $L, H, D \in \text{End}(V)$ are an $\mathfrak{sl}_2$-**triple** if

$$[L, D] = H, [H, L] = 2L, [H, D] = -2D.$$

Example 3.21. Let n be a non-negative integer and put

$$d_k = -k(-n-1+k), \text{ for } k=1,2,\ldots,n, \text{ and } h_i = -n+2i, \text{ for } i = 0,1,\ldots,n.$$

Define $(n+1) \times (n+1)$ matrices L, D, H as follows:

$$L = \begin{pmatrix} 0 & 0 & 0 & \cdots & 0 \\ 1 & 0 & 0 & \cdots & 0 \\ 0 & 1 & 0 & \ddots & \vdots \\ & & \ddots & \ddots & \\ 0 & 0 & 0 & 1 & 0 \end{pmatrix}, \quad D = \begin{pmatrix} 0 & d_1 & 0 & \cdots & 0 \\ 0 & 0 & d_2 & \cdots & 0 \\ \vdots & \vdots & \ddots & \ddots & \vdots \\ & & & \ddots & d_n \\ 0 & 0 & 0 & \cdots & 0 \end{pmatrix}$$

$$H = \begin{pmatrix} h_0 & 0 & 0 & \cdots & 0 \\ 0 & h_1 & 0 & \cdots & 0 \\ 0 & 0 & \ddots & \ddots & \vdots \\ \vdots & \vdots & \ddots & \ddots & 0 \\ 0 & 0 & \cdots & 0 & h_n \end{pmatrix}$$

Then $\{ L, H, D \}$ is an $\mathfrak{sl}_2$-triple.

Example 3.22. Let n be any positive integer. Let $V = K[x,y]_n$ be the homogeneous component of degree n of the polynomial ring in two variables over a field of characteristic zero. Define $L, D, H \in \mathrm{End}_K(V)$ by

$$L = y\frac{\partial}{\partial x}, \quad D = x\frac{\partial}{\partial y}, \quad H = [L, D].$$

Then $\{ L, H, D \}$ is an $\mathfrak{sl}_2$-triple.

Example 3.23. Let $R = K[x]$ be the polynomial ring in one variable over a field of characteristic zero. Let $L \in \mathrm{End}_K(R)$ be the linear map defined by the multiplication by x, i.e., $Lf = xf$. Let $\lambda \in K$ be any element, and set

$$D = -x\frac{d^2}{dx^2} + \lambda\frac{d}{dx}, \quad H = [L, D].$$

Then $\{ L, D, H \} \subset \mathrm{End}_K(R)$ is an $\mathfrak{sl}_2$-triple.

Example 3.24. With the same notation as in the previous example let $2 \le n \in \mathbb{Z}_+$ and $\lambda = n - 1$. Let (x^n) be the ideal of R generated by x^n. Then we have:

1. $L, D, H \in \mathrm{End}_K((x^n))$ form an $\mathfrak{sl}_2$-triple.
2. $L, D, H \in \mathrm{End}_K(R/(x^n))$ form an $\mathfrak{sl}_2$-triple.

In all the examples above the verifications are straightforward.

3.3.2   Irreducible Modules of $\mathfrak{sl}_2$

We intend to prove that for each integer $n \geq 0$, there exists an irreducible $\mathfrak{sl}_2$-module of dimension $n + 1$ unique up to isomorphisms of $\mathfrak{sl}_2$-modules. For the theory of $\mathfrak{sl}_2$, it is usually assumed that the ground field K is algebraically closed. Here we simply assume that K is a field of characteristic zero.

Let V be a finite-dimensional irreducible $\mathfrak{sl}_2$-module, and $v \in V$ is an eigenvector for the action of h, so there is a $c \in K$ with $h \cdot v = cv$. (Note that our ground field K may not be algebraically closed. We are only assuming that there exist $v \in V$ and $c \in K$ such that $h \cdot v = cv$.) An eigenvector for h is called a **weight vector** and its eigenvalue a **weight**, so v is a weight vector of weight c. A weight vector $u \in V$ is called **lowest** if it satisfies in addition the condition $f \cdot u = 0$, and its weight is called a **lowest weight**. A weight vector annihilated by e is called **highest**, and its weight a **highest weight**.

Lemma 3.25. *Let V be a finite dimensional irreducible $\mathfrak{sl}_2$-module and $v \in V$ is a weight vector. Then there exist non-negative integers N, n and an element $\lambda \in K$ such that: if we set $v_0 = f^N \cdot v$, and $v_k = e^k \cdot v_0$ for $k \geq 1$,*

1. $h \cdot v_k = (\lambda + 2k)v_k, k = 0, 1, \ldots, n,$
2. $e \cdot v_k = v_{k+1}, k = 0, 1, \ldots, n-1, e \cdot v_n = 0,$
3. $f \cdot v_0 = 0, f \cdot v_k = k(-\lambda - k + 1)v_{k-1}$ for $k > 0.$

So $v_0, v_1, \ldots, v_n$ are weight vectors; v_0 is a lowest weight vector with lowest weight λ and v_n is a highest weight vector with highest weight $\lambda + 2n$.

Proof. Since $[h, f] = -2f$, we have $h \cdot f \cdot v = ([h, f] + f \cdot h) \cdot v = -2f \cdot v + cf \cdot v = (c-2)f \cdot v$. Repeated use of this identity shows that $h \cdot f^k \cdot v = (c-2k)f^k \cdot v$, for all $k \geq 0$. Hence those elements $f^k \cdot v$ which are non-zero are linearly independent. Finiteness of dimension implies that $f^k \cdot v = 0$ for some $k \geq 0$. Let $N \geq 0$ be the integer such that $f^N \cdot v \neq 0$ and $f^{N+j} \cdot v = 0$ for all $j > 0$. Let $v_0 = f^N \cdot v$ and $\lambda = c - 2N \in K$. Thus v_0 is a lowest weight vector of lowest weight λ. Let $v_k = e^k \cdot v_0 \in V$ for $k \geq 1$. Then, using the fundamental relations, we can verify that $h \cdot v_k = (\lambda + 2k)v_k$, and $f \cdot v_k = k(-\lambda - k + 1)v_{k-1}$ for all $k \geq 1$, (We leave the verification to the reader.) These identities imply that v_k are eigenvectors of distinct eigenvalues unless they are zero, and hence those elements v_k which are non-zero are linearly independent. Again by finiteness of dimension, there exists a non-negative integer n such that $v_n = e^n \cdot v_0 = 0$.   □

In the notation of Lemma 3.25 let W be the non-zero subspace of V spanned by the $n + 1$ weight vectors

$$\{ v_0, v_1, \ldots, v_n \}.$$

Then the lemma shows that W is $\mathfrak{sl}_2$-invariant. The irreducibility of V implies W coincides with V, whence the dimension of V is $n + 1$. Note that if we let $k = n + 1$

in the equation (3) of Lemma 3.25, we get $0 = (n+1)(-\lambda-n)v_n$. Thus $\lambda = -n$, and accordingly, the highest weight $\lambda + 2n$ is equal to n. Furthermore it turned out that all eigenvalues of h are integers. This shows that if V is an irreducible $\mathfrak{sl}_2$-module of dimension $n + 1$, the structure is determined by the rules (1)–(3) of Lemma 3.25 with $\lambda = -n$. (Cf. Example 3.21.)

Conversely, let V be a vector space of dimension $n+1$, and $v_0, v_1, \ldots, v_n$ a basis. The rules in Lemma 3.25 actually define an action of $\mathfrak{sl}_2$ on the vector space V. It is straightforward to verify the fundamental relations of $\mathfrak{sl}_2$.

As a conclusion, we have a unique, up to isomorphism, irreducible $\mathfrak{sl}_2$-module for each dimension, the structure of which is described by the identities in Lemma 3.25:

Proposition 3.26. *Let V be a finite-dimensional irreducible $\mathfrak{sl}_2$-module. Then there exists a basis $v_0, v_1, \ldots, v_n$ such that the action of $\mathfrak{sl}_2$ is described by the identities in Lemma 3.25 with $(\lambda = -n)$. Conversely, the identities give an irreducible $\mathfrak{sl}_2$-module structure on an $(n + 1)$-dimensional vector space V with the basis $v_0, v_1, \ldots, v_n$. The $(n + 1)$-dimensional $\mathfrak{sl}_2$-module $K[x, y]_n$ described in Example 3.23 is an example of an irreducible $\mathfrak{sl}_2$-module of dimension $n + 1$.*

We denote by V_n the unique irreducible $\mathfrak{sl}_2$-module of dimension $n+1$. It follows from the proposition that the set of weights on V_n is

$$\{\, n, n - 2, n - 4, \ldots, -n + 2, -n \,\},$$

each of which occurs exactly once.

Another fundamental fact about $\mathfrak{sl}_2$ is that any finite dimensional $\mathfrak{sl}_2$-module is a direct sum of irreducible modules. This is a consequence of the following result.

Proposition 3.27. *Let V be a finite dimensional $\mathfrak{sl}_2$-module over K. Then any submodule $W \subset V$ has a complementary submodule. In other words there exists a submodule $W' \subset V$ such that $V \cong W \oplus W'$ as $\mathfrak{sl}_2$-modules.*

This will be explained in the next subsection in the context of general of Lie algebras.

3.3.3 Complete Reducibility

In this section we assume that K is an algebraically closed field of characteristic zero, and discuss briefly the complete reducibility of $\mathfrak{sl}_2$ in a more general context. The Lie algebra $\mathfrak{sl}_2$ is a typical example of "simple" Lie algebras. The most important point in our development here concerning simplicity is the fact that any module decomposes into a direct sum of irreducible components. This can be formulated in a more general setting, which is usually called "semisimplicity" of Lie algebras. Namely, any module over a semisimple Lie algebra has such a decomposition. Semisimplicity for a Lie algebra is defined by the vanishing of its "radical", but we will not go further into the definition and its details. What we need here is only the fact that simplicity implies semisimplicity, as defined above, and a module for simple Lie algebra is "completely reducible".

Let $\mathfrak{g}$ be a Lie algebra, and U and V two $\mathfrak{g}$-modules. One can readily see that $\mathfrak{g}$ also acts on the direct sum $U \oplus V$ by $x \cdot (u, v) := (x \cdot u, x \cdot v)$ for $x \in \mathfrak{g}$ and $(u, v) \in U \oplus V$. We can also define the direct sum of any finite number of $\mathfrak{g}$-modules in a similar way. A $\mathfrak{g}$-module V is called **completely reducible** if any $\mathfrak{g}$-submodule W of V has a complementary submodule W' such that $V = W \oplus W'$. When V is finite-dimensional, this is equivalent to the claim that V can be decomposed into a direct sum of irreducible components, which is called an **irreducible decomposition** of V. The following result is known as **Weyl's Theorem**. (See e.g., [60] p. 28.)

Theorem 3.28. *Let $\mathfrak{g}$ be a semisimple Lie algebra. Then any $\mathfrak{g}$-module V is completely reducible.*

We use this theorem as a black box and the only thing we have to know about it is that it can be applied to $\mathfrak{sl}_2$. So any finite dimensional $\mathfrak{sl}_2$-module decomposes as a direct sum of irreducible modules. Let V be a finite-dimensional $\mathfrak{sl}_2$-module. As in the proof of Proposition 3.26, one can reach a lowest weight vector v_0 by making f act successively on a weight vector. If the weight of v_0 is $-n$, then the proposition shows that V contains an irreducible submodule isomorphic to V_n, the irreducible $\mathfrak{sl}_2$-module of dimension $n + 1$. Since V is completely reducible, one can find a complementary submodule W for V_n. Let w_0 be a lowest weight vector which belongs to W with lowest weight $-m$. Then W contains an irreducible $\mathfrak{sl}_2$-module isomorphic to V_m. Since V is finite-dimensional, this process terminates in a finite number of steps, and we will reach an irreducible decomposition of V.

We can also use highest weight vectors to construct an irreducible decomposition of an $\mathfrak{sl}_2$-module. Let v be a weight vector of a finite-dimensional $\mathfrak{sl}_2$-module V. One can find a positive integer N such that $e^N \cdot v \neq 0$ and $e^k \cdot v = 0$ for all $k > N$. Let $v_0 = e^N \cdot v$ and $v_k = f^k \cdot v_0$ for $k \geq 1$. So v_0 is a highest weight vector this time. Since V is finite-dimensional, there exists a positive integer n such that $v_n \neq 0$ and $v_k = 0$ for all $k > n$, and these nonzero v_k are linearly independent, since they have distinct eigenvalues with respect to the action of h. Let W be a subspace of V spanned by these n linearly independent vectors. It is an irreducible submodule of V. Since V is completely reducible, one can find a complementary submodule W' for W. An inductive procedure makes it possible to construct an irreducible decomposition of V.

3.3.4 The Clebsch–Gordan Theorem

Let $\mathfrak{g}$ be a Lie algebra and U, V two $\mathfrak{g}$-modules. The **tensor product** $U \otimes V$ admits a $\mathfrak{g}$-module structure if we let an element $x \in \mathfrak{g}$ act on $u \otimes v$ by $x \cdot (u \otimes v) = (x \cdot u) \otimes v + u \otimes (x \cdot v)$ for any $u \in U$ and $v \in V$. To see that this is well defined, it is enough to check the identity $[x, y] \cdot u \otimes v = x \cdot y \cdot (u \otimes v) + y \cdot x \cdot (u \otimes v)$ for $x, y \in \mathfrak{g}$, $u \in U$, $v \in V$. Let V_m and V_n be irreducible $\mathfrak{sl}_2$-modules of dimension $m + 1$ and $n + 1$ respectively. Then the tensor space $V_m \otimes V_n$ also has an $\mathfrak{sl}_2$-module structure, but it is no longer irreducible in general. The problem here concerns the way $V_m \otimes V_n$ decomposes into a direct sum of irreducible components. The **Clebsch–Gordan theorem** (see e.g., [47]) answers this problem. The method we employ here is based on the procedure outlined at the end of Sect. 3.3.3.

Theorem 3.29. *For $m \geq n$, the tensor product of two irreducible $\mathfrak{sl}_2$-modules V_m and V_n of dimension $m + 1$ and $n + 1$ respectively is isomorphic to a direct sum of $n + 1$ irreducible modules $V_{m+n}, V_{m+n-2}, \ldots, V_{m-n}$:*

$$V_m \otimes V_n \cong V_{m+n} \oplus V_{m+n-2} \oplus \cdots \oplus V_{m-n}.$$

Proof. Let $\{ u_m, u_{m-2}, \ldots, u_{-m} \}$ and $\{ v_n, v_{n-2}, \ldots, v_{-n} \}$ be bases of V_m and V_n respectively, where the subscripts designate the weights of these basis vectors. Then $\{ u_i \otimes v_j \}$ form a basis of $V_m \otimes V_n$, each of which is a weight vector: $h \cdot (u_i \otimes v_j) = (h \cdot u_i) \otimes v_j + u_i \otimes (h \cdot v_j) = (i + j) u_i \otimes v_j$. This shows that the highest weight of $V_m \otimes V_n$ is $m + n$, whose multiplicity is exactly one. The only weight vector which affords the highest weight is $u_m \otimes v_n$. Consider the weight space assigned to the next smaller weight $m + n - 2$. A basis for it consists of the following two weight vectors $u_{m-2} \otimes v_n$ and $u_m \otimes v_{n-2}$, hence the multiplicity of the next weight is two. Thus one can easily check that the multiplicity of a weight increases by one as we proceed down the sequence of the possible weights $m + n, m + n - 2, \ldots$ step by step up to $m - n = m + n - 2n$.

Since $u_m \otimes v_n$ is a highest weight vector of weight $m + n$, the tensor space $V = V_m \otimes V_n$ has an irreducible component W isomorphic to V_{m+n}, and the table $\{ m + n, m + n - 2, \ldots, -m - n \}$ gives the list of all the weights of W, where each appears with multiplicity one. Since W exhausts the weights $m + n, m + n - 2, \ldots, -m - n$ exactly once for each, the multiplicities of the weights of the complementary submodule W' are one less than those for W, hence $m + n$ is not a weight for W' and $m + n - 2$ is a highest weight of W'. In fact there is only one highest weight vector. This shows that the complementary submodule contains V_{m+n-2} as an irreducible component. The process of stripping off the irreducible component containing a highest weight vector reduces the highest weight and will terminate at weight $m - n$. We have constructed $n + 1$ trivially intersecting irreducible submodules of V, isomorphic to $V_{m+n}, V_{m+n-2}, \ldots, V_{m-n}$ respectively, and it is easy to see that the total sum of the dimensions for these components coincides with $\dim(V)$. So they give an irreducible decomposition of $V_m \otimes V_n$. $\qquad\square$

Remark 3.30. A ring-theoretic proof of the Clebsch–Gordan theorem will be given in Sect. 3.5.

We give an example. Let us consider the tensor product of the two irreducible $\mathfrak{sl}_2$-modules V_3 and V_2: $\dim V_3 = 4$, $\dim V_2 = 3$. The dimension of the tensor product $U = V_3 \otimes V_2$ is 12. Let $\{v_3, v_1, v_{-1}, v_{-3}\}$ be a basis of V_3. The subscript designates the weight of each basis element. Recall that the highest weight equals the dimension minus one. Similarly, $\{v_2, v_0, v_{-2}\}$ is a basis for V_2. The corresponding basis of U is

$$v_3 \otimes v_2, \; v_3 \otimes v_0, \; v_3 \otimes v_{-2},$$
$$v_1 \otimes v_2, \; v_1 \otimes v_0, \;\; v_1 \otimes v_{-2},$$
$$v_{-1} \otimes v_2, \; v_{-1} \otimes v_0, \; v_{-1} \otimes v_{-2},$$
$$v_{-3} \otimes v_2, \; v_{-3} \otimes v_0, \;\; v_{-3} \otimes v_{-2},$$

where each column corresponds to a weight space. The weights of U are 5, 3, 1, $-1, -3, -5$ with multiplicities 1, 2, 3, 3, 2, 1 respectively. Each multiplicity equals the dimension of the corresponding weight space. The highest weight of U is 5 and its multiplicity is one, hence U has an irreducible component W isomorphic to V_5. Since W exhaust the weights 5, 3, 1, $-1, -3, -5$ exactly once for each, the set of weights for the complementary submodule W' is $\{3, 1, -1, -3\}$ with multiplicities 1, 2, 2, 1, respectively. The highest weight of W' is three, hence W' has V_2 as an irreducible component. The complementary submodule W'' has the weights 1, -1, hence this is an irreducible $\mathfrak{sl}_2$-module isomorphic to V_1. Thus the tensor product $U = V_3 \otimes V_2$ is isomorphic to $V_5 \oplus V_3 \oplus V_1$.

3.3.5 The SLP and $\mathfrak{sl}_2$

The strong Lefschetz property **in the narrow sense** can be described by utilizing an action of $\mathfrak{sl}_2$, under which the operator representing the element $e \in \mathfrak{sl}_2$ is realized by the multiplication operator corresponding to the Lefschetz element. As a result, we will be able to show that the SLP is closed under taking tensor products of algebras (Theorem 3.34). Throughout this subsection, the symbols e, h, f are reserved for the $\mathfrak{sl}_2$-triple, and the term "SLP" is used only in the narrow sense.

Let $A = \bigoplus_{i=0}^{c} A_i$ be a finite-dimensional graded K-algebra, where A_i denotes the homogeneous component of degree i, $A_0 = K$, and $A_c \neq 0$. Let $a = \dim_K A$. Suppose that the algebra A has the SLP and $l \in A_1$ is a strong Lefschetz element. The symbol E denotes the endomorphism $E : A \to A$, $x \mapsto lx$; the multiplication operator by the Lefschetz element. Since A is finite-dimensional, the linear operator E is nilpotent, hence its eigenvalues are all zero. Let $\mu = (\mu_1, \mu_2, \ldots, \mu_d)$ denote the sequence obtained by ordering the sizes of the Jordan blocks of E in non-increasing order, i.e., μ is a **partition** of the positive integer a, and the integer μ_i is called the (i-th) component of μ. One can depict a partition in another way by using "multiplicity". Let i be a positive integer, and $m_i = m_i(\mu)$ denote the number of the components which are equal to i; one might call it the **multiplicity** of i in μ. Utilizing this, the partition μ can be expressed in the form $(1^{m_1} 2^{m_2} \cdots a^{m_a})$. This means that m_i is the number of blocks of size $i \times i$ in the Jordan block decomposition of E. If $m_i = 0$ for some i, it means that a block of size $i \times i$ does not exist. We want to introduce a notation for a basis for V in which E is represented as a Jordan canonical form. We only assume that $E \in \mathrm{End}(V)$ is a nilpotent endomorphism of the vector space V, where $a = \dim V$, and $(1^{m_1} 2^{m_2} \cdots a^{m_a})$ is the partition of the integer a that describes the Jordan canonical form of E.

Consider the filtration:

$$V \supset \mathrm{Im}\, E \supset \mathrm{Im}\, E^2 \supset \cdots \supset \mathrm{Im}\, E^r = 0.$$

This induces a filtration on the subspace $\mathrm{Ker}\, E \subset V$ by restriction:

$$\mathrm{Ker}\, E \supset \mathrm{Ker}\, E \cap \mathrm{Im}\, E \supset \mathrm{Ker}\, E \cap \mathrm{Im}\, E^2 \supset \cdots \supset \mathrm{Ker}\, E \cap \mathrm{Im}\, E^r = 0.$$

An element $w \in \mathrm{Ker}\, E \cap \mathrm{Im}\, E^j \setminus \mathrm{Ker}\, E \cap \mathrm{Im}\, E^{j+1}$ may be written as $w = E^j v$ for some $v \in V \setminus \mathrm{Im}\, E$. Put $m = m_1 + m_2 + \cdots + m_a$. It is easy to see the following lemma holds.

Lemma 3.31. *There exist m linearly independent elements*

$$S := \bigsqcup_{i=1}^{a} \left\{ v_1^{(i)}, v_2^{(i)}, \ldots, v_{m_i}^{(i)} \right\}$$

which satisfy

1. *S is (a representative of) a basis for $V / \mathrm{Im}\, E$.*
2. *$S' := \bigsqcup_{i=1}^{a} \left\{ E^{i-1} v_1^{(i)}, E^{i-1} v_2^{(i)}, \ldots, E^{i-1} v_{m_i}^{(i)} \right\}$ is a basis for $\mathrm{Ker}\, E$.*
3. *The set of nonzero elements in $S \cup ES \cup E^2 S \cup \cdots$ is a basis for V.*

The lemma simply asserts the existence of a Jordan basis for a nilpotent matrix E. In the context of the SLP, the construction of such a basis is discussed in Lemma 4.13. The set of nonzero elements in $S \cup ES \cup E^2 S \cup \cdots$ is precisely a basis for V in which E is represented as a Jordan canonical form.

We give an example to familiarize the reader with the notations. Suppose that $\dim V = 20$ and the partition for E is $\mu = (1^2 2^1 3^0 4^0 5^2 6^1 8^0 \cdots 20^0)$. Then,

$$S = \left\{ v_1^{(1)}, v_2^{(1)} \right\} \sqcup \left\{ v_1^{(2)} \right\} \sqcup \left\{ v_1^{(5)}, v_2^{(5)} \right\} \sqcup \left\{ v_1^{(6)} \right\}.$$

A Young diagram can be used to show the partition and a basis. The boxes of the Young diagram are labeled by the basis elements of V.

$v_1^{(6)}$	$Ev_1^{(6)}$	$E^2v_1^{(6)}$	$E^3v_1^{(6)}$	$E^4v_1^{(6)}$	$E^5v_1^{(6)}$
$v_2^{(5)}$	$Ev_2^{(5)}$	$E^2v_2^{(5)}$	$E^3v_2^{(5)}$	$E^4v_2^{(5)}$	
$v_1^{(5)}$	$Ev_1^{(5)}$	$E^2v_1^{(5)}$	$E^3v_1^{(5)}$	$E^4v_1^{(5)}$	
$v_1^{(2)}$	$Ev_1^{(2)}$				
$v_2^{(1)}$					
$v_1^{(1)}$					

The elements in the first column are the elements of S. The superscript (i) in $v_j^{(i)}$ indicates that it starts a block of size i.

Now suppose that (A, l) is a K-algebra with the SLP. We apply Lemma 3.31 for $\times l \in \mathrm{End}(A)$. Since l is a homogeneous element, we may choose a set S so that it consists of homogeneous elements and consequently the set $S \cup ES \cup E^2S \cup \cdots$ also consists of homogeneous elements.

Suppose $\times l \in \mathrm{End}(A)$ has the Jordan canonical form $(1^{m_1} 2^{m_2} \cdots a^{m_a})$. Since A has the SLP, it has a unimodal Hilbert function, and we have

$$m_k = \begin{cases} \dim_K A_0 = 1, & \text{for } k = c + 1, \\ \dim_K A_1 - \dim_K A_0, & \text{for } k = c - 1, \\ \dim_K A_2 - \dim_K A_1, & \text{for } k = c - 3, \\ \dim_K A_3 - \dim_K A_2, & \text{for } k = c - 5, \\ \quad\vdots \end{cases}$$

(All other m_j are 0.)

Let $v_j^{(i)} \in S$ and let V be the vector space spanned by $\left\{ E^r v_j^{(i)} \,\middle|\, r = 0, 1, \ldots, i - 1 \right\}$. Then $\dim V = i$ and the endomorphism E is well-defined on V, i.e., $EV \subset V$, and the matrix of the restriction $E_V = E|_V$ is given by the lower $i \times i$ Jordan block matrix

$$\begin{pmatrix} 0 & 0 & \cdots\cdots & 0 \\ 1 & 0 & & \vdots \\ 0 & 1 & \ddots & \vdots \\ \vdots & \ddots & \ddots & 0 & 0 \\ 0 & \cdots & 0 & 1 & 0 \end{pmatrix},$$

the only nonzero entries of which are all 1 arranged along the subdiagonal. Let H_V be the endomorphism on V defined by the $i \times i$ diagonal matrix

$$\mathrm{diag}(-(i-1), -(i-1)+2, -(i-1)+4, \ldots, (i-1)-2, i-1)$$

with respect to the basis, and F_V the nilpotent endomorphism defined by the matrix

$$
\begin{pmatrix}
0 & 1(i-1) & 0 & \cdots & & 0 \\
0 & 0 & 2(i-2) & \ddots & & \vdots \\
\vdots & & \ddots & \ddots & 0 & \\
\vdots & & & & 0 & (i-1)1 \\
0 & \cdots & & \cdots & 0 & 0
\end{pmatrix},
$$

where the only nonzero entries are $s(i-s)$ ($s = 1, 2, \ldots, i-1$) arranged along the superdiagonal. As was shown in Example 3.21, a direct computation shows that

$$[H_V, E_V] = 2E_V, \quad [H_V, F_V] = -2F_V, \quad [E_V, F_V] = H_V,$$

where $[\;,\;]$ denotes the commutator $[A, B] = AB - BA$. These three identities show that the maps $e \mapsto E_V$, $h \mapsto H_V$, $f \mapsto F_V$ make V into an i-dimensional irreducible representation of $\mathfrak{sl}_2$. Thus one can define an $\mathfrak{sl}_2$-module structure on the algebra A. Note also that the unimodal symmetry of the Hilbert series of A forces the dimensions of irreducible components to be either all even or all odd.

Thus we have verified that the SLP makes A into an $\mathfrak{sl}_2$-module satisfying the following conditions:

1. The action of $e \in \mathfrak{sl}_2$ is represented by the multiplication by a linear element l.
2. The grading decomposition of A coincides with the weight space decomposition.

Furthermore the weight vectors are all homogeneous, and if v is a weight vector of weight w, then lv is a weight vector of weight $w + 2$, unless $lv = 0$.

Conversely, let A be a finite-dimensional graded algebra with an $\mathfrak{sl}_2$-action satisfying the conditions (1) and (2). Then the Hilbert function of A is unimodal symmetric and the linear form l is a strong Lefschetz element. Hence A has the SLP in the narrow sense. To summarize:

Theorem 3.32. *Let $A = \bigoplus_{i=0}^{c} A_i$ be a finite-dimensional graded K-algebra, and $l \in A$ a homogeneous element of degree one. Then the following two conditions are equivalent:*

1. *The algebra (A, l) has the SLP in the narrow sense.*
2. *There exists an $\mathfrak{sl}_2$-action on A, with $E = \times l$, such that the weight space decomposition coincides with the grading decomposition.*

In this case, if $v \in A$ is a weight vector then

$$\text{weight } v = 2(\deg v) - c,$$

where c is the maximal socle degree of A.

In the context of Theorem 3.32 the action of the other generators of $\mathfrak{sl}_2$ is described as follows. Let H denote a direct sum $\bigoplus_V H_V$ of H_V's where the sum is over all irreducible components V:

$$H : A = \bigoplus_V V \to A = \bigoplus_V V, \ \alpha = (\alpha_V) \mapsto H\alpha = (H_V \alpha_V).$$

Similarly, we define $F = \bigoplus_V F_V$. Thus H and F are endomorphisms on A satisfying $[H, E] = 2E$, $[H, F] = -2F$ and $[E, F] = H$; therefore the maps $e \mapsto E$, $h \mapsto H$, $f \mapsto F$ define an $\mathfrak{sl}_2$-module structure on A as stated in Theorem 3.32.

Note that S is the set of lowest weight vectors. We illustrate the action of E on A by a simple example.

Example 3.33. Consider $A = K[x_1, x_2, x_3]/(x_1^2, x_2^2, x_3^2)$, $l = x_1 + x_2 + x_3$. (We assume Char $K = 0$.) Then A has the SLP in the narrow sense with $l = x_1 + x_2 + x_3$ as a Lefschetz element. The Jordan block decomposition of the multiplication map

$$E : A \to A, \ x \mapsto lx$$

has three blocks of sizes $4, 2, 2$. Thus the partition for it is $\mu = (4, 2, 2)$. The dimension of A is eight; $\dim A = 4 + 2 + 2$. The multiplicities of μ are all zero apart from $m_2 = 2$ and $m_4 = 1$. Hence, in the second notation of a partition, $\mu = (1^0 2^2 3^0 4^1)$, and the dimension of $\ker F$ is three. Choose a basis $v_1^{(2)}, v_2^{(2)}$ and $v_1^{(4)}$ of $\ker F$. We may assume that $E v_k^{(2)} \neq 0$ and $E^2 v_k^{(2)} = 0$ for $k = 1, 2$, and that $E^j v_1^{(4)} \neq 0$ for $j = 1, 2, 3$ and $E^4 v_1^{(4)} = 0$. Thanks to the SLP, these eight nonzero vectors form a basis of A, and each E-orbit of a $v_k^{(i)}$ spans an irreducible $\mathfrak{sl}_2$-submodule of A:

$$v_1^{(2)} \mapsto E v_1^{(2)}$$
$$v_2^{(2)} \mapsto E v_2^{(2)}$$
$$v_1^{(4)} \mapsto E v_1^{(4)} \mapsto E^2 v_1^{(4)} \mapsto E^3 v_1^{(4)}.$$

As an $\mathfrak{sl}_2$-module, A decomposes as $A \cong V_1 \oplus V_1 \oplus V_3$.

In the rest of this section, we show, as a corollary of Theorem 3.32, that the SLP is closed under taking tensor products of K-algebras.

Theorem 3.34. *Let (A, l) and (A', l') be algebras with the SLP in the narrow sense. Then so is the tensor product $B = A \otimes A'$, with $m = l \otimes 1 + 1 \otimes l'$ as a strong Lefschetz element of B.*

Proof. Let $\{e, f, h\}$ be the $\mathfrak{sl}_2$-triple. Recall that the tensor space $A \otimes A'$ is made into an $\mathfrak{sl}_2$-module with the action of $\{e, f, h\}$ given by (see Sect. 3.3.4)

$$\begin{cases} e \mapsto (e \cdot) \otimes 1 + 1 \otimes (e \cdot) \\ f \mapsto (f \cdot) \otimes 1 + 1 \otimes (f \cdot) \\ h \mapsto (h \cdot) \otimes 1 + 1 \otimes (h \cdot) \end{cases}$$

Thus the multiplication by $m = l \otimes 1 + 1 \otimes l'$ is the same as the action of $e \in \mathfrak{sl}_2$ on $A \otimes A'$. Suppose that $v \in A$ and $v' \in A'$ are homogeneous elements. Then we have

$$\deg(v \otimes v') = \deg v + \deg v'.$$

On the other hand, if v, v' are weight vectors, it is easy to see that $v \otimes v'$ is a weight vector with

$$\text{weight}(v \otimes v') = 2(\deg v \otimes v') - (c + c'),$$

where c, c' are the maximal socle degrees of A and A' respectively. Indeed,

$$h \cdot (v \otimes v') = (h \cdot v) \otimes v' + v \otimes (h \cdot v')$$
$$= ((2 \deg v - c)v) \otimes v' + v \otimes (2 \deg v' - c')v')$$
$$= (2 \deg(v \otimes v') - (c + c'))(v \otimes v')$$

Note that $c + c'$ is the maximal socle degree of B. This shows that the weight space decomposition of B coincides with the grading decomposition of B. Hence by Theorem 3.32, the assertion follows.   □

3.3.6 The SLP with Symmetric Hilbert Function and $\mathfrak{sl}_2$

Theorem 3.34 immediately implies that a monomial complete intersection has the SLP in the narrow sense. This is a very basic fact of the theory of the SLP. We state it as a theorem.

Theorem 3.35. *Let K be a field of characteristic zero. Then every monomial complete intersection*

$$K[x_1, x_2, \ldots, x_n]/(x_1^{a_1}, x_2^{a_2}, \ldots, x_n^{a_n})$$

has the SLP with $x_1 + x_2 + \cdots + x_n$ as a strong Lefschetz element.

Remark 3.36. In the case $K = \mathbb{C}$, R. Stanley [134] showed, among other things, that the Hard Lefschetz Theorem can be applied to prove that a monomial complete intersection has the SLP. He used the fact that a monomial complete intersection is isomorphic to the cohomology ring of the product of projective spaces over the complex number field. J. Watanabe [146] proved the same result using the theory of $\mathfrak{sl}_2$-modules. L. Reid, L. Roberts and M. Roitman proved it purely algebraically for the first time in their paper [119]. Lindsey gave another algebraic proof in [81]. Using combinatorics N. de Bruijn et al. [11] had proved that it has the Sperner property as early as in 1951.

The following result is characteristic free: one of the few of this sort.

Proposition 3.37. *Let* $A = \bigoplus_{i=0}^{c} A_i$ *be a graded ring over a field* $K = A_0$. *Suppose that* A *has the SLP in the narrow sense with* L *as an SL element. Then* $(A/(0 : L^k), \overline{L})$ *has the SLP in the narrow sense, where* $\overline{L}$ *denotes* L mod (L^k).

Proof. In Propositions 3.9 and 3.11 we have already proved that $A/(0 : L^k)$ has the SLP (in the wide sense). Thus it suffices to prove that $A/(0 : L^k)$ has a symmetric Hilbert function. By induction it suffices to prove it for $k = 1$. Let

$$\mathrm{Hilb}(A, t) = \sum_{i=0}^{c} h_i t^i,$$

and let $j = c/2$ or $j = (c - 1)/2$ according as c is even or odd. Notice that

$$A/(0 : L) \cong \mathrm{Im}[\times L : A \to A],$$

with a shift of degrees by -1. So $(A/(0 : L))_i \cong A_i$ for $i < j$ since $\times L : A_i \to A_{i+1}$ is injective and $(A/(0 : L))_i \cong A_{i+1}$ for $i \geq j$ since $\times L : A_i \to A_{i+1}$ is surjective. Thus we have

$$\dim_K (A/(0 : L))_i = \begin{cases} \dim_K A_i & \text{if } i < j, \\ \dim_K A_{i+1} & \text{if } i \geq j. \end{cases}$$

In other words, the Hilbert function of $A/(0 : L)$ is obtained from that of A by deleting the term h_j. Hence it is symmetric. $\square$

Remark 3.38. The algebra $A/(0 : L^k)$ is obtained from A by dualizing k times the element L. (Cf. Remark 3.12.) If A is a Poincaré duality algebra with formal dimension c then Proposition 3.37 leads to a cofiltration of A:

$$A \twoheadrightarrow A/(0 : L) \twoheadrightarrow A/(0 : L^2) \twoheadrightarrow \cdots \twoheadrightarrow A/(0 : L^{c-1}) \twoheadrightarrow K,$$

which decreases the formal dimension by one at each step. This seems to give a very useful mechanism for studying the SLP for small formal dimensions.

Problem 3.39 (A. Iarrobino). Let $P(A, k)$ be the Jordan partition of multiplication by a generic element in $\mathfrak{m}^k$ ($\mathfrak{m}$ the maximal ideal of A) and $\mathbf{P}(A, k)$ the set of Jordan partitions of multiplication by elements in $\mathfrak{m}^k$. Broaden the study of weak Lefschetz/strong Lefschetz to study the properties of $\mathbf{P}(A, k)$, $P(A, k)$ under deformation and algebra operation such as linkage and products and products with dual (Remark 5.18). Study also as an invariant of A. For example if A has the SLP, then $P(A, 1)$ is the dual to the Hilbert function of A. Same questions for modules over A.

If A is a graded ring over a field K, an element $f \in A_k$ is written as a linear combination of basis elements for A with coefficients in K. These coefficients can be regarded as an element in the affine space $\mathbb{A}^d$, where $d = \dim_K A_k$. By choosing the coefficients in a suitable open set in $\mathbb{A}^d$, we can avoid an exceptional behavior of the linear homomorphism induced by the multiplication $\times f : A \to A$, $x \mapsto fx$. We say that an element $f \in A_k$ is **sufficiently general** if the coefficients of f are in an open dense set $U \subset \mathbb{A}^d$ such that for any $f' \in A_k$ with coefficients in U, the multiplication maps $\times f$ and $\times f'$ behave in the same way.

Corollary 3.40. *Let $A = \bigoplus_{i=0}^c A_i$ be a graded ring with the SLP in the narrow sense and suppose that L is a strong Lefschetz element. Then for a sufficiently general homogeneous element $f \in A_k$, for $0 \le k \le c$, the quotient algebra $(A/(0 : f), \overline{L})$ has the SLP.*

Proof. In Proposition 3.11 we proved that $(A/L^k, L)$ has the SLP. Reflecting on the proof, the SLP of $(A/(0 : f), L)$ would follow if $A/(0 : f)$ and $A/(0 : fL)$ have the same Hilbert functions as those of $A/(0 : L^k)$ and $A/(0 : L^{k+1})$ respectively. This indeed is the case if $f \in A_k$ is general enough. $\qquad\square$

Proposition 3.41. *Suppose that A is a graded ring with a symmetric Hilbert function, suppose that B is a graded subring of A with the same socle degree as that of A. Suppose that B has a symmetric Hilbert function and suppose that there exists a linear form $L \in B_1$ which is an SL element for A. Then B has the SLP with L as an SL element.*

Proof. The assertion follows immediately from the diagram

$$
\begin{array}{ccc}
A_i & \xrightarrow{\times L^{c-2i}} & A_{c-i} \\
\uparrow & & \uparrow \\
B_i & \xrightarrow{\times L^{c-2i}} & B_{c-i}
\end{array}
$$

where c is the socle degree of A and B and the vertical maps are natural injection.
$\qquad\square$

The unimodality of the Hilbert function is not preserved by tensor product. Hence the SLP in the wide sense is not preserved by tensor product. The following is such an example.

Example 3.42. Let $R = K[x, y, z]$, $I = (x^2, xy, y^2, xz, yz, z^5)$, and $A = R/I$. Then

$$\mathrm{Hilb}(A, t) = 1 + 3t + t^2 + t^3 + t^4.$$

Obviously A has the SLP. It is an example of an algebra with SLP that is not a Poincaré duality algebra. (See Definition 2.78.) Let $B = A \otimes_K A$. Then

$$\mathrm{Hilb}(B, t) = 1 + 6t + 11t^2 + 8t^3 + 9t^4 + 8t^5 + 3t^6 + 2t^7 + t^8.$$

Since it is not unimodal, B cannot have the SLP. It is easy to compute that $\dim_K (A \otimes A/((x + y + z) \otimes 1 + 1 \otimes (x + y + z))) = 13$, and $\mu(I + J) = 13$, where

$$I = (x \otimes x, y \otimes x, z \otimes x, x \otimes y, y \otimes y, z \otimes y, x \otimes z, y \otimes z),$$
$$J = (z^4 \otimes 1, z^3 \otimes z, z^2 \otimes z^2, z \otimes z^3, 1 \otimes z^4).$$

Thus $d(A) = r(A) = 13 > \mathrm{Sperner}(A) = 11$.

The following are special cases of Theorem 5.6 (2) which we prove later:

Proposition 3.43 ([50]). *Let A be an Artinian graded K-algebra with a symmetric Hilbert function, and let ξ be an indeterminate over K.*

1. *A has the SLP if and only if $A \otimes_K K[\xi]/(\xi^n)$ has the WLP for all $n = 1, 2, \ldots, c + 1$, where c is the socle degree of A.*
2. *If g is a strong Lefschetz element of A, then $g \otimes 1 + 1 \otimes \bar{\xi}$ is a weak Lefschetz element of $A \otimes_K K[\xi]/(\xi^n)$ for every n.*

Proposition 3.44. *Let C be the family of all Artinian complete intersections with standard grading over a field K of characteristic zero. Then the following assertions are equivalent.*

1. *All members of C have the weak Lefschetz property.*
2. *All members of C have the strong Lefschetz property.*

Proof. (2) $\implies$ (1) is clear. (1) $\implies$ (2): Let $A \in C$. Then $A[\xi]/(\xi^n)$ is also a complete intersection, i.e., $A[\xi]/(\xi^n) \in C$. Hence, since $A[\xi]/(\xi^n)$ has the WLP, it follows by Proposition 3.43 that A has the SLP. $\qquad\square$

Remark 3.45. Proposition 3.44 was proved in [119] by L. Reid, L. Roberts and M. Roitman. Later Harima and Watanabe [50] proved it independently.

Conjecture 3.46. The assertions (1) and (2) in Proposition 3.44 are true. Therefore all complete intersections over a field of characteristic zero enjoy the strong Lefschetz property.

3.4   The WLP and SLP in Low Codimensions

In T. Harima et al. [55] it was proved that an Artinian complete intersection A in embedding codimension three has the WLP over a field of characteristic zero. When we think of other cases where the WLP and/or SLP can be proved, it seems encouraging to conjecture that every complete intersection in any codimension in characteristic zero has the strong Lefschetz property. In this section we give an outline of a proof for the case emb. dim $A = 3$, which uses a theorem about vector bundles. Before proceeding, however, we would like to make some remarks on the SLP and WLP for codimension two.

3.4.1   The WLP and SLP in Codimension Two

A proof of the WLP in codimension two is easy as we show in the next proposition.

Proposition 3.47. *Let* $A = \bigoplus_{i=0}^{c} A_i$ *be a standard graded Artinian algebra of embedding codimension two over an infinite field* $K = A_0$ *in any characteristic. Then* A *has the WLP.*

Proof. Let $A = R/I$, where I is a homogeneous ideal of the polynomial ring $R = K[x, y]$. Let $d = \min\{\deg f \mid f \in I\}$. Then $\dim A_{d-1} = \dim R_{d-1} = d \geq \dim A_j$ for $j \geq d$. Thus d is equal to the Sperner number of A. Since K is infinite there exists a linear form $l \in A$ such that $A/lA \cong K[x]/(x^d)$. Hence A has the WLP. □

If A is a complete intersection and Char $K = 0$ or Char $K > c$, where c is the socle degree of A, then any algebra $A = K[x, y]/I$, for a homogeneous ideal I, has the SLP in the wide sense, but a lot of argument is involved to prove this. (See Proposition 3.15.) For complete intersections this was first proved in Iarrobino [64] in a more general setup as was mentioned earlier.

To prove the SLP only for the monomial complete intersection

$$K[x, y]/(x^r, y^s)$$

one uses the theory of the linear Lie algebra $\mathfrak{sl}_2$. An alternate proof is to compute the determinants of the matrix for the linear map $(x + y)^{c-2i} : A_i \to A_{c-i}$, where $c = r + s - 2$. See [74, 80, 120].

It is easy to see that A fails to have the WLP if K is finite, and even if K is infinite it may fail to have the SLP in positive characteristic.

3.4.2 The WLP in Codimension Three

Theorem 3.48. *Let* $R = K[x_1, x_2, x_3]$ *be the polynomial ring over a field* K *of characteristic zero. Let* f_1, f_2, f_3 *be a homogeneous regular sequence of degrees* d_1, d_2, d_3 *respectively. Then* $A := R/(f_1, f_2, f_3)$ *has the weak Lefschetz property.*

Proof. We may assume $2 \le d_1 \le d_2 \le d_3$. Let Hilb$(A, t)$ be the Hilbert series of A. We know that

$$\text{Hilb}(A, t) = \prod_{i=1}^{3}(1 + t + t^2 + \cdots + t^{d_i - 1})$$

First assume $d_3 > d_1 + d_2 - 2$. Rewrite Hilb(A, t) as

$$\text{Hilb}(A, t) = (1 + t + t^2 + \cdots + t^{d_3 - 1})H(t),$$

where

$$H(t) = (1 + t + t^2 + \cdots + t^{d_1 - 1})(1 + t + t^2 + \cdots + t^{d_2 - 1}).$$

Since deg $H(t) = d_1 + d_2 - 2$, we see that the maximum of the coefficients of the polynomial Hilb(A, t) occurs in degree at $d_3 - 1$ and higher, and this coefficient is equal to the sum of all the coefficients of the polynomial $H(t)$. It turns out that this is just $d_1 d_2$, so we have Sperner$(A) = d_1 d_2$. Let $l \in A$ be a general linear form. Then we have $A/lA = R/(f_1, f_2, f_3, l) = R/(f_1, f_2, l)$, since the socle degree of $R/(f_1, f_2, l)$ is $d_1 + d_2 - 2$ and deg $f_3 > d_1 + d_2 - 2$. So $\mu(I + (l)/(l)) = 2$.

Hence we have $r(A) = d_1 d_2 = $ Sperner(A), so the WLP holds for A. Next assume that $d_3 \le d_1 + d_2 - 2$. Let $l \in R$ be any linear element, $\overline{R} = R/lR$, and $\overline{f}_1, \overline{f}_2, \overline{f}_3$ their natural images in $\overline{R}$. Then there exist positive integers e_1, e_2, such that

$$0 \to \overline{R}(-e_1) \oplus \overline{R}(-e_2) \to \overline{R}(-d_1) \oplus \overline{R}(-d_2) \oplus \overline{R}(-d_3) \to \overline{R}(0)$$

$$\to \overline{R}/(\overline{f}_1, \overline{f}_2, \overline{f}_3) \to 0$$

is exact, where $e_1 + e_2 = d_1 + d_2 + d_3$. Thus we have

$$\text{Hilb}(A/lA, t) = \frac{1 - t^{d_1} - t^{d_2} - t^{d_3} + t^{e_1} + t^{e_2}}{(1 - t)^2}. \tag{3.3}$$

This enables us to express the number $\dim(A/lA)$ in terms of the binomial coefficients as the sum of the coefficients of Hilb$(A/lA, t)$. On the other hand, the minimal free resolution of $A = R/(f_1, f_2, f_3)$ has the form

$$0 \to F_3 \to F_2 \to F_1 \to R \to A \to 0, \tag{3.4}$$

where

$$F_3 = R(-d_1 - d_2 - d_3),$$
$$F_2 = R(-d_2 - d_3) \oplus R(-d_1 - d_3) \oplus R(-d_1 - d_2),$$
$$F_1 = R(-d_1) \oplus R(-d_2) \oplus R(-d_3).$$

Thus we may write the Hilbert function $\text{Hilb}(A, t)$ in terms of binomial coefficients. By the Key lemma below, if l is a general element, then $|e_1 - e_2| \leq 1$. (Note that in the following argument we may assume without loss of generality that the ground field K is algebraically closed.) It is straightforward, although lengthy, to show that the sum of terms with positive coefficients of $\text{Hilb}(A, t) - \text{Hilb}(A, t - 1)$ is equal to $\text{Hilb}(A/lA, t)$ under the condition $|e_2 - e_1| \leq 1$. (LHS can be computed by (3.4) and RHS by (3.3).) Evaluating both polynomials at $t = 1$, we have $s(A) = r(A)$.

If one wishes to avoid computation, one may argue as follows: The Key lemma says that $\dim A/lA$ for a general linear form l depends only on the degrees d_1, d_2, d_3 and not on the particular choice of elements f_1, f_2, f_3. Therefore, for each triple of degrees d_1, d_2, d_3 we know at least one instance for which the WLP holds (i.e., the monomial complete intersection). Thus for all complete intersections A, it follows that $s(A) = r(A)$. □

Lemma 3.49 (Key lemma). *Let K be an algebraically closed field of characteristic 0. Let $I \subset R = K[x, y, z]$ be a complete intersection ideal generated by homogeneous elements f_1, f_2, f_3 of degrees d_1, d_2, d_3 with $d_3 \geq d_2 \geq d_1$. Let E be the kernel of the map*

$$\begin{bmatrix} f_1 \\ f_2 \\ f_3 \end{bmatrix} : R(-d_1) \oplus R(-d_2) \oplus R(-d_3) \rightarrow R(0),$$

and let $\mathscr{E}$ be the sheafification of the module E. Note that $\mathscr{E}$ is a rank 2 locally free sheaf over $\mathscr{O}_{\mathbb{P}^2}$, and there is an exact sequence:

$$0 \rightarrow \mathscr{E} \rightarrow \mathscr{O}_{\mathbb{P}^2}(-d_1) \oplus \mathscr{O}_{\mathbb{P}^2}(-d_2) \oplus \mathscr{O}_{\mathbb{P}^2}(-d_3) \rightarrow \mathscr{O}_{\mathbb{P}^2} \rightarrow 0.$$

In this situation the following hold.

1. *If $d_3 \leq d_1 + d_2 + 1$, then $\mathscr{E}$ is semi-stable.*
2. *(Theorem of Grauert–Mülich) If (the rank two vector bundle) $\mathscr{E}$ is semi-stable and if $\mathscr{E}$ is restricted to a general line L in $\mathbb{P}^2$, then $\mathscr{E}|_L$ splits as $\mathscr{E}|_L \cong \mathscr{O}_{\mathbb{P}^1}(-e_1) \oplus \mathscr{O}_{\mathbb{P}^1}(-e_2)$, with $|e_1 - e_2| \leq 1$.*

Proof. See e.g., [10, 63, 113]. □

Lemma 3.49 gives us a further consequence on complete intersection ideals as follows:

Corollary 3.50. *With the same* $R = K[x, y, z]$, $I = (f_1, f_2, f_3)$ *as in Theorem 3.48, assume that* $2 \leq \deg f_1 \leq \deg f_2 \leq \deg f_3$. *Let* $l \in R$ *be a general linear element. Then* $\mu((f_1, f_2, f_3, l)/(l)) = 3$ *if and only if* $\deg f_3 \leq \deg f_1 + \deg f_2 - 2$.

Proof. Put $d_1 = \deg f_1$, $d_2 = \deg f_2$, $d_3 = \deg f_3$. Assume $d_3 > d_1 + d_2 - 2$. Then we have that $(f_1, f_2, f_3, l)/(l) = (f_1, f_2, l)/(l)$ for the degree reasons. Conversely, assume that $d_3 \leq d_1 + d_2 - 2$. Then the rank two vector bundle $\mathscr{E}$ as defined in the Key lemma is semi-stable. Hence for a general element l, the free resolution of $I + (l)/(l)$ over $\overline{R} = R/(l)$ takes the form

$$0 \to \overline{R}(-e_1) \oplus \overline{R}(-e_2) \to \overline{R}(-d_1) \oplus \overline{R}(-d_2) \oplus \overline{R}(-d_3) \to I + (l)/(l),$$

where $|e_1 - e_2| \leq 1$ and $e_1 + e_2 = d_1 + d_2 + d_3$. This means that if $d := d_1 + d_2 + d_3$ is even, then $e_1 = e_2 = (d_1 + d_2 + d_3)/2$, and if d is odd, then $\{e_1, e_2\} = \{(d-1)/2, (d+1)/2\}$. In any case none of e_i coincides with any of d_1, d_2, d_3, which shows that $\mu(I + (l)/(l)) = 3$.   □

Remark 3.51. The result in Corollary 3.50 was proved in [148] without using the Grauert–Mülich theorem. This can be generalized to complete intersections of any codimension, and it can be used to prove the WLP for complete intersections in some special cases. This is shown below.

Theorem 3.52. *Let* K *be a field of characteristic* 0, $I \subset R = K[x_1, \ldots, x_n]$ *an ideal of height* n *minimally generated by* n *homogeneous elements of degrees* $d_1, \ldots, d_n$ *with* $2 \leq d_1 \leq \cdots \leq d_n$. *Let* l *be a general linear form in* R. *Then* $\mu(I + lR/lR) = n - 1$ *if and only if* $d_n > d_1 + \cdots + d_{n-1} - (n-1)$.

Proof. Let $f_1, \ldots, f_n$ minimally generate I. Suppose that $d_n > d_1 + \cdots + d_{n-1} - (n-1)$. Choose l general enough so that $l, f_1, \ldots, f_{n-1}$ is a regular sequence. Then $A := R/(l, f_1, \ldots, f_{n-1})$ is a graded complete intersection whose socle degree is $d_1 + \cdots + d_{n-1} - (n-1)$. Hence the natural image of f_n in the quotient ring A is zero for degree reasons. Hence we have $\mu((f_1, \ldots, f_n, l)/(l)) = n - 1$. The hard part is the other implication. This was proved in [148] for the case $n = 3$. For the general case see [150].   □

Remark 3.53. Theorem 3.52 can also be proved by the "Socle Lemma" in [62].

Corollary 3.54. *Let* $R = K[x_1, \ldots, x_n]$ *be the polynomial ring over a field* K *of characteristic zero. Let* $A = R/(f_1, \ldots, f_n)$ *be a standard graded complete intersection with* $d_i = \deg f_i$, *and* $2 \leq d_1 \leq \cdots \leq d_n$. *If* $d_n \geq d_1 + \cdots + d_{n-1} - n$, *then* A *has the WLP.*

Proof. We have to show that the Sperner number Sperner(A) equals the Rees number $r(A)$, which is, by definition, $\dim A/lA$ for a general linear form. Put $\delta = d_1 + d_2 + \cdots + d_{n-1} - (n-1)$ and $m = d_1 d_2 \cdots d_{n-1}$. Then we have

$$\text{Sperner}(A) = \begin{cases} m & \text{if } d_n > \delta, \\ m-1 & \text{if } d_n = \delta, \\ m-2 & \text{if } d_n = \delta - 1. \end{cases}$$

To see this, note that the Hilbert function of A is given by the polynomial

$$\text{Hilb}(A,t) = (1 + t + \cdots + t^{d_n-1}) \times \prod_{i=1}^{n-1}(1 + t + \cdots + t^{d_i-i}).$$

δ is equal to the degree of the second factor. If $d_n > \delta$, it is easy to see that Sperner(A) is equal to the sum of the coefficients of the second factor, which is equal to m. If $d_n = \delta$, then Sperner(A) is one less than that of the case $d_n > \delta$, and if $d_n = \delta - 1$, then Sperner(A) is two less than that of the case $d_n > \delta$; this is shown by a simple computation.

On the other hand we have

$$r(A) = \begin{cases} m & \text{if } d_n > \delta, \\ m-1 & \text{if } d_n = \delta, \\ m-2 & \text{if } d_n = \delta - 1. \end{cases}$$

To verify this, assume first that $d_n > \delta$. Then we have $A/lA = R/(f_1, \ldots, f_n, l) = R/(f_1, \ldots, f_{n-1}, l)$ for a general linear form l. Thus $r(A) = \dim A/lA = m$, as is claimed. If $d_n = \delta$, then the element f_n generates the socle of the algebra $B :=$ $R/(f_1, \ldots, f_{n-1}, l)$, and if $d_n = \delta - 1$, then f_n generates an ideal of B of dimension 2. Thus all three cases are justified and Sperner$(A) = r(A)$ for $d_n \geq \delta - 1$. □

Remark 3.55. In the proof of Corollary 3.54, we used Theorem 3.52 only for the last two cases $(d_n = \delta$ or $\delta - 1)$, but not for the first case where $d_n > \delta$. So, if $d_n > \delta$, then the assertion is valid for an infinite field K of positive characteristic also. If K is finite, it will affect the equality $\dim_K A/lA = m$ for some linear form l.

3.4.3 The WLP of Almost Complete Intersection in Codimension Three

We continue to assume Char $K = 0$ to the end of this section. The theorem of Grauert–Mülich can be used further to determine the almost complete intersections in codimension three which have the WLP. (An $\mathfrak{m}$-primary ideal is an almost complete intersection if it is generated by $n + 1$ elements in $K[x_1, x_2, \ldots, x_n]$.) We state the result of H. Brenner and A. Kaid [10] and their examples without proof.

Let $R = K[x, y, z]$. Let I be an ideal $I = (f_1, f_2, f_3, f_4) \subset R$ of finite colength minimally generated by four homogeneous elements and $A = R/I$. Let $\mathscr{E}$ be the "syzygy bundle" defined by (f_1, f_2, f_3, f_4). Namely $\mathscr{E}$ is defined by the exact sequence

$$0 \to \mathscr{E} \to \mathscr{O}_{\mathbb{P}^2}(-d_1) \oplus \mathscr{O}_{\mathbb{P}^2}(-d_2) \oplus \mathscr{O}_{\mathbb{P}^2}(-d_3) \oplus \mathscr{O}_{\mathbb{P}^2}(-d_4) \to \mathscr{O}_{\mathbb{P}^2} \to 0,$$

where $d_i = \deg f_i$, and the surjective map is

$$\begin{pmatrix} f_1 \\ f_2 \\ f_2 \\ f_4 \end{pmatrix}.$$

We have the following results in this context.

Theorem 3.56 ([10]). *Let $L \subset \mathbb{P}^2$ be a general line and suppose that $\mathscr{E}|_L \cong \mathscr{O}_{\mathbb{P}^1}(-a_1) \oplus \mathscr{O}_{\mathbb{P}^1}(-a_2) \oplus \mathscr{O}_{\mathbb{P}^1}(-a_3)$ and $a_1 \geq a_2 \geq a_3$.*

1. *If $\mathscr{E}$ is semi-stable, then A has the WLP if and only if $0 \leq a_1 - a_3 \leq 1$.*
2. *If $\mathscr{E}$ is not semi-stable, then A has the WLP.*

Example 3.57 ([10]).

1. Let $A = K[x, y, z]/(x^3, y^3, z^3, xyz)$. Then $\mathscr{E}$ is semi-stable, and A does not have the WLP.
2. $A = K[x, y, z]/(x^9, y^9, z^9, x^3 y^3 z^3)$. Then $\mathscr{E}$ is semi-stable, and A does not have the WLP.
3. $A = K[x, y, z]/(x^4, y^4, z^4, x^3 y)$. Then $\mathscr{E}$ is not semi-stable, and A has the WLP.

Remark 3.58. The stability and semi-stability of vector bundles were defined by D. Mumford and D. Giesecker for compactification of moduli spaces of vector bundles. The interested reader is referred to Mumford [105] and R. Pandharipande [114].

3.5 Jordan Decompositions and Tensor Products

In Sect. 1.4.3 we introduced the notation $(n_1, n_2, \ldots, n_r)$ for a partition of a positive integer n. It is a non-increasing sequence of non-negative integers such that $\sum_{i=1}^{r} n_i = n$. In this Sect. 3.5, we change slightly the notation for a partition and write

$$n_1 \oplus n_2 \oplus \cdots \oplus n_r$$

if $\sum_{i=1}^r n_i = n$ and if $n_i \geq 0$, for all i. (We do not assume the entries are in the non-increasing order.) Two such expressions $n_1 \oplus n_2 \oplus \cdots \oplus n_r$ and $n_1' \oplus n_2' \oplus \cdots \oplus n_{r'}'$ are regarded as the same if they differ only by permutation of terms disregarding the terms which are zero. Thus $n_1 \oplus n_2 \oplus \cdots \oplus n_r$ is another way to express a partition of a positive integer.

We are interested in the Jordan canonical forms of nilpotent matrices. Let $(A, \mathfrak{m}, K)$ be an Artinian local ring which contains the residue field $K = A/\mathfrak{m}$, and let $y \in \mathfrak{m}$. Consider the map $\times y : A \longrightarrow A$ defined by $a \mapsto ya$. Since A is Artinian, the linear map $\times y \in \mathrm{End}(A)$ is nilpotent, and hence the Jordan canonical form $J_A(\times y)$ of $\times y$ is a matrix of the following form:

$$J_A(\times y) = \begin{pmatrix} J(n_1) & & & \\ & J(n_2) & & \huge 0 \\ & & \ddots & \\ \huge 0 & & & J(n_r) \end{pmatrix},$$

where $J(m)$ is a Jordan block of size $m \times m$

$$J(m) = \begin{pmatrix} 0 & & & \\ 1 & 0 & & \huge 0 \\ & \ddots & \ddots & \\ \huge 0 & & 1 & 0 \end{pmatrix}.$$

We denote this **Jordan decomposition** of $\times y$ by writing

$$P_A(\times y) = n_1 \oplus n_2 \oplus \cdots \oplus n_r.$$

We note that

$$r = \dim_K A/yA \quad \text{and} \quad \dim_K A = \sum_{i=1}^r n_i.$$

Assume that $n_1 \geq n_2 \geq \cdots \geq n_r$ and let $(f_1, f_2, \ldots, f_s)$ be the finest subsequence of $(n_1, n_2, \ldots, n_r)$ such that $f_1 > f_2 > \cdots > f_s$. Then we may rewrite the same Jordan decomposition $P_A(\times y)$ as

$$P_A(\times y) = n_1 \oplus n_2 \oplus \cdots \oplus n_r$$
$$= \underbrace{f_1 \oplus f_1 \oplus \cdots \oplus f_1}_{m_{f_1}} \oplus \underbrace{f_2 \oplus f_2 \oplus \cdots \oplus f_2}_{m_{f_2}} \oplus \cdots \oplus \underbrace{f_s \oplus f_s \oplus \cdots \oplus f_s}_{m_{f_s}}$$
$$= f_1^{m_{f_1}} \oplus f_2^{m_{f_2}} \oplus \cdots \oplus f_s^{m_{f_s}},$$

so m_{f_i} is the number of times a Jordan block of size f_i occurs.

For a partition $P = n_1 \oplus n_2 \oplus \cdots \oplus n_r$ of a positive integer n, consider the following polynomial

$$Y(P, \lambda) = \sum_{i=1}^{r} (1 + \lambda + \lambda^2 + \cdots + \lambda^{n_i - 1}),$$

and let d_i be the coefficient of λ^i in $Y(P, \lambda)$ for all $i = 0, 1, \ldots, c$, where $c = \max_i \{ n_i - 1 \}$. Then we call $\hat{P} = d_0 \oplus d_1 \oplus \cdots \oplus d_c$ the **dual partition** of P.

Notation 3.59. *Let $(A, \mathfrak{m}, K)$ be an Artinian local ring, $y \in \mathfrak{m}$ and $P_A(\times y) = n_1 \oplus n_2 \oplus \cdots \oplus n_r$ the Jordan decomposition of the linear map $\times y : A \to A$. Take subspaces $\mathscr{V}_k$ of A corresponding to the Jordan blocks of sizes $n_k \times n_k$. Namely these are the subspaces $\mathscr{V}_1, \mathscr{V}_2, \ldots, \mathscr{V}_r$ of A with $A = \bigoplus_{k=1}^{r} \mathscr{V}_k$ which satisfy the following conditions for all $k = 1, 2, \ldots, r$:*

1. *$y\mathscr{V}_k \subset \mathscr{V}_k$,*
2. *The Jordan canonical form of the map $\times y : \mathscr{V}_k \to \mathscr{V}_k$ is a single Jordan block of size n_k,*
3. *$\dim_K \mathscr{V}_k = n_k$.*

Furthermore we choose elements $v_1, v_2, \ldots, v_r$ of A such that $\{ v_k, v_k y, v_k y^2, \ldots, v_k y^{n_k - 1} \}$ is a basis of $\mathscr{V}_k$ for all $k = 1, 2, \ldots, r$.

Lemma 3.60. *Let $(A, \mathfrak{m})$ be an Artinian local ring which contains the residue field $K = A/\mathfrak{m}$. Let $y \in \mathfrak{m}$ and set $e_i = \dim_K (0 : y^{i+1})/(0 : y^i)$ for all $i = 0, 1, 2, \ldots, b$, where $b = \max \{ j \mid y^j \neq 0 \}$. Then $P(\times y)$ is equal to the dual partition of $e_0 \oplus e_1 \oplus \cdots \oplus e_b$.*

Proof. We use Notation 3.59, and prove this by induction on $n = \dim_K A$. Since

$$\left\{ v_1 y^{n_1 - 1}, v_2 y^{n_2 - 1}, \ldots, v_r y^{n_r - 1} \right\}$$

is a basis of $(0 : y)$ we have $e_0 = r$. Write $\bar{a}$ for the image of $a \in A$ in $A/(0 : y)$. Then

$$\left\{ \overline{v_k y^{m_k}} \mid k = 1, 2, \ldots, r; \ n_k \geq 2; \ m_k = 0, 1, 2, \ldots, n_k - 2 \right\}$$

is a basis of $A/(0 : y)$ so we have

$$P_{A/(0:y)}(\times \bar{y}) = \bigoplus_{1 \leq k \leq r, n_k \geq 2} (n_k - 1),$$

and hence $\hat{P}_A(\times y) = e_0 \oplus \hat{P}_{A/(0:y)}(\times \bar{y})$ by the definition of the dual partition. Therefore our desired equality is immediately proved by using the induction hypothesis that $P_{A/(0:y)}(\times \bar{y})$ is equal to the dual partition of $e_1 \oplus e_2 \oplus \cdots \oplus e_b$. $\square$

Remark 3.61. Lemma 3.60 also gives that

$$\hat{P}_{A/(0:y^i)}(\times \overline{y}) = e_i \oplus e_{i+1} \oplus \cdots \oplus e_b$$

for all $i = 1, 2, \ldots, b$.

Definition 3.62. Let $P = n_1 \oplus n_2 \oplus \cdots \oplus n_r$ and $Q = m_1 \oplus m_2 \oplus \cdots \oplus m_s$ be two partitions of a positive integer n, where we assume that $n_1 \geq n_2 \geq \cdots \geq n_r$ and $m_1 \geq m_2 \geq \cdots \geq m_s$. Then we will write $P \succ Q$ and say Q is less than P if and only if (i) $r < s$ or (ii) $r = s$ and $n_i = m_i$ for $i = 1, 2, \ldots, j - 1$ and $n_j > m_j$.

Remark 3.63. Let A be a standard graded Artinian K-algebra with a unimodal Hilbert function and y a homogeneous element of A. Generally speaking, suppose that N is a nilpotent matrix. Then we have that $\dim_K \mathrm{Ker}(N)$ is equal to the number of blocks in the Jordan decomposition of N. Keeping this in mind it is easy to see the following:

1. The number of Jordan blocks of $\times y$ is greater than or equal to the Sperner number of A.
2. Suppose that y is a linear form. Then y is a weak Lefschetz element of A if and only if the number of Jordan blocks of $\times y$ is equal to the Sperner number of A.

The following is a characterization of strong Lefschetz elements for a standard graded Artinian K-algebra.

Proposition 3.64. *Let $A = \bigoplus_{i=0}^{c} A_i$ be a standard graded Artinian K-algebra with $A_c \neq (0)$, and let $y \in \mathfrak{m}$ be a homogeneous element.*

1. *Suppose that the Hilbert function $\{ h_i \}$ of A is unimodal. Then $P_A(\times y)$ is less than or equal to the dual partition of $h_0 \oplus h_1 \oplus \cdots \oplus h_c$.*
2. *Suppose that $y \in A_1$ is a linear form. Then the following conditions are equivalent.*

   a. *y is a strong Lefschetz element of A.*
   b. *the Hilbert function of A is unimodal and $P_A(\times y)$ is equal to the dual partition of $h_0 \oplus h_1 \oplus \cdots \oplus h_c$.*

We precede the proof with a lemma.

Lemma 3.65. *With the same notation as Proposition 3.64, suppose that y is a weak Lefschetz element of A and $\overline{y}$, the image of y in $A/(0 : y)$, is a strong Lefschetz element of $A/(0 : y)$. Then y is a strong Lefschetz element of A.*

Proof. Let h_t be a term of the Hilbert function of A which is equal to the Sperner number of A. We would like to show that the linear map $\times y^k : A_i \to A_{i+k}$ has full rank for all k and i.

We first consider the case where $i + k \leq t$. Consider the commutative diagram:

$$A_i \xrightarrow{\times y^k} A_{i+k}$$

$$(A/(0:y))_i \xrightarrow{\times \overline{y}^k} (A/(0:y))_{i+k}$$

In this case the vertical maps are bijective and the map $\times\overline{y}^k$ has full rank by our assumption, and hence the map $\times y^k$ also has full rank.

Secondly we consider the case where $i < t < i + k$. There is the commutative diagram:

$$A_i \xrightarrow{\times y^{k-1}} A_{i+k-1} \xrightarrow{\times y} A_{i+k}$$

$$(A/(0:y))_i \xrightarrow{\times \overline{y}^{k-1}} (A/(0:y))_{i+k-1} \xrightarrow{\times \overline{y}} A_{i+k}$$

In this case the map $\times\overline{y}^{k-1}$ has full rank and the map $\times\overline{y}$ is bijective by our assumption, and hence the composite $(A/(0:y))_i \to A_{i+k}$ also has full rank. Therefore the map $\times y^k: A_i \to A_{i+k}$ has full rank, since the first vertical map is bijective.

Finally we consider the case where $t \le i$. In this case the map $\times y: A_j \to A_{j+1}$ is surjective for all $j \ge i$ by our assumption, and hence the composite $\times y^k: A_i \to A_{i+k}$ has full rank.   □

Proof of Proposition 3.64. (1) Let $\hat{P}$ be the dual partition of $P = h_0 \oplus h_1 \oplus \cdots \oplus h_c$. Set $\hat{P} = m_1 \oplus m_2 \oplus \cdots \oplus m_s$ and $P_A(\times y) = n_1 \oplus n_2 \oplus \cdots \oplus n_r$, where $m_1 \ge m_2 \ge \cdots \ge m_s$ and $n_1 \ge n_2 \ge \cdots \ge n_r$. Since the Hilbert function of A is unimodal, we have

$$\sum_{i=0}^{c-1} \max\{h_i - h_{i+1}, 0\} = \text{Sperner}(A),$$

and hence

$$r = \dim_K \text{Ker}(\times y: A \to A)$$

$$= \sum_{k=0}^{c-1} \dim_K \text{Ker}(\times y: A_k \to A_{k+1})$$

$$\ge \sum_{k=0}^{c-1} \max\{h_k - h_{k+1}, 0\}$$

$$= \text{Sperner}(A)$$

$$= s.$$

If $r > s$ then $P_A(\times y) \prec \hat{P}$. So, consider the case $r = s$. The first term m_1 of $\hat{P}$ is equal to $c+1$. If $\deg y \geq 2$ then the first term n_1 of $P(\times y)$ is less than $(c+1)/2$, and hence $P(\times y) \preceq \hat{P}$. Let $\deg y = 1$. We would like to prove the following inequality

$$n_1 \oplus n_2 \oplus \cdots \oplus n_s \preceq m_1 \oplus m_2 \oplus \cdots \oplus m_s.$$

Since $\hat{P}$ is the dual partition of $P = h_0 \oplus h_1 \oplus \cdots \oplus h_c$, it is easy to show that

1. $n_1 \leq c + 1$, i.e., $n_1 \leq m_1$
2. $m_1 = n_1, m_2 = n_2, \ldots, m_i = n_i \implies m_{i+1} \geq n_{i+1}$

and this gives our desired inequality.

(2) Assume (2a) holds. Since A has the SLP, the Hilbert function of A is unimodal. Set $e_i = \dim_K (0 : y^{i+1})/(0 : y^i)$ for all $i = 0, 1, 2, \ldots, c$. Lemma 3.60 says that the dual partition of $P(\times y)$ is the set of positive integers $\{ e_i \mid i = 0, 1, 2, \ldots, c \}$ with multiplicity counted. Furthermore, we have

$$\hat{P}_{A/(0:y)}(\times \overline{y}) = e_1 \oplus e_2 \oplus \cdots \oplus e_c \tag{3.5}$$

by Remark 3.61. We would like to show that if the sequence $h_0, h_1, \ldots, h_c$ is arranged in decreasing order it coincides with the sequence $e_0, e_1, \ldots, e_c$. Let h_t be a term of $\{ h_i \}$ which is equal to the Sperner number of A. Then, since y is a weak Lefschetz element, we get the following isomorphism as K-vector spaces,

$$A/(0 : y) \cong A_0 \oplus A_1 \oplus \cdots \oplus A_{t-1} \oplus A_{t+1} \oplus \cdots \oplus A_c,$$

and the Hilbert function of $A/(0 : y)$ is equal to

$$h_0, h_1, \ldots, h_{t-1}, h_{t+1}, \ldots, h_c. \tag{3.6}$$

Hence, since e_0 is the Sperner number of A, the proof is complete by induction on c by using (3.5) and (3.6) above.

Assume (2b) holds. We use induction on c. The condition (2b) says that the dimension of $(0 : y)$ is equal to the Sperner number of A by Lemma 3.60, and hence y is a weak Lefschetz element of A. Furthermore using the above (3.5) and (3.6) again, the induction hypothesis implies that $\overline{y}$ is a strong Lefschetz element of $A/(0 : y)$. Hence (2a) follows immediately from Lemma 3.65.

Proposition 3.66. *Let A and B be Artinian local rings which contain the same residue field K of characteristic zero. Let $z \in A$ and $w \in B$ be non-unit elements. Set $P_A(\times z) = d_1 \oplus d_2 \oplus \cdots \oplus d_r$ and $P_B(\times w) = f_1 \oplus f_2 \oplus \cdots \oplus f_s$.*

1. $P_{A \otimes_K B}(\times(z \otimes 1 + 1 \otimes w)) = \bigoplus_{i,j} \bigoplus_{k=1}^{\min\{d_i, f_j\}} (d_i + f_j + 1 - 2k)$.
2. $\dim_K \mathrm{Ker}\,(\times(z \otimes 1 + 1 \otimes w)) = \sum_{i,j} \min \{ d_i, f_j \}$.

The proof is postponed until after Lemma 3.70.

Proposition 3.67 (Ikeda). *Let K be a field of characteristic $p \geq 0$. Let $B = \bigoplus_{i=0}^{c} B_i$, with $B_c \neq 0$ be a standard graded K-algebra. Assume that $p = 0$, or $p > c + 1$. Let $A = B[z]/(z^2)$. Then A has the SLP in the narrow sense if and only if B has the SLP in the narrow sense.*

Proof. We give a proof after H. Ikeda [66]. Notice that $\mathrm{Hilb}(B,t)$ is symmetric if and only if $\mathrm{Hilb}(A,t) = \mathrm{Hilb}(B,t)(1 + t)$ is symmetric. Thus for the rest of proof we assume that both A and B have symmetric Hilbert functions. Assume that B has the SLP with $y \in B_1$ an SL element. We are going to show that $\times(y + z)^{c+1-2i}: A_i \rightarrow A_{c+1-i}$ is injective, for $i = 0, 1, 2, \ldots, [(c + 1)/2]$. Assume that $(y + z)^{c+1-2i}(a + bz) = 0$, where $a + bz \in A_i = B_i \oplus B_{i-1}z$. Then, since

$$(y + z)^{c+1-2i}(a + bz) = (y^{c+1-2i} + (c + 1 - 2i)y^{c-2i}z)(a + bz)$$
$$= y^{c+1-2i}a + y^{c-2i}(yb + (c + 1 - 2i)a)z$$
$$= 0,$$

we have

$$y^{c+1-2i}a = 0, \tag{3.7}$$

$$y^{c-2i}(yb + (c + 1 - 2i))a = 0. \tag{3.8}$$

With (3.7), (3.8) implies $y^{c-2(i-1)}b = 0$. Since B has the SLP, we have $b = 0$. By (3.8), we get $(c + 1 - 2i)y^{c-2i}a = 0$. Since $c + 1 - 2i \neq 0$, again by the SLP of B, we have $a = 0$. Thus we have shown that $\times(y + z)^{c+1-2i}: A_i \rightarrow A_{c+1-i}$ is injective, for $i = 0, 1, 2, \ldots, [(c + 1)/2]$. Since the Hilbert function of A is symmetric, this shows that all these maps are bijective. Conversely assume that A has the SLP. Suppose that $y \in B_1$ and $y + z \in A_1$ is an SL element for A. Let $b \in B_i$ and $y^{c-2i}b = 0$. Then $(y + z)^{c+1-2i}b = y^{c+1-2i}b + (c + 1 - 2i)y^{c-2i}zb = 0$. Hence by the SLP for A, we have $b = 0$. This show that y is an SL element for B. $\qquad\square$

Remark 3.68. Theorem 4.10 is a generalization of Ikeda's result.

Corollary 3.69. *Let K be a field of characteristic $p \geq 0$. Let n be a positive integer. Then $A = K[x_1, x_2, \ldots, x_n]/(x_1^2, x_2^2 \ldots, x_n^2)$ has the SLP, if $p = 0$ or $p > n$, and in this case $\overline{x_1} + \overline{x_2} + \cdots + \overline{x_n}$ is an SL element.*

Proof. Immediate by Proposition 3.67. $\qquad\square$

Lemma 3.70. *Let $A = K[x, y]/(x^d, y^f)$, where $K[x, y]$ is the polynomial ring in two variables over a field K of characteristic zero.*

1. *$x + y$ is a strong Lefschetz element of A.*
2. *$P_A(\times(x + y)) = \underbrace{(d + f - 1) \oplus (d + f - 3) \oplus \cdots \oplus (|d - f| + 1)}_{\min\{d, f\}}.$*

3. *$\dim_K \mathrm{Ker}(\times(x + y)) = \min\{d, f\}.$*

Proof. (1) By Corollary 3.69 the Artinian algebra

$$A := K[z_1, z_2, \ldots, z_n]/(z_1^2, z_2^2, \ldots, z_n^2)$$

has the strong Lefschetz property with a strong Lefschetz element $\bar{z}_1 + \bar{z}_2 + \cdots + \bar{z}_n$. Choose n so that $n = d + f - 2$ and consider the ring homomorphism

$$\phi \colon K[X, Y] \to A$$

defined by $X \mapsto \bar{z}_1 + \bar{z}_2 + \cdots + \bar{z}_{d-1}$ and $Y \mapsto \bar{z}_d + \bar{z}_{d+1} + \cdots + \bar{z}_n$. Put $J = \operatorname{Ker} \phi$. We claim that $J = (X^d, Y^f)$. It is easy to see the inclusion

$$J \supset (X^d, Y^f). \tag{3.9}$$

We have to show that it is an equality. By way of contradiction assume that J is strictly larger than (X^d, Y^f). Then the homogeneous part of degree $d + f - 2$ of R/J is (0), since $R/(X^d, Y^f)$ has a unique minimal ideal, the socle, at that degree. On the other hand we have

$$\phi(X^{d-1} Y^{f-1}) = (d-1)! \, (f-1)! \, \overline{z_1 z_2 \cdots z_n} \neq 0.$$

This is a contradiction. Thus the inclusion (3.9) is in fact the equality.

We have just shown that the subring $B := K[\phi(X), \phi(Y)]$ of A is isomorphic to

$$K[X, Y]/(X^d, Y^f).$$

Note that B has a symmetric Hilbert function and moreover it shares the same socle with A. Then, since the Lefschetz element $\bar{z}_1 + \bar{z}_2 + \cdots + \bar{z}_n$ of A lies in B, it is also a Lefschetz element of B.

(2) Assume that $d \leq f$. Then the Hilbert function of A is as follows:

$$\underbrace{1, 2, 3, \ldots, d-2, d-1, \overbrace{d, d, \ldots, d}^{|d-f|+1}, d-1, d-2, \ldots, 3, 2, 1}_{d+f-1}$$

(The case $d > f$ is similar.) Hence, the assertion (2) follows from Proposition 3.64 (2), since $x + y$ is a strong Lefschetz element of A.

The assertion (3) follows from Proposition 3.5, since $x + y$ is a weak Lefschetz element of A and $\operatorname{Sperner}(A) = \min \{ d, f \}$. $\qquad \square$

Proof (Proof of Proposition 3.66).

Case 1. Let $r = s = 1$. Set $d = d_1$ and $f = f_1$. We would like to show that

$$P(\times(z \otimes 1 + 1 \otimes w)) = \underbrace{(d + f - 1) \oplus (d + f - 3) \oplus \cdots \oplus (|d - f| + 1)}_{\min\{d, f\}}$$

and $\dim_K \mathrm{Ker}(\times(z \otimes 1 + 1 \otimes w)) = \min\{d, f\}$. Consider the algebras $K[x]/(x^d)$ and $K[y]/(y^f)$. The linear map $\times x \colon K[x]/(x^d) \to K[x]/(x^d)$ may be identified with $\times z \colon A \to A$ and similarly for $\times y$ and $\times w$. Thus the multiplication map

$$\times(x + y) \colon K[x, y]/(x^d, y^f) \to K[x, y]/(x^d, y^f)$$

is essentially the same as $\times(z \otimes 1 + 1 \otimes w)$. Hence our assertion follows from Lemma 3.70.

Case 2. Next we consider the case where $r > 1$ and $s > 1$. Let $\mathcal{V}_1, \mathcal{V}_2, \ldots, \mathcal{V}_r$ be the subspaces of A corresponding to Jordan blocks for $\times z \in \mathrm{End}(A)$. Namely $\{\mathcal{V}_i\}$ are vector subspaces of A with $A = \bigoplus_{i=1}^r \mathcal{V}_i$ which satisfy the following conditions for each $i = 1, 2, \ldots, r$.

1. $z\mathcal{V}_i \subset \mathcal{V}_i$.
2. The Jordan canonical form of $\times z \in \mathrm{End}(\mathcal{V}_i)$ is a single Jordan block.
3. $\dim_K \mathcal{V}_i = d_i$

Similarly, there are vector subspaces $\mathcal{W}_1, \mathcal{W}_2, \ldots, \mathcal{W}_r$ of B with $B = \bigoplus_{i=1}^s \mathcal{W}_i$ which satisfy the similar conditions. Then $A \otimes B = \bigoplus_{i,j}(\mathcal{V}_i \otimes \mathcal{W}_j)$, and $\mathcal{V}_i \otimes \mathcal{W}_j$ is closed under the multiplication by $\times(z \otimes 1 + 1 \otimes w) \colon A \otimes B \to A \otimes B$ for all i and j. Hence our assertion is proved by applying the result of Case 1 for each $\mathcal{V}_i \otimes \mathcal{W}_j$. $\square$

Remark 3.71. Lemma 3.70 is equivalent to Clebsch–Gordan Theorem. Yet another proof can be found in Martsinkovsky and Vlassov [91].

The following is a good property of strong Lefschetz elements.

Proposition 3.72. *Let* $A = \bigoplus_{i=0}^c A_i$ *be a standard graded Artinian K-algebra with the SLP, g a strong Lefschetz element of A and k a positive integer. Then* $P(\times g^k) \succeq P(\times y^k)$ *for all linear forms* $y \in A_1$.

Proof. (Step 1) Let $\{h_i\}$ be the Hilbert function of A. Set $P(\times g) = m_1 \oplus m_2 \oplus \cdots \oplus m_s$ and $P(\times y) = n_1 \oplus n_2 \oplus \cdots \oplus n_r$, where $m_1 \geq m_2 \geq \cdots \geq m_s$ and $n_1 \geq n_2 \geq \cdots \geq n_r$. Then, since g is a strong Lefschetz element, it follows that $m_1 = c + 1$ and $m_1 \geq n_1$. Hence, using that $P(\times g)$ is the dual partition of $h_0 \oplus h_1 \oplus \cdots \oplus h_c$ by Proposition 3.64 (2), we have the following:

$$m_1 = n_1, m_2 = n_2, \ldots, m_i = n_i \implies m_{i+1} \geq n_{i+1}.$$

Note that this is clear by Definition of the order "$\succ$" in the case where $r = s$.

(Step 2) Set $P(\times g^k) = f_1 \oplus f_2 \oplus \cdots \oplus f_t$ and $P(\times y^k) = f_1' \oplus f_2' \oplus \cdots \oplus f_u'$. Then, since g is a strong Lefschetz element, we have

$$\dim_K \mathrm{Ker}(\times g^k \colon A_i \to A_{i+k}) = \max\{h_i - h_{i+k}, 0\}.$$

Hence, it follows that

$$u = \dim_K (0 : y^k)$$

$$= \dim_K \text{Ker}(\times y^k : A \to A)$$

$$= \sum \dim_K \text{Ker}(\times y^k : A_i \to A_{i+k})$$

$$\geq \sum \max\{ h_i - h_{i+k}, 0 \}$$

$$= \sum \dim_K \text{Ker}(\times g^k : A_i \to A_{i+k})$$

$$= \dim_K (0 : g^k)$$

$$= t.$$

Thus, we see that $P(\times g^k) \succ P(\times y^k)$ if $u > t$.

(Step 3) Next we consider the case where $u = t$. Chose non-negative integers q_i, r_i, q'_j and r'_j satisfying the following conditions for all $1 \leq i \leq s$ and $1 \leq j \leq r$; $m_i = q_i \times k + r_i$ $(0 \leq r_i < k)$ and $n_j = q'_j \times k + r'_j$ $(0 \leq r'_j < k)$. Then we have

$$P(\times g^k) = \bigoplus_{i=1}^{r} (\underbrace{(q_i + 1) \oplus (q_i + 1) \oplus \cdots \oplus (q_i + 1)}_{r_i} \oplus \underbrace{q_i \oplus q_i \oplus \cdots \oplus q_i}_{k-r_i})$$

and

$$P(\times y^k) = \bigoplus_{j=1}^{s} (\underbrace{(q'_j + 1) \oplus (q'_j + 1) \oplus \cdots \oplus (q'_j + 1)}_{r'_j} \oplus \underbrace{q'_j \oplus q'_j \oplus \cdots \oplus q'_j}_{k-r'_j}).$$

Hence, using Step 1, it is easy to show that $P(\times g^k) \succeq P(\times y^k)$.   □

Question 3.73. Let A be a standard graded Artinian K-algebra, g a strong Lefschetz element for A and k a positive integer. Does the inequality $P(\times g^k) \succeq P(\times z)$ hold for all $z \in A_k$?

3.6   SLP for Artinian Gorenstein Algebras and Hessians

In this section we assume that char. $K = 0$. Let us regard the polynomial algebra $R := K[x_1, x_2, \ldots, x_n]$ as a module over the algebra $Q := K[X_1, X_2, \ldots, X_n]$ via the identification $X_i = \partial/\partial x_i$.

Lemma 3.74. *Let $A = \bigoplus_{i=0}^{c} A_i$, $A_c \neq 0$, be a graded Artinian Gorenstein K-algebra with the standard grading. Fix an isomorphism $[\ ]: A_c \overset{\sim}{\to} K$. For a K-linear basis $\beta_1, \beta_2, \ldots, \beta_n$ of A_1, define the polynomial*

$$F(x_1, x_2, \ldots, x_n) := [(x_1 \beta_1 + x_2 \beta_2 + \cdots + x_n \beta_n)^c]$$

in the variables $x_1, x_2, \ldots, x_n$ *taking values in* K. *Then we have the following presentation of* A:

$$A \cong Q / \operatorname{Ann}_Q(F).$$

Proof. Extend the linear map $[\ ]: A_c \xrightarrow{\sim} K$ by setting $[\alpha] = 0$ for homogeneous elements α with $\deg \alpha < c$. Note that

$$F(x_1, x_2, \ldots, x_n) = c! \, [\exp(x_1 \beta_1 + x_2 \beta_2 + \cdots + x_n \beta_n)].$$

Let us consider the algebra homomorphism $\varphi: Q \to A$ defined by $X_i \mapsto \beta_i$. For $f(X) \in \operatorname{Ker} \varphi$, it is clear that

$$f(X) F(x) = c! \, [f(\beta) \cdot \exp(x_1 \beta_1 + x_2 \beta_2 + \cdots + x_n \beta_n)] = 0.$$

This means that $\operatorname{Ker} \varphi \subset \operatorname{Ann}_Q(F)$. Conversely, for $f(X) \in \operatorname{Ann}_Q(F)$, we have

$$[\alpha \cdot f(\beta) \cdot \exp(x_1 \beta_1 + x_2 \beta_2 + \cdots + x_n \beta_n)] = 0$$

for arbitrary element $\alpha \in A$. After the substitution $x_1 = x_2 = \cdots = x_n = 0$, we obtain that $[\alpha \cdot f(\beta)] = 0$. Since A is a Poincaré duality algebra, $f(\beta)$ must be zero. Hence we have $\operatorname{Ann}_Q(F) \subset \operatorname{Ker} \varphi$. $\qquad\square$

Definition 3.75. Let G be a polynomial in $R = K[x_1, x_2, \ldots, x_n]$. If a family $\boldsymbol{B}_d = \{\alpha_i^{(d)}\}_i$ of homogeneous polynomials of degree $d > 0$ in Q is given, we call the polynomial

$$\det\left((\alpha_i^{(d)}(X) \, \alpha_j^{(d)}(X) \, G(x))|_{i,j=1}^{|\boldsymbol{B}_d|} \right) \in K[x_1, x_2, \ldots, x_n]$$

the d-th **Hessian** of G with respect to $\boldsymbol{B}_d$, and denote it by $\operatorname{Hess}_{\boldsymbol{B}_d}^{(d)} G$. We also denote the d-th Hessian simply by $\operatorname{Hess}^{(d)} G$ if the choice of $\boldsymbol{B}_d$ is clear.

Note that if $d = 1$ and $\boldsymbol{B}_1 = \{X_1, \ldots, X_n\}$, then $\operatorname{Hess}_{\boldsymbol{B}_1}^{(1)} G$ coincides with the usual Hessian

$$\operatorname{Hess}(G) = \det\left(\frac{\partial^2 G}{\partial x_i \, \partial x_j} \right)_{ij}.$$

Let $A = K[X_1, X_2, \ldots, X_n] / \operatorname{Ann}_Q(F)$ be a graded Artinian Gorenstein algebra with the standard grading and socle degree c.

Theorem 3.76 ([89, 149]). *Fix an arbitrary* K-*linear basis* $\boldsymbol{B}_d$ *of* A_d *for* $d = 1, 2, \ldots, [c/2]$. *An element* $L = a_1 X_1 + a_2 X_2 + \cdots + a_n X_n \in A_1$ *is a strong Lefschetz element of* $A = Q / \operatorname{Ann}_Q(F)$ *if and only if* $F(a_1, a_2, \ldots, a_n) \neq 0$ *and*

$$(\mathrm{Hess}_{B_d}^{(d)} F)(a_1, a_2, \ldots, a_n) \neq 0$$

for $d = 1, 2, \ldots, [c/2]$.

Proof. Fix the identification [ ]: $A_c \xrightarrow{\sim} K$ by $[\omega(X)] := \omega(X)F(x)$ for any $\omega(X) \in A_c$. Note that $\omega(X)F(x) \in K$, because $\deg(\omega) = \deg(F) = c$. Since A is a Poincaré duality algebra, the necessary and sufficient condition for $L = a_1 X_1 + a_2 X_2 + \cdots + a_n X_n \in A_1$ to be a strong Lefschetz element is that the bilinear pairing

$$
\begin{array}{ccccc}
A_d \times A_d & \to & A_c & \cong & K \\
(\xi, \eta) & \mapsto & L^{c-2d}\xi\eta & \mapsto & [L^{c-2d}\xi\eta]
\end{array}
$$

is non-degenerate for $d = 0, 1, 2, \ldots, [c/2]$. Therefore L is a Lefschetz element if and only if the matrix

$$(L^{c-2d} \alpha_i^{(d)}(X) \alpha_j^{(d)}(X) F(x))_{ij}$$

has nonzero determinant. For a homogeneous polynomial $G(x_1, x_2, \ldots, x_n) \in K[x_1, x_2, \ldots, x_n]$ of degree d, we have the formula

$$(a_1 X_1 + a_2 X_2 + \cdots + a_n X_n)^d G(x_1, x_2, \ldots, x_n) = d!\, G(a_1, a_2, \ldots, a_n),$$

so

$$L^{c-2d} \alpha_i^{(d)}(X)\alpha_j^{(d)}(X)F(x) = (c-2d)!\, \alpha_i^{(d)}(X)\alpha_j^{(d)}(X)F(x)|_{(x_1, x_2, \ldots, x_n) = (a_1, a_2, \ldots, a_n)}.$$

This completes the proof.   □

Proposition 3.77. *Let* $R := K[x_1, x_2, \ldots, x_n]$ *and* $R' := K[y_1, y_2, \ldots, y_m]$ *be the polynomial rings. Denote by* $Q = K[\partial/\partial x_1, \partial/\partial x_2, \ldots, \partial/\partial x_n]$ *and* $Q' = K[\partial/\partial y_1, \partial/\partial y_2, \ldots, \partial/\partial y_m]$ *the rings of differential polynomials over* R *and* R' *respectively.*

1. For $F \in R$ *and* $G \in R'$, *we have*

$$(Q \otimes Q')/\mathrm{Ann}_{Q \otimes Q'}(F \cdot G) \cong (Q/\mathrm{Ann}_Q F) \otimes (Q'/\mathrm{Ann}_{Q'} G).$$

2. Assume that $F \in R$, $G \in R'$ *and* $\deg(F) = \deg(G)$. *The algebra* $(Q \otimes Q')/\mathrm{Ann}_{Q \otimes Q'}(F + G)$ *has the SLP if and only if* $Q/\mathrm{Ann}_Q(F)$ *and* $Q'/\mathrm{Ann}_{Q'}(G)$ *have the SLP.*

Proof. (1) Let I be the ideal generated by $\mathrm{Ann}_Q(F)$ and $\mathrm{Ann}_{Q'}(G)$ in $Q \otimes Q'$. We will show that $\mathrm{Ann}_{Q \otimes Q'}(F \cdot G) = I$. It is easy to see that $I \subset \mathrm{Ann}_{Q \otimes Q'}(F \cdot G)$. Hence we have the surjective homomorphism

$$\varphi: (Q \otimes Q')/I \to (Q \otimes Q')/\mathrm{Ann}_{Q \otimes Q'}(F \cdot G).$$

Since the tensor product of two Poincaré duality algebras is also a Poincaré duality algebra, the algebra

$$(Q \otimes Q')/I \cong (Q/\operatorname{Ann}_Q F) \otimes (Q'/\operatorname{Ann}_{Q'} G)$$

is Gorenstein. Note that $(Q \otimes Q')/I$ and $(Q \otimes Q')/\operatorname{Ann}_{Q \otimes Q'}(F \cdot G)$ have the same socle degree $c := \deg(F) + \deg(G)$. So the homomorphism φ induces an isomorphism between the socle of $(Q \otimes Q')/I$ and that of $(Q \otimes Q')/\operatorname{Ann}_{Q \otimes Q'}(F \cdot G)$. Choose a polynomial $\omega \in Q \otimes Q'$ that gives the socle generators of $Q \otimes Q'/I$ and of $Q \otimes Q'/\operatorname{Ann}_{Q \otimes Q'}(F \cdot G)$. Let $f \in Q \otimes Q' \setminus I$ be a homogeneous polynomial with $\deg(f) < c$. Since $Q \otimes Q'/I$ is a Poincaré duality algebra, there exists a polynomial $g \in Q \otimes Q'$ such that $fg = \omega \bmod I$. This means that $\varphi(fg) = \omega$ in $(Q \otimes Q')/\operatorname{Ann}_{Q \otimes Q'}(F \cdot G)$. In particular, we have $f \neq 0$ in $(Q \otimes Q')/\operatorname{Ann}_{Q \otimes Q'}(F \cdot G)$. Hence we have the equality $\operatorname{Ann}_{Q \otimes Q'}(F \cdot G) = I$.

(2) Since we have

$$\frac{\partial}{\partial x_i}\frac{\partial}{\partial y_j}(F + G) = 0$$

for $i = 1, 2, \ldots, n$ and $j = 1, 2, \ldots, m$, the Hessians $\operatorname{Hess}^{(k)}(F + G)$ factor into the products of $\operatorname{Hess}^{(k)}(F)$ and $\operatorname{Hess}^{(k)}(G)$ for $k = 1, 2, \ldots, [(\deg F)/2]$. The claim follows from Theorem 3.76. $\square$

Example 3.78. In [41], Gordan and Noether proved that if a polynomial $f = f(x_1, x_2, x_3, x_4)$ in four variables has zero Hessian, then a variable can be eliminated from f by using a linear transformation of variables. In other words, we can find an invertible matrix $(a_{ij}) \in GL(4, K)$ and a polynomial $g = g(y_1, y_2, y_3, y_4)$ such that $f(x_1, x_2, x_3, x_4) = g(y_1, y_2, y_3, 0)$ under the change of variables $y_j = \sum_{i=1}^{4} a_{ij} x_i$, $j = 1, 2, 3, 4$. The polynomial $F = u^2 x + uvy + v^2 z \in K[u, v, x, y, z]$ is the simplest example in five variables that has zero Hessian, but none of the variables can be eliminated by a linear transformation. Gordan-Noether [41] described all polynomials in five variables with zero Hessian (Cf. Lossen [82] and Watanabe [151]).

For the polynomial $F = u^2 x + uvy + v^2 z \in Q = K[u, v, x, y, z]$, the algebra $A = Q/\operatorname{Ann}_Q(F)$ does not have the strong Lefschetz property by Theorem 3.76. The algebra A can also be constructed as follows. Let us consider the algebra $B := K[u, v]/(u, v)^3$ and its canonical module E. Then the algebra A is isomorphic to the idealization $B \ltimes E$. The Hilbert series of A and B are given by $\operatorname{Hilb}(A) = (1, 5, 5, 1)$ and $\operatorname{Hilb}(B) = (1, 2, 3)$.

Note that the subring $K[u, v, F]$ can be characterized as the space of solutions in $K[u, v, x, y, z]$ of the following system of differential equations:

$$\left(-v\frac{\partial}{\partial x} + u\frac{\partial}{\partial y} + 0\frac{\partial}{\partial z}\right) F = 0,$$

$$\left(0\frac{\partial}{\partial x} - v\frac{\partial}{\partial y} + u\frac{\partial}{\partial z}\right) F = 0.$$

These equations imply that

$$\left(v^2\frac{\partial}{\partial x} - 2uv\frac{\partial}{\partial y} + u^2\frac{\partial}{\partial z}\right) F = 0.$$

Hence any element of $K[u, v, F]$ has zero Hessian, because the vector $(0, 0, v^2, -2uv, u^2)$ is in the kernel of the Hessian matrix

$$\left(\frac{\partial^2 F}{\partial x_i \partial x_j}\right)_{i,j}.$$

The space of polynomials with zero Hessian is discussed also in (Maeno and Watanabe, The theory of gordan-noether on homogeneous forms with zero hessian, unpublished).

Example 3.79 (Ikeda's Example). Define the polynomial $F \in K[w, x, y, z]$ by $F = w^3xy + wx^3z + y^3z^2$. The Hessian is given as follows:

$$\text{Hess}(F) = 8(3w^7xy^4 + 8w^6x^6 - 27w^5x^3y^3z + 27w^4y^6z^2$$

$$- 45w^3x^5y^2z^2 - 54w^2x^2y^5z^3 + 9wx^7yz^3 + 27x^4y^4z^4).$$

On the other hand, we have $\text{Hess}^{(2)}(F) = 0$, so $A := Q/\text{Ann}_Q(F)$ does not have the strong Lefschetz property. The Hilbert series is given by $\text{Hilb}(A) = (1, 4, 10, 10, 4, 1)$.

Problem 3.80. It is still an open problem whether there exist any Artinian Gorenstein graded algebras over a field of characteristic zero with dim $A_1 = 3$ without the strong (or weak) Lefschetz property. Equivalently, an example of a homogeneous polynomial F in three variables such that $\text{Hess}(F) \neq 0$ but $\text{Hess}^{(d)}(F) = 0$ for some $d > 1$ is not known.

Example 3.81 (Hesse cubic). Let us consider the Hesse cubic $F = x^3 + y^3 + z^3 - 6sxyz$, where s is a parameter. Then the ideal $\text{Ann}_Q(F)$ is given as follows according to the choice of s:

$$\text{Ann}_Q(F) = \begin{cases} (sX^2 + YZ, sY^2 + XZ, sZ^2 + XY) & s^3 \neq 0, 1, \\ (X^3 - Y^3, X^3 - Z^3, XY, YZ, XZ) & s = 0, \\ (sX^2 + YZ, sY^2 + XZ, sZ^2 + XY, XZ^2, YZ^2) & s^3 = 1. \end{cases}$$

In any cases, we have $\mathrm{Hilb}(A) = (1, 3, 3, 1)$. The set of *non*-Lefschetz elements in A_1 is the union of the loci given by

$$a^3 + b^3 + c^3 - 6s \cdot abc = 0$$

and

$$s^2 a^3 + s^2 b^3 + s^2 c^3 - (1 - 2s^3)abc = 0.$$

These equations describes two elliptic curves in the projective plane $\mathbb{P}A_1 \cong \mathbb{P}^2$ intersecting at the inflection points of each other.

Chapter 4
Complete Intersections with the SLP

The main result of this chapter is Theorem 4.10. This may be regarded as a generalization of Theorem 3.34 which states that the SLP is preserved by tensor products. Using the main theorem, we give some examples of complete intersections with the strong Lefschetz property.

4.1 Central Simple Modules

In order to prove the main theorem, we introduce the notion of central simple modules for a graded Artinian K-algebra, and review the definition of the SLP for modules and a characterization of the SLP in terms of central simple modules. Central simple modules are a useful tool to study the SLP. For details we refer the reader to [52,53] and [54].

Definition 4.1. Let A be a graded Artinian K-algebra, z a linear form of A and $P(\times z) = f_1^{m_{f_1}} \oplus f_2^{m_{f_2}} \oplus \cdots \oplus f_s^{m_{f_s}}$ the Jordan decomposition of $\times z \in \mathrm{End}(A)$, where $f_1 > f_2 > \cdots > f_s$. The graded A-module

$$U_i = \frac{(0 : z^{f_i}) + (z)}{(0 : z^{f_i+1}) + (z)}$$

is called the i-th **central simple module** of (A, z), with $1 \leq i \leq s$ and the convention $f_{s+1} = 0$. Note that these are defined for a pair consisting of the algebra A and a linear form $z \in A$.

By the definition, it is easy to see that the modules $U_1, U_2, \ldots, U_s$ are the non-zero terms of the successive quotients of the descending chain of ideals

$$A = (0 : z^{f_1}) + (z) \supset (0 : z^{f_1-1}) + (z) \supset \cdots \supset (0 : z) + (z) \supset (z).$$

T. Harima et al., *The Lefschetz Properties*, Lecture Notes in Mathematics 2080, DOI 10.1007/978-3-642-38206-2_4, © Springer-Verlag Berlin Heidelberg 2013

Remark 4.2. We indicate a property of central simple modules of graded Artinian K-algebras with the SLP. Let $A = \oplus_{i=0}^{c} A_i$ be a graded Artinian K-algebra with the SLP and z a strong Lefschetz element of A. Set

$$A/(z) = \oplus_{i=0}^{c'} \overline{A}_i,$$

where $\overline{A}_i = A_i/zA_{i-1}$ and c' is the largest integer such that $(A/(z))_{c'} \neq 0$. Since A has the SLP, one sees easily that (A, z) has $c' + 1$ central simple modules $U_1, \ldots, U_{c'+1}$ and that $U_i \cong \overline{A}_{i-1}$ as graded K-vector spaces. This shows that U_i has only one non-trivial graded piece concentrated at the degree $i - 1$. Hence U_i has the SLP for trivial reasons.

In analogy to a graded module over a graded ring, we define the Hilbert series for a graded vector space as follows. Let $V = \oplus_{i=a}^{b} V_i$ be a graded vector space over a field K, where $V_a \neq 0$, $V_b \neq 0$ and $V_i = 0$ if $i < a$ or $i > b$. We call a the **initial degree** of V and b the **end degree** of V. Then the **Hilbert function** of V is the map $i \mapsto \dim_K V_i$, which we denote by $h(V, i)$, and its **Hilbert series** is the polynomial

$$\mathrm{Hilb}(V, t) = \sum_{i=a}^{b} (\dim_K V_i) t^i.$$

We define the **Sperner number** of V by

$$\mathrm{Sperner}(V) = \max \{ \dim_K V_a, \dim_K V_{a+1}, \ldots, \dim_K V_b \}$$

and **CoSperner number**

$$\mathrm{CoSperner}(V) = \sum_{i=a}^{b-1} \min \{ \dim_K V_i, \dim_K V_{i+1} \}.$$

The Hilbert function $h(V, i)$ of $V = \oplus_{i=a}^{b} V_i$ (where $V_a \neq (0)$ and $V_b \neq (0)$) is **symmetric** if $\dim_K V_{a+i} = \dim_K V_{b-i}$ for all $i = 0, 1, \ldots, [(b-a)/2]$, and we call the half integer $(a + b)/2$ the **reflecting degree** of $h(V, i)$. The Hilbert function of V is **unimodal** if there exists an integer m ($a \leq m \leq b$) such that

$$\dim_K V_a \leq \dim_K V_{a+1} \leq \cdots \leq \dim_K V_m \geq \dim_K V_{m+1} \geq \cdots \geq \dim_K V_b.$$

Definition 4.3. Let $A = \oplus_{i=0}^{c} A_i$ be a graded Artinian K-algebra. Suppose that $V = \oplus_{i=a}^{b} V_i$ is a finite graded A-module with $V_a \neq (0)$ and $V_b \neq (0)$.

1. V has the **weak Lefschetz property** as an A-module if there is a linear form $g \in A_1$ such that the multiplication map $\times g: V_i \to V_{i+1}$ has full rank for all $a \leq i \leq b - 1$.

2. V has the **strong Lefschetz property** as an A-module if there is a linear form $g \in A_1$ such that the multiplication map $\times g^d : V_i \to V_{i+d}$ has full rank for all $1 \le d \le b - a$ and $a \le i \le b - d$.
3. V has the **strong Lefschetz property in the narrow sense** as an A-module if there is a linear form $g \in A_1$ such that the multiplication map $\times g^{b-a-2i} : V_{a+i} \to V_{b-i}$ is bijective for all $0 \le i \le [(b-a)/2]$.

Remark 4.4. 1. If V has a unimodal Hilbert function, then $\mathrm{CoSperner}(V) = \dim_K V - \mathrm{Sperner}(V)$.

2. If V has the SLP in the narrow sense, then the Hilbert function of V is symmetric and unimodal.

The implication (2) $\Rightarrow$ (1) in the following theorem plays an important role in our proof of Theorem 4.10. The converse (1) $\Rightarrow$ (2) easily follows from Remark 4.2. We omit the proof of Theorem 4.5, which can be found in [53].

Theorem 4.5. *Let K be a field of characteristic zero and let A be a graded Artinian K-algebra. Then the following conditions are equivalent.*

1. *A has the SLP in the narrow sense.*
2. *There exists a linear form z of A such that all the central simple modules U_i of (A, z) have the SLP in the narrow sense and the reflecting degree of the Hilbert function of $\tilde{U}_i = U_i \otimes_K K[t]/(t^{f_i})$ coincides with that of A for $i = 1, 2, \ldots, s$, where $P(\times z) = f_1^{m_{f_1}} \oplus f_2^{m_{f_2}} \oplus \cdots \oplus f_s^{m_{f_s}}$.*

Example 4.6. (See Example 26 in [54] for more details.) Let

$$A = K[w, x, y, z]/(w^2, wx, x^3, xy, y^3, yz, z^3).$$

We use the same letters w, z to denote their images in A. The Jordan decomposition for $\times z$ is given by $P(\times z) = 3^4 \oplus 1^4$. Hence (A, z) has two central simple modules U_1 and U_2 such that

$$\begin{cases} U_1 = A/((0 : z) + (z)) \cong K[w, x, y, z]/(w^2, wx, x^3, y, z), \\ U_2 = ((0 : z) + (z)/(z)) \cong K[w, x, y]/(w^2, x, y^2). \end{cases}$$

Both U_1 and U_2 have the SLP, and they have the same Hilbert function $(1, 2, 1)$. Hence A has the SLP by Theorem 4.5.

We next consider the central simple modules of (A, w). The Jordan decomposition for $\times w$ is given by $P(\times w) = 2^5 \oplus 1^6$. Hence since there are again two central simple modules we have

$$\begin{cases} U_1 = A/((0 : w) + (w)) \cong K[w, x, y, z]/(w, x, y^3, yz, z^3), \\ U_2 = ((0 : w) + (w))/(w) \cong K[x, y, z]/(x^2, y, z^3). \end{cases}$$

Since U_1 has the Hilbert function $(1, 2, 2)$ it does not have a symmetric Hilbert function, we can not apply Theorem 4.5 to prove A has the SLP.

We next state a proposition on the Hilbert functions of central simple modules for a Gorenstein K-algebra.

Proposition 4.7. *Let K be any field, A a graded Artinian Gorenstein K-algebra and z a linear form of A. (We use the notation of Theorem 4.5.)*

1. *Every central simple module U_i of (A, z) has a symmetric Hilbert function.*
2. *Every module $\tilde{U}_i$ has a symmetric unimodal Hilbert function with the same reflecting degree as that of A.*

Again, we omit the proof, which can be found in [53].

Remark 4.8. These ideas were used much earlier than [53] in A. Iarrobino [64] Theorem 1.7. Related results can be found in Lindsey [81].

As an immediate consequence of Theorem 4.5 and Proposition 4.7, we have the following result which allows one to check for the SLP in sort of divide and conquer manner.

Theorem 4.9. *Let K be a field of characteristic zero and let A be a graded Artinian Gorenstein K-algebra. Then the following conditions are equivalent.*

1. *A has the SLP.*
2. *There exists a linear form z of A such that all the central simple modules U_i of (A, z) have the SLP.*

4.2   Finite Free Extensions of a Graded K-Algebra

We now state the main theorem in this chapter.

Theorem 4.10 ([53, Theorem 6.1]). *Let K be a field of characteristic zero, B a graded Artinian K-algebra and A a finite flat algebra over B such that the algebra map $\varphi \colon B \to A$ preserves grading. Suppose that both B and $A/\mathfrak{m}A$ have the SLP in the narrow sense, where $\mathfrak{m}$ is the maximal ideal of B. Then A has the SLP in the narrow sense.*

In order to prove this theorem, we need a few lemmas.

Lemma 4.11. *We use the same notation as Theorem 4.10. Let z' be any linear form of B and $z = \varphi(z')$. Let U_i' and U_i be the i-th central simple modules of (B, z') and (A, z), respectively. Then $U_i' \otimes_B A \cong U_i$ as A-modules.*

Proof. By assumption, we can write $A \cong Be_1 \oplus Be_2 \oplus \cdots \oplus Be_k$ for some homogeneous elements $e_i \in A$. Let

$$P(\times z') = f_1^{m_{f_1}} \oplus f_2^{m_{f_2}} \oplus \cdots \oplus f_s^{m_{f_s}}$$

be the Jordan decomposition of $\times z' \in \mathrm{End}(B)$. Then it follows immediately that

$$P(\times z) = \underbrace{f_1 \oplus f_1 \oplus \cdots \oplus f_1}_{m_{f_1} \times k} \oplus \underbrace{f_2 \oplus f_2 \oplus \cdots \oplus f_2}_{m_{f_2} \times k} \oplus \cdots \oplus \underbrace{f_s \oplus f_s \oplus \cdots \oplus f_s}_{m_{f_s} \times k}$$

$$= f_1^{m_{f_1} \times k} \oplus f_2^{m_{f_2} \times k} \oplus \cdots \oplus f_s^{m_{f_s} \times k}.$$

Therefore, noting that

$$(0 :_A z^m) + zA =$$

$$\big((0 :_B (z')^m) + z'B\big)e_1 \oplus \big((0 :_B (z')^m) + z'B\big)e_2 \oplus \cdots \oplus \big((0 :_B (z')^m) + z'B\big)e_k$$

for all integers $m > 0$, we have that

$$U_i' \otimes_B A \cong \left(\big((0 :_B (z')^{f_i}) + z'B\big)/\big((0 :_B (z')^{f_i+1}) + z'B\big)\right) \otimes_B \left(\bigoplus_{i=1}^k Be_i\right)$$

$$\cong \bigoplus_{i=1}^k \left(\big((0 :_B (z')^{f_i}) + z'B\big)/\big((0 :_B (z')^{f_i+1}) + z'B\big)\right)e_i$$

$$\cong \big((0 :_A z^{f_i}) + zA\}/\big((0 :_A z^{f_i+1}) + zA\big)$$

$$\cong U_i$$

for all $i = 1, 2, \ldots, s$. $\qquad\qquad\qquad\qquad\qquad\qquad\qquad\qquad\qquad\qquad\qquad\qquad\square$

The following is an analogue of Theorem 3.34.

Proposition 4.12. *Let K be a field of characteristic zero. Let A and B be graded Artinian K-algebras, let V be a finite graded A-module and W a finite graded B-module. If V and W have the SLP in the narrow sense, then $V \otimes_K W$ has the SLP in the narrow sense as an $A \otimes_K B$-module.*

We postpone the proof of Proposition 4.12 until after Lemma 4.14.

Lemma 4.13. *Let A be a graded Artinian K-algebra and let $V = \bigoplus_{i=a}^b V_i$ be a finite graded A-module, where $V_a \neq 0$ and $V_b \neq 0$. Put $r = \mathrm{Sperner}(V)$. Let g be a linear form of A. Then the following conditions are equivalent.*

1. *g is a strong Lefschetz element for V in the narrow sense.*
2. *There are graded vector subspaces $\mathcal{V}_1, \mathcal{V}_2, \ldots, \mathcal{V}_r$ of V with $V = \bigoplus_{i=1}^r \mathcal{V}_i$ which satisfy the following conditions for each $i = 1, 2, \ldots, r$.*

   a. *$g\mathcal{V}_i \subset \mathcal{V}_i$.*
   b. *The Jordan canonical form of $\times g \in \mathrm{End}(\mathcal{V}_i)$ is a single Jordan block.*
   c. *The reflecting degree of the Hilbert function of $\mathcal{V}_i$ is equal to that of V.*

In this case $P(\times g) = \dim_K \mathcal{V}_1 \oplus \dim_K \mathcal{V}_2 \oplus \cdots \oplus \dim_K \mathcal{V}_r$.

Proof. (1) $\implies$ (2): Assume g is a strong Lefschetz element for V. Let $h_i = h(V, i)$ be the Hilbert function of V. A basis for the Jordan decomposition of $\times g \in \mathrm{End}(V)$ is obtained as follows. Let $v_1, v_2, \ldots, v_{h_a}$ be a basis of V_a. By the SLP of V, the elements

$$\left\{ g^k v_j \mid 1 \le j \le h_a, 0 \le k \le b - a \right\},$$

being linearly independent, will be a part of the basis. Next let $\{v'_{h_a+1}, v'_{h_a+2}, \ldots, v'_{h_{a+1}}\}$ be a basis of $\mathrm{Ker}[V_{b-1} \overset{\times g}{\to} V_b]$. By the SLP of V, there exist elements $\{v_{h_a+1}, v_{h_a+2}, \ldots, v_{h_{a+1}}\}$ of V_{a+1} such that $g^{b-a-2}v_j = v'_j$ for all $j = h_a + 1, h_a + 2, \ldots, h_{a+1}$. Then the elements

$$\left\{ g^k v_j \mid h_a + 1 \le j \le h_{a+1}, 0 \le k \le b - a - 2 \right\}$$

have the property that none of these is dependent on the previously chosen basis elements, so will be another part of the basis. We repeat this procedure to extend basis elements. We may continue this process to decompose $\times g$ into Jordan blocks. Since

$$h_a + (h_{a+1} - h_a) + (h_{a+2} - h_{a+1}) + \cdots + (h_{[(b-a)/2]} - h_{[(b-a)/2]-1}) = r,$$

one sees that there are r Jordan blocks.

Now let $\mathcal{V}_i$ be the subspace of V spanned by $\{v_i, g v_i, \ldots, g^{b-a-2(d_i-a)} v_i\}$ for $i = 1, 2, \ldots, r$, where $d_i = \deg(v_i)$. Then it is easy to verify the conditions stated in (2) are satisfied.

(2) $\implies$ (1): By (2c) it suffices to prove this for each $\mathcal{V}_i$, which is obvious. $\qquad\square$

Lemma 4.14. *With the same notation as Proposition 4.12, let $g \in A$ be any linear form and let $\mathcal{V}$ be a graded subspace of V such that $g\mathcal{V} \subset \mathcal{V}$. Similarly let $h \in B$ be a linear form and $\mathcal{W} \subset W$ a graded subspace such that $h\mathcal{W} \subset \mathcal{W}$. Assume that $\times g \in \mathrm{End}(\mathcal{V})$ is a single Jordan block and the same for $\times h \in \mathrm{End}(\mathcal{W})$. Moreover let $m = \dim_K \mathcal{V}$, $n = \dim_K \mathcal{W}$ and $s = \mathrm{Sperner}(\mathcal{V} \otimes \mathcal{W})$.*

Then there exist graded vector subspaces $\mathcal{U}_1, \mathcal{U}_2, \ldots, \mathcal{U}_s$ of $\mathcal{V} \otimes \mathcal{W}$ such that $\mathcal{V} \otimes \mathcal{W} = \bigoplus_{i=1}^{s} \mathcal{U}_i$, which satisfy the following conditions for $i = 1, 2, \ldots, s$.

1. *$(g \otimes 1 + 1 \otimes h) \mathcal{U}_i \subset \mathcal{U}_i$.*
2. *The Jordan canonical form of $\times(g \otimes 1 + 1 \otimes h) \in \mathrm{End}(\mathcal{U}_i)$ is a single Jordan block.*
3. *The reflecting degree of the Hilbert function of $\mathcal{U}_i$ is equal to that of $\mathcal{V} \otimes \mathcal{W}$.*

Proof. By Lemma 4.13, it is enough to show that $g \otimes 1 + 1 \otimes h$ is a strong Lefschetz element for $\mathcal{V} \otimes \mathcal{W}$. Let d and e be the initial degrees of $\mathcal{V}$ and $\mathcal{W}$ respectively. Then the Hilbert series of $\mathcal{V}$ and $\mathcal{W}$ are:

$$\begin{cases} \mathrm{Hilb}(\mathscr{V},t) = t^d + t^{d+1} + \cdots + t^{d+m-1}, \\ \mathrm{Hilb}(\mathscr{W},t) = t^e + t^{e+1} + \cdots + t^{e+n-1}. \end{cases}$$

Let $K[x]$ be the polynomial ring in one variable. As a graded vector space we may choose an isomorphisms $\mathscr{V} \to K[x]/(x^m)(-d)$ so that we have the commutative diagram:

$$\begin{array}{ccc} \mathscr{V} & \xrightarrow{\times g} & \mathscr{V} \\ \downarrow & & \downarrow \\ (K[x]/(x^m))(-d) & \xrightarrow{\times x} & (K[x]/(x^m))(-d). \end{array}$$

Likewise we may choose an isomorphism $\mathscr{W} \to K[y]/(y^n)(-e)$ to get a similar diagram for $\mathscr{W}$.

These diagrams give rise the following commutative diagram:

$$\begin{array}{ccc} \mathscr{V} \otimes \mathscr{W} & \xrightarrow{\times(g\otimes1+1\otimes h)} & \mathscr{V} \otimes \mathscr{W} \\ \downarrow & & \downarrow \\ (K[x,y]/(x^m,y^n))(-(d+e)) & \xrightarrow{\times(x+y)} & (K[x,y]/(x^m,y^n))(-(d+e)), \end{array}$$

where the vertical maps are isomorphisms as graded vector spaces. Since the characteristic of K is zero, it follows by Lemma 3.70 that $x+y$ is a strong Lefschetz element for $K[x,y]/(x^m,y^n)$. Hence $g \otimes 1 + 1 \otimes h$ is a strong Lefschetz element for $\mathscr{V} \otimes \mathscr{W}$. □

Proof (Proof of Proposition 4.12). Let $V = \bigoplus_{i=1}^r \mathscr{V}_i$ and $W = \bigoplus_{i=j}^u \mathscr{W}_j$ be the direct sum decompositions of V and W constructed in Lemma 4.13 with respect to the strong Lefschetz elements $g \in A_1$ and $h \in B_1$ for V and W, where $r = \mathrm{Sperner}(V)$ and $u = \mathrm{Sperner}(W)$. Then we see that $V \otimes W = \bigoplus_{i,j}(\mathscr{V}_i \otimes \mathscr{W}_j)$ and $\mathscr{V}_i \otimes \mathscr{W}_j$ is closed under the multiplication $\times(g \otimes 1 + 1 \otimes h): V \otimes W \to V \otimes W$ for all i and j. Noting that

$$\begin{cases} \mathrm{Hilb}(V \otimes W,t) = \mathrm{Hilb}(V,t)\mathrm{Hilb}(W,t) = (\sum_{i=1}^r \mathrm{Hilb}(\mathscr{V}_i,t))(\sum_{j=1}^u \mathrm{Hilb}(\mathscr{W}_j,t)), \\ \mathrm{Hilb}(\mathscr{V}_i \otimes \mathscr{W}_j,t) = \mathrm{Hilb}(\mathscr{V}_i,t)\mathrm{Hilb}(\mathscr{W}_j,t), \end{cases}$$

we see that the reflecting degree of the Hilbert function of $\mathscr{V}_i \otimes \mathscr{W}_j$ is equal to that of $V \otimes W$. By Lemma 4.14, for each pair (i,j), there are subspaces $\mathscr{U}_k^{(ij)}$ satisfying the conditions (1), (2) and (3) of Lemma 4.14. We observe that the Sperner number of $V \otimes W$ is equal to the sum of those of $\mathscr{V}_i \otimes \mathscr{W}_j$. Hence, by Lemma 4.13, the linear form $g \otimes 1 + 1 \otimes h$ is a strong Lefschetz element for $V \otimes W$. □

Remark 4.15. The converse of Proposition 4.12 is also true. For details see Theorem 3.10 in [53].

Let us prove the main theorem.

Proof (Proof of Theorem 4.10). Let z' be a strong Lefschetz element of B and let $z = \varphi(z')$. Since every central simple module U_i' of (B, z') has only one non-trivial graded piece, U_i' has the SLP in the narrow sense. Also, since U_i' is annihilated by $\mathfrak{m}$, it follows by Lemma 4.11 that $U_i' \otimes_K A/\mathfrak{m}A \cong U_i$, where U_i is the i-th central simple module of (A, z). Hence, using our assumption that $A/\mathfrak{m}A$ has the SLP in the narrow sense, we have by Proposition 4.12 that every central simple module U_i of (A, z) has the SLP in the narrow sense.

To complete the proof, we use the same notation as the proof of Lemma 4.11. Fix i such that $1 \le i \le s$ and set

$$\widetilde{U}_i = U_i \otimes_K K[t]/(t^{f_i}) \quad \text{and} \quad \widetilde{U}_i' = U_i' \otimes_K K[t]/(t^{f_i}).$$

Let a, b be the initial and end degrees of the Hilbert series of $\widetilde{U}_i'$ so that

$$\mathrm{Hilb}(\widetilde{U}_i', t) = h_a t^a + \text{(mid terms)} + h_b t^b.$$

Similarly let

$$\mathrm{Hilb}(B, t) = 1 + \text{(mid terms)} + t^c, \quad \mathrm{Hilb}(A/\mathfrak{m}A, t) = 1 + \text{(mid terms)} + t^d$$

be the Hilbert series of B and $A/\mathfrak{m}A$. Since $\mathrm{Hilb}(\widetilde{U}_i, t) = \mathrm{Hilb}(\widetilde{U}_i', t)$ $\mathrm{Hilb}(A/\mathfrak{m}A, t)$ and $\mathrm{Hilb}(A, t) = \mathrm{Hilb}(B, t) \mathrm{Hilb}(A/\mathfrak{m}A, t)$, the Hilbert series of $\widetilde{U}_i$ and A are of the following form:

$$\mathrm{Hilb}(\widetilde{U}_i, t) = h_a t^a + \text{(mid terms)} + h_b t^{b+d},$$

$$\mathrm{Hilb}(A, t) = 1 + \text{(mid terms)} + t^{c+d}.$$

Since the reflecting degree of the Hilbert series of $\widetilde{U}_i'$ coincides with that of B, we have $(a + b)/2 = c/2$, and $(a + b + d)/2 = (c + d)/2$. This means that the reflecting degree of the Hilbert function of $\widetilde{U}_i$ coincides with that of A independent of i. Therefore by Theorem 4.5 the proof is finished.   □

Remark 4.16. In general the converse of Theorem 4.10 is not true. To give such an example, consider $A = K[x_1, x_2]/(e_1^2, e_2^2)$ and $B = K[e_1, e_2]/(e_1^2, e_2^2)$, where $e_1 = x_1 + x_2$ and $e_2 = x_1 x_2$. Note that A has the standard grading, but B does not. By Lemma 4.21 below, A is a finite free module over B. Furthermore, by Proposition 3.15, both A and $A/\mathfrak{m}A$ have the SLP. However, B does not have the SLP. In fact e_1 is the only candidate for a strong Lefschetz element for B while we have $\mathrm{Hilb}(B, t) = 1 + t + t^2 + t^3$ and $e_1^2 = 0$ in B.

Corollary 4.17 ([50, Corollary 29], [57, Theorem]). *Let K be a field of characteristic zero, and B a graded Artinian K-algebra with the SLP in the narrow sense. Let $A = B[z]/(h)$, where h is a monic homogeneous polynomial in the variable z with coefficients in B. Then A has the SLP in the narrow sense.*

Proof. Theorem 4.10 applies since the ring A is a flat extension of B and $A/\mathfrak{m}A \cong K[z]/(z^r)$ for some r, where $\mathfrak{m}$ is the maximal ideal of B. □

4.3 Power Sums of Consecutive Degrees

As an application of Theorem 4.10, we give an interesting class of complete intersections with the SLP.

Proposition 4.18 ([53, Proposition 7.1]). *Let K be a field of characteristic zero, and let $R = K[x_1, x_2, \ldots, x_n]$ be the polynomial ring over K with the standard grading, i.e., $\deg(x_i) = 1$ for all i. Fix a positive integer a, and let*

$$I = (p_a(x_1, x_2, \ldots, x_n), p_{a+1}(x_1, x_2, \ldots, x_n), \ldots, p_{a+n-1}(x_1, x_2, \ldots, x_n)),$$

*where $p_d = x_1^d + x_2^d + \cdots + x_n^d$ is the **power sum** of degree d. Then $A = R/I$ is a complete intersection and has the SLP.*

The proof is given at the end of this section after a series of lemmas.

Remark 4.19 ([139]). Let K be a field of characteristic zero and $R = K[x_1, x_2, \ldots, x_n]$ the polynomial ring over K with $\deg(x_i) = 1$ for all i. Let $e_i = e_i(x_1, x_2, \ldots, x_n)$ be the elementary symmetric polynomial of degree i in R for all $i = 1, 2, \ldots, n$. We denote by S the subring $S = K[e_1, e_2, \ldots, e_n]$ of R. Put $S_j = S \cap R_j$. Then we have $S = \bigoplus_{j \geq 0} S_j$, which we regard as defining the grading of S, so that the natural injection $S \subset R$ is a grade-preserving algebra map. Let $\mathscr{H}$ be the set of harmonic functions in R. Namely, $\mathscr{H}$ is the vector space spanned by the partial derivatives of the alternating polynomial $\prod_{i<j}(x_i - x_j)$. (For a construction of a standard linear basis of $\mathscr{H}$, see also Sect. 8.2.1.) The following are well known.

1. The ring S contains all symmetric polynomials of R.
2. $\mathscr{H}$ is isomorphic to $R/(e_1, e_2, \ldots, e_n)$ as a graded vector space.
3. The map

$$\mathscr{H} \otimes_K S \ni h \otimes e \mapsto he \in R$$

is an isomorphism as graded vector spaces, and hence R is a finite free module over S.

Lemma 4.20. *We use the same notation as Proposition 4.18. Then the ideal I is a complete intersection.*

Proof. Recall the Newton formula [85],

$$p_m = -\sum_{j=1}^{n}(-1)^j e_j\, p_{m-j}.$$

This implies that $p_m \in I$ for all $m > a + n - 1$, so that $(p_m, p_{2m}, \ldots, p_{nm}) \subset I$. It is easy to see that $(p_m, p_{2m}, \ldots, p_{nm})$ is a complete intersection, and hence so is I. $\square$

Lemma 4.21. *With the same notation as Remark 4.19, let $f_1, f_2, \ldots, f_n \in S$ be homogeneous polynomials. Let $I = (f_1, f_2, \ldots, f_n)R$ and $J = (f_1, f_2, \ldots, f_n)S$. Then R/I is a finite free module over S/J.*

Proof. This follows immediately from the last statement of Remark 4.19. $\square$

Lemma 4.22. *The Artinian complete intersection*

$$B = K[e_1, e_2, \ldots, e_n]/(p_a, p_{a+1}, \ldots, p_{a+n-1})$$

has the SLP, where p_d is the power sum of degree d. We assume $a > 1$, so $B \neq K$.

Proof. Since

$$\mathrm{Hilb}(B,t) = \frac{(1 - t^a)(1 - t^{a+1})\cdots(1 - t^{a+n-1})}{(1 - t)(1 - t^2)\cdots(1 - t^n)} \quad \text{and} \quad t^{an-n}\,\mathrm{Hilb}(B,t^{-1}) = \mathrm{Hilb}(B,t),$$

B has a symmetric Hilbert function and the maximal socle degree of B is equal to $an - n$. Set

$$C = K[x_1, x_2, \ldots, x_n]/(x_1^a, x_2^a, \ldots, x_n^a).$$

One notices that B and C have the same maximal socle degree.

Next we show that B is naturally a graded subring of C. The symmetric group $G = S_n$ acts on C by permutation of the variables. Consider the exact sequence

$$0 \rightarrow (x_1^a, x_2^a, \ldots, x_n^a) \rightarrow R \rightarrow C \rightarrow 0.$$

Since $\mathrm{Char}(K) = 0$, we have the exact sequence

$$0 \rightarrow (x_1^a, x_2^a, \ldots, x_n^a)^G \rightarrow R^G \rightarrow C^G \rightarrow 0,$$

where M^G denotes the invariant subspace for a G-module M. Since $R^G = S$, we see $C^G \cong S/((x_1^a, x_2^a, \ldots, x_n^a) \cap S)$. We would like to prove that

$$(x_1^a, x_2^a, \ldots, x_n^a) \cap S = (p_a, p_{a+1}, \ldots, p_{a+n-1}),$$

or equivalently, the natural surjection

$$\psi \colon B \to S/((x_1^a, x_2^a, \ldots, x_n^a) \cap S) \cong C^G$$

is an isomorphisms. By way of contradiction assume that $\mathrm{Ker}(\psi) \neq (0)$. Since B is Gorenstein, B_{an-n} is the unique minimal ideal of B and would be contained in $\mathrm{Ker}(\psi)$. Hence the maximal socle degree of $B/\mathrm{Ker}(\psi)$ is less than $an-n$. However, since the element $(x_1 x_2 \cdots x_n)^{a-1} \in C$ lies in C^G, the maximal socle degree of C^G is equal to $an - n$. This is a contradiction, so we have proved $\mathrm{Ker}(\psi) = (0)$.

It is known that the image of e_1 in C is a strong Lefschetz element for C (cf. Corollary 3.5 in [146]). Thus, by Proposition 3.41, e_1 is a strong Lefschetz element for B.   □

Lemma 4.23. *Let* $e_i = e_i(x_1, x_2, \ldots, x_n)$ *be the elementary symmetric polynomial of degree* i *in* $R = K[x_1, x_2, \ldots, x_n]$. *Then* $A = R/(e_1, e_2, \ldots, e_n)$ *has the SLP.*

Proof. We use induction on n. It is obvious that both $K[x_1]/(x_1)$ and $K[x_1, x_2]/(x_1+x_2, x_1 x_2)$ have the SLP. Let $n \geq 3$, and set $S = R/(e_1, e_2, \ldots, e_{n-1})$ and $T = K[x_n]$. We first show that

$$(e_1, e_2, \ldots, e_{n-1}, e_n) = (e_1, e_2, \ldots, e_{n-1}, x_n^n).$$

Consider the following polynomial

$$F(X) = (X - x_1)(X - x_2) \cdots (X - x_n)$$
$$= X^n - e_1 X^{n-1} + e_2 X^{n-2} - \cdots + (-1)^n e_n.$$

Since $F(x_n) = 0$, we have

$$x_n^n = e_1 x_n^{n-1} - e_2 x_n^{n-1} + \cdots - (-1)^n e_n,$$

and this gives the equality above. Thus $\{e_1, e_2, \ldots, e_{n-1}, x_n\}$ is a regular sequence, so that by Exercise 22.2 in [93], the natural map $T \to S$ is a free extension. This is also a finite extension, since x_n is a regular element mod $e_1, e_2, \ldots, e_{n-1}$. Hence $A \simeq S \otimes_{K[x_n]} K[x_n]/(x_n^n)$ is also a finite free extension over $B = K[x_n]/(x_n^n)$, and its fiber $A \otimes_B K[x_n]/(x_n)$ is isomorphic to $K[x_1, x_2, \ldots, x_{n-1}]/(e_1', e_2', \ldots, e_{n-1}')$ where e_i' is the elementary symmetric polynomial of degree i in $K[x_1, x_2, \ldots, x_{n-1}]$. By the induction hypothesis, the fiber ring has the SLP. Therefore A has the SLP by Theorem 4.10.   □

Proof (Proof of Proposition 4.18). The first assertion is proved in Lemma 4.20. Let $\mathfrak{m}$ be the maximal ideal of $K[e_1, e_2, \ldots, e_n]/(p_a, p_{a+1}, \ldots, p_{a+n-1})$. By Lemma 4.23, the algebra

$$A/\mathfrak{m}A = K[x_1, x_2, \ldots, x_n]/(e_1, e_2, \ldots, e_n)$$

has the SLP. Therefore the second assertion follows from Lemmas 4.21, 4.22 and Theorem 4.10.   □

Question 4.24. Let p_{d_i} be the power sum of degree d_i in $R = K[x_1, x_2, \ldots, x_n]$ and set $I = (p_{d_1}, p_{d_2}, \ldots, p_{d_n})$, where $d_1 < d_2 < \cdots < d_n$.

1. Find a necessary and sufficient condition for I to be complete intersection.
2. Does R/I have the SLP if I is complete intersection?

Let $n = 3$ and suppose that $\gcd(d_1, d_2, d_3) = 1$. Conca, Krattenthaler and Watanabe conjectured in [24] that $I = (p_{d_1}, p_{d_2}, p_{d_3})$ is complete intersection in $R = K[x_1, x_2, x_3]$ if and only if $d_1 d_2 d_3 \equiv 0 \pmod{6}$, and provided evidence for the conjecture by proving it in several special instances.

4.4   More Applications of Finite Free Extensions

Throughout this section let K be a field of characteristic zero.

Proposition 4.25. *1. Let $R = K[x_1, x_2, \ldots, x_n]$ be the polynomial ring and let $I \subset R$ be a homogeneous complete intersection ideal of height n. Suppose that a power of a linear element z of R can be a member of a minimal generating set for I. If $R/I + (z)$ has the SLP, so does R/I.*
2. Let R be the polynomial ring in three variables over K, and let $I \subset R$ be a homogeneous complete intersection ideal of height three. If a power of a linear form can be an element of a minimal generating set of I, then R/I has the SLP.
3. Let I be a complete intersection ideal of $R = K[x_1, x_2, \ldots, x_n]$ generated by $\{ f, g, \ell_3^{d_3}, \ell_3^{d_4}, \ldots, \ell_n^{d_n} \}$, where f, g are any homogeneous forms and ℓ_j are any linear forms. Then R/I has the SLP.

Proof. (1) Suppose that z^r is one of the members of a minimal generating set for I. Put $B = K[z]/(z^r)$. One notices that $(I : z^s) = I + (z^{r-s})$ for all $1 \le s \le r$, which shows that $A = R/I$ is a flat extension of B, because the injection $0 \to (z^s) \to B$ remains exact after taking the tensor with A over B. So one may apply Theorem 4.10.
(2) Let z be a linear form such that $I = (f, g, z^d)$ for some d. Then $R/I + (z)$ has the SLP by Proposition 3.15. Hence the SLP of R/I follows from (1).
    Using (1) and (2) the assertion (3) can be proved by induction on n.   □

Proposition 4.26. *Let* $R = K[x_1, x_2, \ldots, x_n]$ *be the polynomial ring and let* $e_i(x_1, x_2, \ldots, x_n)$ *be the elementary symmetric polynomial of degree* i *for* $i = 1, 2, \ldots, n$. *For positive integers* r *and* s, *set*

$$\begin{cases} f_i = e_i(x_1^r, x_2^r, \ldots, x_n^r), & \text{for } i = 1, 2, \ldots, n - 1, \\ f_n = e_n(x_1^s, x_2^s, \ldots, x_n^s), \end{cases}$$

and $I = (f_1, f_2, \ldots, f_n)$. *Suppose that* s *is a multiple of* r *or less than* r. *Then the complete intersection* $A = R/I$ *has the SLP.*

Proof. Case 1. Suppose that s is a multiple of r. Noting the relation

$$(-1)^{n+1} e_n + (-1)^n x_n e_{n-1} + (-1)^{n-1} x_n^2 e_{n-2} + \cdots + (-1)^2 x_n^{n-1} e_1 = x_n^n, \quad (4.1)$$

we see that there exist polynomials $P_1, \ldots, P_n \in R$ such that

$$x_n^{rn} = P_1 f_1 + \cdots + P_{n-1} f_{n-1} + P_n x_1^r \cdots x_n^r.$$

Since

$$(x_n^{rn})^{s/r} = (P_1 f_1 + \cdots + P_{n-1} f_{n-1} + P_n x_1^r \cdots x_n^r)^{s/r},$$

it follows that $x_n^{sn} \in I = (f_1, f_2, \ldots, f_{n-1}, f_n)$. Thus we obtain that

$$(f_1, f_2, \ldots, f_{n-1}, x_n^{sn}) \subset I.$$

Next, noting that $R/(f_1, f_2, \ldots, f_{n-1}, x_n^{sn})$ and A are complete intersections with the same Hilbert function, we have

$$I = (f_1, f_2, \ldots, f_{n-1}, x_n^{sn}).$$

Let z be the image of x_n in A. Then we see that

$$A/(z) \cong K[x_1, x_2, \ldots, x_{n-1}]/(e_1'(x_1^r, x_2^r, \ldots, x_{n-1}^r), \ldots, e_{n-1}'(x_1^r, x_2^r, \ldots, x_{n-1}^r)),$$

where $e_i'(x_1, x_2, \ldots, x_{n-1})$ is the elementary symmetric function of degree i in the variables $x_1, x_2, \ldots, x_{n-1}$ for all $1 \leq i \leq n - 1$. We use induction on n. The induction hypothesis implies that $A/(z)$ has the SLP. So the SLP of A follows from Proposition 4.25 (1).

Case 2. Suppose that $r > s$. We use induction on n. By Theorem 4.9 and the induction hypothesis, it is sufficient to show that (A, z) has only two central simple modules U_1 and U_2 of the following forms:

$$U_1 \cong K[x_1, x_2, \ldots, x_{n-1}]/(\overline{f}_1, \overline{f}_2, \ldots, \overline{f}_{n-2}, (x_1 \cdots x_{n-1})^s),$$

$$U_2 \cong K[x_1, x_2, \ldots, x_{n-1}]/(\overline{f}_1, \overline{f}_2, \ldots, \overline{f}_{n-2}, (x_1 \cdots x_{n-1})^{r-s}),$$

where $\overline{f}_j$ is the image of f_j in $K[x_1, x_2, \ldots, x_{n-1}]$ (note that $\overline{f}_j = e'_j(x_1^r, \ldots, x_{n-1}^r)$).

We first show that

1. $(I : x_n^k) = (f_1, f_2, \ldots, f_{n-1}, x_1^s \cdots x_{n-1}^s x_n^{s-k})$ for all $0 \le k \le s$,
2. $(I : x_n^k) = (f_1, f_2, \ldots, f_{n-2}, x_n^{(n-1)r-(k-s)}, x_1^s \cdots x_{n-1}^s)$ for all $s < k < (n-1)$ $r + s$,
3. $(I : x_n^k) = R$ for all $k \ge (n-1)r + s$.

(1) and (3) are easy. So we give a proof for (2). Using (4.1) we get the following relation

$$(-1)^{n+1} x_1^r \cdots x_n^r + (-1)^n x_n^r f_{n-1} + (-1)^{n-1} x_n^{2r} f_{n-2} + \cdots + (-1)^2 x_n^{(n-1)r} f_1 = x_n^{nr}.$$

Hence we have

$$(-1)^{n+1} x_1^r x_2^r \cdots x_{n-1}^r + (-1)^n f_{n-1} + (-1)^{n-1} x_n^r f_{n-2} + \cdots + (-1)^2 x_n^{(n-2)r} f_1 = x_n^{(n-1)r}, \tag{4.2}$$

and

$$I : x_n^s = (f_1, f_2, \ldots, f_{n-1}, x_1^s \cdots x_{n-1}^s) = (f_1, f_2, \ldots, f_{n-2}, x_n^{(n-1)r}, x_1^s \cdots x_{n-1}^s).$$

Therefore, for all $s < k \le (n-1)r + s$, it follows that

$$I : x_n^k = (I : x_n^s) : x_n^{k-s} = (f_1, \ldots, f_{n-2}, x_n^{(n-1)r-(k-s)}, x_1^s \cdots x_{n-1}^s).$$

We next calculate the central simple modules of (A, z). From (1), (2) and (3), we have

$$(I : x_n^k) + (x_n) = \begin{cases} (f_1, f_2, \ldots, f_{n-1}, x_n) & k = 0, 1, \ldots, s-1, \\ (f_1, f_2, \ldots, f_{n-1}, x_1^s \cdots x_{n-1}^s, x_n) & k = s, \\ (f_1, f_2, \ldots, f_{n-2}, x_1^s \cdots x_{n-1}^s, x_n) & k = s+1, s+2, \ldots, (n-1)r+s-1, \\ R & k = (n-1)r + s, \ldots. \end{cases}$$

Here, noting the equality (4.2), we see that

$$(f_1, f_2, \ldots, f_{n-1}, x_1^s \cdots x_{n-1}^s, x_n) = (f_1, f_2, \ldots, f_{n-2}, x_1^s \cdots x_{n-1}^s, x_n).$$

Hence we have

$$(0 : z^k) + (z) = \begin{cases} (z) & k = 0, 1, \ldots, s-1, \\ (\overline{x}_1^s \cdots \overline{x}_{n-1}^s, z) & k = s, s+1, \ldots, (n-1)r+s-1, \\ A & k = (n-1)r + s, \ldots, \end{cases}$$

where $\overline{x}_j$ is the image of x_j in A. Thus we obtain that

$$U_1 = A/(\overline{x}_1^s \cdots \overline{x}_{n-1}^s, z) \cong K[x_1, x_2, \ldots, x_{n-1}]/(\overline{f}_1, \overline{f}_2, \ldots, \overline{f}_{n-2}, (x_1 \cdots x_{n-1})^s)$$

and

$$U_2 = (\overline{x}_1^s \cdots \overline{x}_{n-1}^s, z)/(z)$$

$$\cong \frac{A/(z)}{(0 : \overline{x}_1^s \cdots \overline{x}_{n-1}^s) + (z)/(z)}$$

$$\cong R/(I + (x_n)) : x_1^s \cdots x_{n-1}^s$$

$$\cong K[x_1, x_2, \ldots, x_{n-1}]/(\overline{f}_1, \overline{f}_2, \ldots, \overline{f}_{n-2}, (x_1 \cdots x_{n-1})^{r-s}). \qquad \Box$$

In the rest of this section we fix $R = K[x_1, x_2, \ldots, x_n]$, $S = K[e_1, e_2, \ldots, e_n]$ with the same assumption as in Remark 4.19, i.e., K has characteristic zero.

Remark 4.27. Suppose that $f_1, f_2, \ldots, f_n$ is a regular sequence of S. We use the notation $B = S/(f_1, f_2, \ldots, f_n)S$ and $A = R/(f_1, f_2, \ldots, f_n)R$. Let $\mathfrak{m}$ be the maximal ideal of B. Then we have (1) A is finite flat over B by Lemma 4.21 and (2) $A/\mathfrak{m}A$ has the SLP by Lemma 4.23. Therefore, by Theorem 4.10, if B has the SLP, then A has the SLP. This is the idea behind the proof of Proposition 4.18.

Proposition 4.28. *Let $f \in S$ be a homogeneous element of degree d, and let e_i be the elementary symmetric polynomial of degree i. Suppose that $(e_2, e_3, \ldots, e_n, f)$ is a complete intersection in S. Then $R/(e_2, e_3, \ldots, e_n, f)R$ has the SLP.*

Proof. Set $A = R/(e_2, e_3, \ldots, e_n, f)R$ and $B = S/(e_2, e_3, \ldots, e_n, f)$. Note that B has embedding dimension one. Hence it has the SLP, as one easily checks. Thus the assertion is proved by Remark 4.27. $\qquad \Box$

Lemma 4.29. *Suppose that*

$$f_2, f_3, \ldots, f_n, f_d$$

is a regular sequence in S such that $\deg f_i = i$ for $i = 2, 3, \ldots, n, d$ with $d > n$. Set $B = S/(f_2, f_3, \ldots, f_n, f_d) = \bigoplus_{i \geq 0} B_i$. Then the following conditions are equivalent.

1. B has the SLP.
2. The embedding dimension of B is one.

Proof. Since B is a complete intersection, we have

$$\mathrm{Hilb}(B, t) = 1 + t + \cdots + t^{d-1}.$$

Assume (2). Then as a K-algebra, B is generated by $\overline{e_1}$, the image of e_1. Hence B is isomorphic to $K[X]/(X^d)$, so that (1) follows. Conversely, if we assume (1), then the maximal ideal of B is generated by $\overline{e_1}, \overline{e_2}, \ldots, \overline{e_n}$, the images of $e_1, \ldots, e_n$. Since B has the SLP, $\overline{e_k}$ is a constant multiple of $\overline{e_1}^k$. Therefore the maximal ideal is generated by a single element. $\qquad\qquad\qquad\qquad\qquad\qquad\qquad\qquad\qquad\qquad\qquad$ □

Proposition 4.30. *With the same notation as Proposition 4.28, assume that d is a prime number.*

1. $B = S/(f_2, f_3, \ldots, f_n, f_d)S$ *has the SLP.*
2. $R/(f_2, f_3, \ldots, f_n, f_d)R$ *has the SLP.*

Proof. (2) follows from (1) by Remark 4.27. To prove (1) it suffices to show that the embedding dimension of B is one by Lemma 4.29. Write $f_2 = \alpha e_1^2 + \beta e_2$, with $\alpha, \beta \in K$. Assume $\beta = 0$. Then since $(e_1^2, f_3, \ldots, f_n, f_d)$ is a complete intersection, so is $(e_1, f_3, \ldots, f_n, f_d)$. Hence, if we put $B' = S/(e_1, f_3, \ldots, f_n, f_d)$, then

$$\text{Hilb}(B', t) = \frac{(1-t)(1-t^3)\cdots(1-t^n)(1-t^d)}{(1-t)(1-t^2)\cdots(1-t^n)} = \frac{1+t+\cdots+t^{d-1}}{1+t}.$$

This forces the numerator of the last function to be divisible by $1+t$, a contradiction, since d is a prime number. This means that $\beta \neq 0$. Thus we may replace e_2 by f_2 as a generator of the algebra B. Hence, by modifying the elements $f_3, \ldots, f_n, f_d$ suitably, we have

$$B \cong K[e_1, e_3, \ldots, e_n]/(f_3, f_4, \ldots, f_n, f_d).$$

Next suppose that $f_3 = \alpha e_1^3 + \beta e_3$. As with the preceding case, $\beta = 0$ does not occur. Hence f_3 may replace e_3 as a generator of the algebra, so that

$$B \cong K[e_1, e_4, \ldots, e_n]/(f_4, f_5, \ldots, f_n, f_d),$$

with modification of the generators $f_4, f_5, \ldots, f_n, f_d$. We may repeat the same argument to obtain

$$B \cong K[e_1]/(e_1^d). \qquad\qquad\qquad\qquad\qquad$$ □

Chapter 5
A Generalization of Lefschetz Elements

In this chapter we would like to discuss a generalization of Lefschetz elements for an Artinian local ring to study the Jordan decomposition of a general element. The point of departure for us is Theorem 5.1 due to D. Rees. Several results from Chap. 6 (e.g., stable ideals, Borel fixed ideals, $\text{gin}(I)$, etc) are needed at a few points in Chap. 5.

5.1 Weak Rees Elements

Theorem 5.1. *Let* $(A, \mathfrak{m})$ *be a local ring with the residue field* $K = A/\mathfrak{m}$, $y_1, y_2, \ldots, y_t \in \mathfrak{m}$ *and* $y = x_1 y_1 + x_2 y_2 + \cdots + x_t y_t$ *with* $x_1, x_2, \ldots, x_t \in A$. *Let* $X_1, X_2, \ldots, X_t$ *be indeterminates over* A, $Y = X_1 y_1 + X_2 y_2 + \cdots + X_t y_t \in A[X_1, X_2, \ldots, X_t]$ *and* $A(X) = A(X_1, X_2, \ldots, X_t)$ *the polynomial ring* $A[X_1, X_2, \ldots, X_t]$ *localized at* $\mathfrak{m}A[X_1, X_2, \ldots, X_t]$. *Finally let* $\mathfrak{a}$ *be an* $\mathfrak{m}$-*primary ideal of* A.

1. $\text{length}_{A(X)}(A(X)/\mathfrak{a}A(X) + YA(X)) \leq \text{length}_A(A/\mathfrak{a} + yA)$.
2. *Further, suppose that* K *is infinite. Then there exists a non-zero radical ideal* $\mathfrak{b}$ *of* $K[X_1, X_2, \ldots, X_t]$ *such that the above inequality becomes an equality if the ideal* $(X_1 - \overline{x_1}, X_2 - \overline{x_2}, \ldots, X_t - \overline{x_t})$ *does not contain* $\mathfrak{b}$, *where* $\overline{x_i}$ *denotes the image of* x_i *in* K.

We first prove the following proposition.

Proposition 5.2. *Let* $(B, \mathfrak{n}, K)$ *be an Artinian local ring,* X *an indeterminate over* B, *and* $B(X) = B[X]_{\mathfrak{n}[X]}$. *Then*

$$\text{length}_{B(X)}(B(X)/(Xz_1 + z_2)B(X)) \leq \text{length}_B(B/(xz_1 + z_2)B)$$

for any $x \in B$ *and* $z_1, z_2 \in \mathfrak{n}$.

T. Harima et al., *The Lefschetz Properties*, Lecture Notes in Mathematics 2080, DOI 10.1007/978-3-642-38206-2_5, © Springer-Verlag Berlin Heidelberg 2013

Proof. Set $n = \text{length}_{B(X)}(B(X)/(Xz_1 + z_2)B(X))$ and let

$$B(X) = J_0 \underset{\neq}{\supseteq} J_1 \underset{\neq}{\supseteq} \cdots \underset{\neq}{\supseteq} J_n = (Xz_1 + z_2)B(X) \underset{\neq}{\supseteq} \cdots \underset{\neq}{\supseteq} J_{n'} = (0)$$

be a composition series of $B(X)$. Consider the natural homomorphisms $\varphi_1 \colon B[X] \longrightarrow B(X)$ and $\varphi_2 \colon B[X] \longrightarrow B[X]/(X - x) \cong B$. Let $I_k = \varphi_1^{-1}(J_k)$ for all $0 \leq k \leq n$ and let $\overline{I}_k$ be the image of I_k in B. Note that $I_n \supset (Xz_1 + z_2)B[X]$ and $\overline{I}_n \supset (xz_1 + z_2)B$. To prove the desired inequality, it suffices to show that $\overline{I}_0 \underset{\neq}{\supseteq} \overline{I}_1 \underset{\neq}{\supseteq} \cdots \underset{\neq}{\supseteq} \overline{I}_n$.

Step 1.   Between two ideals I_k and I_{k+1}, we can take a chain of ideals $I_k = \mathfrak{a}_0 \underset{\neq}{\supseteq} \mathfrak{a}_1 \underset{\neq}{\supseteq} \cdots \underset{\neq}{\supseteq} \mathfrak{a}_{s_k} = I_{k+1}$ such that, for any $j = 0, 1, \ldots, s_k - 1$, $\mathfrak{a}_j/\mathfrak{a}_{j+1} \cong B[X]/\mathfrak{p}_j$ for some $\mathfrak{p}_j \in \text{Spec}(B[X])$. We show that

$$\mathfrak{a}_j/\mathfrak{a}_{j+1} \cong B[X]/\mathfrak{n}[X] \cong K[X]$$

for some j. Note that $\mathfrak{n}[X]$ is the unique minimal prime ideal of $B[X]$; in fact, $\mathfrak{n}[X]$ is nilpotent and $B[X]/\mathfrak{n}[X] \cong K[X]$. We show this by way of contradiction, for suppose that $\mathfrak{n}[X] \underset{\neq}{\subsetneq} \mathfrak{p}_j$ for some j. Then, $B(X)/\mathfrak{p}_j B(X) = (0)$, and hence

$$\mathfrak{a}_j B(X)/\mathfrak{a}_{j+1}B(X) \cong \mathfrak{a}_j/\mathfrak{a}_{j+1} \otimes_{B[X]} B(X)$$

$$\cong B[X]/\mathfrak{p}_j \otimes_{B[X]} B(X)$$

$$\cong B(X)/\mathfrak{p}_j B(X) = (0).$$

This implies that $I_k B(X) = I_{k+1}B(X)$. Therefore we have that $J_k = J_{k+1}$, so this is a contradiction.

Step 2.   We would like to prove that $\overline{I}_k \underset{\neq}{\supseteq} \overline{I}_{k+1}$ under the assumption that $I_k/I_{k+1} \cong B[X]/\mathfrak{n}[X] \cong K[X]$. First we show that $(I_k : (X - x)) = I_k$ for all k. Let $f \in (I_k : (X - x))$. Then $\dfrac{X - x}{1} \cdot \dfrac{f}{1} \in I_k B(X)$. Noting that $(X - x)$ is a unit in $B(X)$ and $I_k B(X) = J_k$, we have $\dfrac{f}{1} \in J_k$, and hence $f \in I_k$. The converse is clear. Note our assumption says that there is an isomorphism $\psi \colon I_k/I_{k+1} \longrightarrow K[X]$ as $B[X]$-modules. By way of contradiction, assume that $\overline{I}_k = \overline{I}_{k+1}$ for some k. Then $I_k + (X - x) = I_{k+1} + (X - x)$ in $B[X]$, so that for any $f \in I_k$ there are $g \in I_{k+1}$ and $h \in B[X]$ such that $f = g + (X - x)h$. Since $(X - x)h = f - g \in I_k$, it follows that $h \in I_k$ by the equality $(I_k : (X - x)) = I_k$. Therefore we have

$$\psi(\overline{f}) = \psi(\overline{g} + (X - x)\overline{h})$$

$$= \psi(\overline{g}) + (X - \overline{x})\psi(\overline{h})$$

$$= (X - \overline{x})\psi(\overline{h}).$$

Since ψ is bijective, this equality implies that any polynomial in $K[X]$ is divisible by $X - \overline{x}$. This is a contradiction, and completes the proof.   □

Proof (Proof of Theorem 5.1). To prove this theorem we may assume that A is Artinian and $\mathfrak{a} = (0)$.

(1) First note that $A(X_1, X_2, \ldots, X_{i+1}) = A(X_1, X_2, \ldots, X_i)(X_{i+1})$ for all $i = 1, 2, \ldots, t - 1$. Applying Proposition 5.2 to the case where $B = A(X_1, X_2, \ldots, X_{t-1})$, $z_1 = y_t$ and $z_2 = X_1 y_1 + X_2 y_2 + \cdots + X_{t-1} y_{t-1}$, we have

$$\text{length}_{A(X_1, X_2, \ldots, X_t)}(A(X_1, X_2, \ldots, X_t)/(X_1 y_1 + X_2 y_2 + \cdots + X_t y_t))$$

$$\leq \text{length}_{A(X_1, X_2, \ldots, X_{t-1})}(A(X_1, X_2, \ldots, X_{t-1})/(X_1 y_1 + X_2 y_2 + \cdots + X_{t-1} y_{t-1} + x_t y_t)).$$

Furthermore, applying Proposition 5.2 to the case where $B = A(X_1, X_2, \ldots, X_{t-2})$, $z_1 = y_{t-1}$ and $z_2 = X_1 y_1 + X_2 y_2 + \cdots + X_{t-2} y_{t-2} + x_t y_t$, we have

$$\text{length}_{A(X_1, X_2, \ldots, X_t)} \left(\frac{A(X_1, X_2, \ldots, X_t)}{(X_1 y_1 + X_2 y_2 + \cdots + X_t y_t)} \right)$$

$$\leq \text{length}_{A(X_1, X_2, \ldots, X_{t-1})} \left(\frac{A(X_1, X_2, \ldots, X_{t-1})}{(X_1 y_1 + X_2 y_2 + \cdots + X_{t-1} y_{t-1} + x_t y_t)} \right)$$

$$\leq \text{length}_{A(X_1, X_2, \ldots, X_{t-2})} \left(\frac{A(X_1, X_2, \ldots, X_{t-2})}{(X_1 y_1 + X_2 y_2 + \cdots + X_{t-2} y_{t-2} + x_{t-1} y_{t-1} + x_t y_t)} \right).$$

Therefore our desired inequality can be obtained by repeating this argument.

(2) The proof is divided into three steps.

Step 1.   We fix a chain of ideals of $S = A[X_1, X_2, \ldots, X_t]$:

$$S = I_0 \supsetneq I_1 \supsetneq \cdots \supsetneq I_n = YS,$$

such that $I_k/I_{k+1} \cong S/P_k$ for some $P_k \in \text{Spec}(S)$. Since any prime ideal of S contains $\mathfrak{m}S$, we may suppose $S/P_k \cong K[X_1, X_2, \ldots, X_t]/\mathfrak{p}_k$ for prime ideals $\mathfrak{p}_k$ of $K[X_1, X_2, \ldots, X_t]$. From the exact sequence

$$0 \longrightarrow I_k/I_{k+1} \longrightarrow S/I_{k+1} \longrightarrow S/I_k \to 0,$$

we get the following exact sequence

$$0 \longrightarrow (I_k/I_{k+1}) \otimes_S A(X) \longrightarrow S/I_{k+1} \otimes_S A(X) \longrightarrow S/I_k \otimes_S A(X) \longrightarrow 0.$$

Hence we have the exact sequence

$$0 \longrightarrow (S/P_k) \otimes_S A(X) \longrightarrow A(X)/J_{k+1} \longrightarrow A(X)/J_k \longrightarrow 0,$$

where $J_k = I_k A(X)$. Thus, since

$$A(X) = J_0 \supset J_1 \supset \cdots \supset J_n = YA(X),$$

the isomorphisms

$$(S/P_k) \otimes_S A(X) \cong \begin{cases} K(X_1, X_2, \ldots, X_t) & \text{if } P_k = \mathfrak{m}S, \text{ i.e., } \mathfrak{p}_k = (0) \\ (0) & \text{if } P_k \supsetneqq \mathfrak{m}S, \text{ i.e., } \mathfrak{p}_k \neq (0) \end{cases}$$

imply

$$\operatorname{length}_{A(X)}(A(X)/YA(X)) = \left| \{ k \mid \mathfrak{p}_k = (0) \} \right|.$$

Step 2.   Next we consider $\operatorname{length}_A(A/yA)$. Tensoring the exact sequence

$$0 \longrightarrow I_k/I_{k+1} \xrightarrow{h_k} S/I_{k+1} \longrightarrow S/I_k \longrightarrow 0$$

through with $S/(X_1 - x_1, X_2 - x_2, \ldots, X_t - x_t) \cong A$, we obtain the exact sequence

$$(I_k/I_{k+1}) \otimes_S A \xrightarrow{h_k \otimes 1_A} (S/I_{k+1}) \otimes_S A \longrightarrow (S/I_k) \otimes_S A \longrightarrow 0.$$

Hence we have the exact sequence

$$(I_k/I_{k+1}) \otimes_S A \xrightarrow{h_k \otimes 1_A} A/\overline{I_{k+1}} \longrightarrow A/\overline{I_k} \longrightarrow 0,$$

where $\overline{I_k}$ is the image of I_k in $S/(X_1 - x_1, X_2 - x_2, \ldots, X_t - x_t) \cong A$. Since

$$(I_k/I_{k+1}) \otimes_S A \cong (S/P_k) \otimes_S A$$
$$\cong S/(P_k + (X_1 - x_1, X_2 - x_2, \ldots, X_t - x_t))$$
$$\cong K[X_1, X_2, \ldots, X_t]/(\mathfrak{p}_k + (X_1 - \overline{x_1}, X_2 - \overline{x_2}, \ldots, X_t - \overline{x_t})),$$

it follows that

$$(I_k/I_{k+1}) \otimes_S A \cong \begin{cases} K & \text{if } \mathfrak{p}_k \subset (X_1 - \overline{x_1}, X_2 - \overline{x_2}, \ldots, X_t - \overline{x_t}), \\ (0) & \text{otherwise.} \end{cases}$$

Hence, noting that $\overline{I_k}/\overline{I_{k+1}}$ is the image of $h_k \otimes 1_A$, we obtain the following isomorphisms

$$\overline{I_k}/\overline{I_{k+1}} \cong \begin{cases} K & \text{if } \mathfrak{p}_k \subset (X_1 - \overline{x_1}, X_2 - \overline{x_2}, \ldots, X_t - \overline{x_t}) \text{ and } h_k \otimes 1_A \neq 0, \\ (0) & \text{otherwise.} \end{cases}$$

Since

$$A = \overline{I_0} \supset \overline{I_1} \supset \cdots \supset \overline{I_n} = yA,$$

the isomorphisms above imply

$$\text{length}_A(A/yA) = \left| \left\{ k \;\middle|\; \begin{array}{l} (a)\ \mathfrak{p}_k \subset (X_1 - \overline{x_1}, X_2 - \overline{x_2}, \ldots, X_t - \overline{x_t}) \\ (b)\ h_k \otimes 1_A \text{ is not the zero-map} \end{array} \right\} \right|.$$

Step 3. Set

$$\mathfrak{b} = \bigcap_{\mathfrak{p}_k \neq (0)} \mathfrak{p}_k,$$

which is a non-zero radical ideal of $K[X_1, X_2, \ldots, X_t]$. Let $\{x_1, x_2, \ldots, x_t\}$ be a set of elements of A such that $(X_1 - \overline{x_1}, X_2 - \overline{x_2}, \ldots, X_t - \overline{x_t}) \not\supseteq \mathfrak{b}$. If $\mathfrak{p}_k \neq (0)$ and $\mathfrak{p}_k \subset (X_1 - \overline{x_1}, X_2 - \overline{x_2}, \ldots, X_t - \overline{x_t})$ then $\mathfrak{b} \subset \mathfrak{p}_k \subset (X_1 - \overline{x_1}, X_2 - \overline{x_2}, \ldots, X_t - \overline{x_t})$. This is a contradiction. Hence if $\mathfrak{p}_k \neq (0)$ then $\mathfrak{p}_k \not\subset (X_1 - \overline{x_1}, X_2 - \overline{x_2}, \ldots, X_t - \overline{x_t})$, so therefore the set $\{ \mathfrak{p}_k \mid (a)'\ \mathfrak{p}_k = (0) \}$ contains the set $\{ \mathfrak{p}_k \mid (a)\ \mathfrak{p}_k \subset (X_1 - \overline{x_1}, X_2 - \overline{x_2}, \ldots, X_t - \overline{x_t}) \}$. Consequently

$$\text{length}_A(A/yA) \leq \left| \left\{ k \;\middle|\; \begin{array}{l} (a)'\ \mathfrak{p}_k = (0) \\ (b)\ h_k \otimes 1_A \text{ is not the zero-map} \end{array} \right\} \right|$$

$$\leq \left| \{ k \mid \mathfrak{p}_k = (0) \} \right|$$

$$= \text{length}_{A(X)}(A(X)/YA(X)).$$

The opposite inequality was proved in (1). We have thus proved the theorem. □

The following is an immediate corollary of Theorem 5.1.

Theorem 5.3. *With the same notation as in Theorem 5.1, suppose that A is Artinian, $\mathfrak{a} = (0)$ and $\{y_1, y_2, \ldots, y_t\}$ is a set of generator of $\mathfrak{m}$.*

1. $\text{length}_{A(X)}(A(X)/YA(X)) \leq \text{length}_A(A/yA)$ for all $y \in \mathfrak{m}$.
2. Further, if K is infinite, then there exists a non-zero radical ideal $\mathfrak{b}$ of the ring $K[X_1, X_2, \ldots, X_t]$ such that the inequality in (1) becomes an equality if the ideal

$(X_1 - \overline{x_1}, X_2 - \overline{x_2}, \ldots, X_t - \overline{x_t})$ *does not contain* $\mathfrak{b}$*, where* $\overline{x_i}$ *denotes the image of* x_i *in* K.

Definition 5.4. We keep the same notation as in Theorem 5.3. We call an element $g \in \mathfrak{m}$ a **weak Rees element** for an Artinian local ring A if

$$\mathrm{length}_A(A/gA) = \mathrm{length}_{A(X)}(A(X)/YA(X)).$$

Remark 5.5. 1. The assertion (2) of Theorem 5.3 says that the set of weak Rees elements of A contains a non-empty Zariski open subset of the affine space of linear forms.
2. Proposition 3.5 and Theorem 5.3 say that the notion of weak Rees elements for an Artinian local ring is a generalization of weak Lefschetz elements for a graded Artinian K-algebra with the WLP.

5.2   Strong Rees Elements

Theorem 5.6. *Let* $(A, \mathfrak{m})$ *be an Artinian local ring which contains the residue field* $K = A/\mathfrak{m}$ *of characteristic zero. Let* $\{y_1, y_2, \ldots, y_t\}$ *be a set of generators of* $\mathfrak{m}$ *and let* $X_1, X_2, \ldots, X_t$ *be indeterminates over* A. *Let* $A(X) = A(X_1, X_2, \ldots, X_t)$ *denote* $A[X_1, X_2, \ldots, X_t]$ *localized at* $\mathfrak{m}[X_1, X_2, \ldots, X_t]$ *and let* $Y = X_1 y_1 + X_2 y_2 + \cdots + X_t y_t \in A[X]$. *Then we have the following.*

1. $P_{A(X)}(\times Y) \succeq P_A(\times y)$ *for all* $y \in \mathfrak{m}$ *(see Definition 3.62).*
2. Set $c = \max \{ i \mid \mathfrak{m}^i \neq (0) \}$ *and let* $g \in \mathfrak{m}$. *Then* $P_A(\times g) = P_{A(X)}(\times Y)$ *if and only if* $g + \overline{z}$ *is a weak Rees element of* $A[z]/(z^p)$ *for every* $p = 1, 2, \ldots, c + 1$, *where* z *is a new indeterminate over* A.
3. There exists an element $g \in \mathfrak{m}$ *such that* $P_A(\times g) = P_{A(X)}(\times Y)$.

Proof. First we fix some notation. Let X_{t+1} be a new indeterminate and let $A(X)' = A(X_1, X_2, \ldots, X_t, X_{t+1})$ denote $A[X_1, X_2, \ldots, X_t, X_{t+1}]$ localized at $\mathfrak{m}[X_1, X_2, \ldots, X_t, X_{t+1}]$. Further set $B = A[z]/(z^p)$, let $\mathfrak{n}$ be the maximal ideal of B, let $B(X) = B(X_1, X_2, \ldots, X_t, X_{t+1})$ denote $B[X_1, X_2, \ldots, X_t, X_{t+1}]$ localized at $\mathfrak{n}[X_1, X_2, \ldots, X_t, X_{t+1}]$ and let $Y' = X_1 y_1 + X_2 y_2 + \cdots + X_t y_t + X_{t+1} \overline{z} \in B[X_1, X_2, \ldots, X_t, X_{t+1}]$. Note that

$$A(X)'[z]/(z^p) \cong (A \otimes_K K(X_1, X_2, \ldots, X_t, X_{t+1})) \otimes_K K[z]/(z^p)$$

$$\cong (A \otimes_K K[z]/(z^p)) \otimes_K K(X_1, X_2, \ldots, X_t, X_{t+1})$$

$$\cong B \otimes_K K(X_1, X_2, \ldots, X_t, X_{t+1})$$

$$\cong B(X).$$

We take up the three points one at a time.

(1) Let g be an element of A such that $P(\times g)$ is the maximum element in $\{ P(\times y) \mid y \in \mathfrak{m} \}$. In Theorem 5.1 it was proved that

$$\dim_K A/gA = \dim_{K(X_1, X_2, \dots, X_t)} A(X)/YA(X).$$

Let $r = \dim_K A/gA$,

$$P(\times g) = d_1 \oplus d_2 \oplus \cdots \oplus d_r \text{ and } P(\times Y) = d'_1 \oplus d'_2 \oplus \cdots \oplus d'_r,$$

where we assume that $d_1 \geq d_2 \geq \cdots \geq d_r$ and $d'_1 \geq d'_2 \geq \cdots \geq d'_r$. (Both $\times g$ and $\times Y$ have the same number r of Jordan blocks.) By way of contradiction assume that $P(\times g) \succ P(\times Y)$. This implies that there exists an integer $k > 0$ such that $d_i = d'_i$ for $i < k$ and $d_k > d'_k$. Put $p = d'_k$. Compare the numbers of Jordan blocks of $P(\times(g + \bar{z})) \in \mathrm{End}_K(B)$ and $P(\times(Y + X_{t+1}\bar{z})) \in \mathrm{End}_L(B(X))$, where $L = B(X)/\mathfrak{n}B(X)$. By Proposition 3.66 we have

$$\dim_K \mathrm{Ker}(\times(g + \bar{z})) = \sum_{i=1}^{r} \min\{d_i, p\} = pk + \sum_{i=k+1}^{r} \min\{d_i, p\}$$

$$\leq pk + \sum_{i=k+1}^{r} d_i$$

and

$$\dim_L \mathrm{Ker}(\times(Y + X_{t+1}\bar{z})) = \sum_{i=1}^{r} \min\{d'_i, p\} = pk + \sum_{i=k+1}^{r} \min\{d'_i, p\}$$

$$= pk + \sum_{i=k+1}^{r} d'_i.$$

Hence, since $\sum_{i=1}^{r} d_i = \sum_{i=1}^{r} d'_i$ and $\sum_{i=k+1}^{r} d_i < \sum_{i=k+1}^{r} d'_i$, it follows that

$$\dim_K \mathrm{Ker}(\times(g + \bar{z})) < \dim_L \mathrm{Ker}(\times(Y + X_{t+1}\bar{z})).$$

That is,

$$\mathrm{length}_B(B/(y + \bar{z})B) < \mathrm{length}_{B(X)}(B(X)/(Y + X_{t+1}\bar{z})B(X)).$$

Since the opposite inequality was proved in Theorem 5.1, we get a contradiction.

(2) The only if part: We show that $\mathrm{length}(B/(g + \bar{z})B) = \mathrm{length}(B(X)/Y'B(X))$, i.e., $\dim_K \mathrm{Ker}(\times(g + \bar{z})) = \dim_L \mathrm{Ker}(\times Y')$. Let $P(\times y) = P(\times Y) = d_1 \oplus d_2 \oplus \cdots \oplus d_r$. Since $B = A \otimes_K K[z]/(z^p)$ and $y + \bar{z} = y \otimes 1 + 1 \otimes \bar{z}$, it follows by Proposition 3.66 that $\dim_K \mathrm{Ker}(\times(g + \bar{z})) = \sum_{i=1}^{r} \min\{d_i, p\}$. Also,

since $B(X) \cong A(X)'[z]/(z^p) \cong A(X)' \otimes_L L[z]/(z^p)$ and $Y' = Y \otimes 1 + 1 \otimes X_{t+1}\bar{z}$, it follows by Proposition 3.66 again that $\dim_L \text{Ker}(\times Y') = \sum_{i=1}^r \min\{d_i, p\} = \dim_K \text{Ker}(\times(g + \bar{z}))$.

The if part: Since g is a weak Rees element of A by our assumption, both linear maps $\times g$ and $\times Y$ have the same number r of Jordan blocks. So let

$$P(\times g) = d_1 \oplus d_2 \oplus \cdots \oplus d_r \text{ and } P(\times Y) = d_1' \oplus d_2' \oplus \cdots \oplus d_r',$$

where we assume that $d_1 \geq d_2 \geq \cdots \geq d_r$ and $d_1' \geq d_2' \geq \cdots \geq d_r'$. By way of contradiction assume that $P(\times g) \prec P(\times Y)$. This implies that there exists an integer $k > 0$ such that $d_i = d_i'$ for $i < k$ and $d_k < d_k'$. Put $p = d_k$. Compare the numbers of blocks of $P(\times(g + \bar{z})) \in \text{End}_K(B)$ and $P(\times(Y + X_{t+1}\bar{z})) \in \text{End}_L(B(X))$. By Proposition 3.66 we have

$$\dim_K \text{Ker}(\times(g + \bar{z})) = \sum_{i=1}^r \min\{d_i, p\} = pk + \sum_{i=k+1}^r \min\{d_i, p\}$$

and

$$\dim_L \text{Ker}(\times(Y + X_{t+1}\bar{z})) = \sum_{i=1}^r \min\{d_i', p\} = pk + \sum_{i=k+1}^r \min\{d_i', p\}.$$

Hence, since $\sum_{i=k+1}^r d_i > \sum_{i=k+1}^r d_i'$, it follows that

$$\text{length}_B(B/(y + \bar{z})B) > \text{length}_{B(X)}(B(X)/(Y + X_{t+1}\bar{z})B(X)).$$

But, by our assumption that $y + \bar{z}$ is a weak Rees element of B, so we have the equality

$$\text{length}(B/(y + \bar{z})B) = \text{length}(B(X)/(Y + X_{t+1}\bar{z})B(X)).$$

Thus we get a contradiction.

The assertion (3) is easily proven by combining (1) of the preceding Theorem 5.1 and (2) of this theorem. $\square$

Definition 5.7. We keep the same notation as Theorem 5.6. We call an element $g \in \mathfrak{m}$ a **strong Rees element** of an Artinian local ring A if $P_A(\times g) = P_{A(X)}(\times Y)$.

Remark 5.8. 1. A strong Rees element is a weak Rees element.
2. Let $g \in \mathfrak{m}$ be a strong Rees element of A. Then $P_A(\times g)$ is the maximum element in $\{ P_A(\times y) \mid y \in \mathfrak{m} \}$. (Hence we say that $P_A(\times g)$ is the **strongest Jordan decomposition** of A.)
3. By Theorem 5.3 (2) and Theorem 5.6 (2), the set of strong Rees elements of A contains a non-empty Zariski open subset of the affine space of linear forms.

4. Proposition 3.64 and Theorem 5.6 say that the notion of strong Rees elements for an Artinian local ring is a generalization of strong Lefschetz elements for a graded Artinian K-algebra with the SLP.

Example 5.9. Let I be a strongly stable ideal (see Definition 6.4) of $R = K[x_1, x_2, \ldots, x_n]$ whose associated quotient ring $A = R/I$ is Artinian. Then the last variable x_n mod I is a strong Rees element of A.

To verify this we need to make use of some results from the next section and Chap. 6. Let $\text{gin}(I)$ be the generic initial ideal of I with respect to the graded reverse lexicographic order with $x_1 > x_2 > \cdots > x_n$ (see Definition 6.12). Let g be a general linear form of A which is in the Zariski open set given by Remark 5.8 (3). Since I is a strongly stable ideal, we have $\text{gin}(I) = I$. Furthermore, by the proof of Proposition 6.15, $R/(I, g^k)$ and $R/(I, x_n^k)$ have the same Hilbert function for every $k = 1, 2, \ldots$, and hence, by Lemma 5.12 (2), we have $P_{R/I}(\times g) = P_{R/\text{gin}(I)}(\times x_n)$. This implies that x_n is a strong Rees element of A as claimed.

5.3 Some Properties of Strong Rees Elements

In this section we prove some propositions which indicate the good properties of strong Rees elements.

Theorem 5.10. *Let $(A, \mathfrak{m})$ and $(B, \mathfrak{n})$ be Artinian local rings which contain the same residue field K of characteristic zero. Let g and h be strong Rees elements of A and B, respectively. Then $g \otimes 1 + 1 \otimes h$ is also a strong Rees element of $A \otimes_K B$.*

Proof. Let $\{y_1, y_2, \ldots, y_t\}$ be a set of generators of $\mathfrak{m}$ and let $\{z_1, z_2, \ldots, z_u\}$ be a set of generators of $\mathfrak{n}$. First we show that $A \otimes_K B$ is a local ring and its maximal ideal is generated by $\{y_1 \otimes 1, y_2 \otimes 1, \ldots, y_t \otimes 1, 1 \otimes z_1, 1 \otimes z_2, \ldots, 1 \otimes z_u\}$. Let $\mathfrak{a}$ be the ideal generated by $\{y_1 \otimes 1, y_2 \otimes 1, \ldots, y_t \otimes 1\}$ and $\mathfrak{b}$ the ideal generated by $\{1 \otimes z_1, 1 \otimes z_2, \ldots, 1 \otimes z_u\}$. Then $\mathfrak{a} = \mathfrak{m}(A \otimes_K B)$, $\mathfrak{b} = \mathfrak{n}(A \otimes_K B)$ and

$$(A \otimes_K B)/\mathfrak{b} \cong (A \otimes_K B)/\mathfrak{n}(A \otimes_K B)$$

$$\cong (A \otimes_K B) \otimes_B (B/\mathfrak{n})$$

$$\cong A.$$

Hence $(A \otimes_K B)/(\mathfrak{a} + \mathfrak{b}) \cong K$, and $\mathfrak{a} + \mathfrak{b}$ is a maximal ideal. Furthermore, since $\mathfrak{a} + \mathfrak{b}$ is nilpotent, any prime ideal contains $\mathfrak{a} + \mathfrak{b}$. This implies that $\mathfrak{a} + \mathfrak{b}$ is the unique maximal ideal.

Let $X_1, X_2, \ldots, X_t$ be indeterminates over A and let $X'_1, X'_2, \ldots, X'_u$ be indeterminates over B. Set $Y = X_1 y_1 + X_2 y_2 + \cdots + X_t y_t \in A(X_1, X_2, \ldots, X_t)$ and $Y' = X'_1 z_1 + X'_2 z_2 + \cdots + X'_u z_u \in B(X'_1, X'_2, \ldots, X'_u)$. Since g and h are strong Rees elements of A and B, we have $P_{A}(\times g) = P_{A(X)}(\times Y)$ and

$P_B(\times h) = P_{B(X')}(\times Y')$. Hence, noting that $P_{A(X)}(\times Y) = P_{A(X)\otimes_{K(X)}K(X,X')}(\times Y)$
and $P_{B(X')}(\times Y') = P_{B(X')\otimes_{K(X')}K(X,X')}(\times Y')$, it follows by Proposition 3.66 that
the Jordan decomposition

$$P_{A\otimes_K B}(\times(g \otimes 1 + 1 \otimes h))$$

is the same as that of

$$\times(Y \otimes 1 + 1 \otimes Y')$$

as an endomorphism of the tensor space

$$\left(A(X) \otimes_{K(X)} K(X, X')\right) \otimes \left(B(X') \otimes_{K(X')} K(X, X')\right).$$

Here the tensor product in the middle is over $K(X, X')$. Since this tensor space is
the same as $(A \otimes_K B) \otimes_K K(X, X')$, we obtain the desired equality:

$$P_{A\otimes_K B}(\times(g \otimes 1 + 1 \otimes h)) = P_{(A\otimes_K B)\otimes_K K(X,X')}(\times(Y \otimes 1 + 1 \otimes Y')).$$

This implies that $g \otimes 1 + 1 \otimes h$ is a strong Rees element of $A \otimes_K B$.   □

In the rest of this section, let $R = K[x_1, x_2, \ldots, x_n]$ be the polynomial ring over
a field K of characteristic zero with the standard grading. The following proposition
says that any strong Rees element is a linear form.

Proposition 5.11. *Let I be a homogeneous ideal of R whose quotient ring $A = R/I$ is Artinian, set $A = \bigoplus_{i=0}^{c} A_i$ with $A_c \neq (0)$ and $\mathfrak{m} = \bigoplus_{i=1}^{c} A_i$. Let g be a strong Rees element of A and $P(\times g) = d_1 \oplus d_2 \oplus \cdots \oplus d_r$ where $d_1 \geq d_2 \geq \cdots \geq d_r$.*

1. *The maximum size of the Jordan blocks of $P(\times g)$ is equal to $c + 1$, i.e., $d_1 = c + 1$.*
2. *$g \notin \mathfrak{m}^2$. In particular, if g is homogeneous, then g is a linear form.*
3. *The set of all strong Rees elements of A_1 contains a Zariski open subset of the affine space A_1.*

Before we prove this proposition, we introduce two lemmas.

Lemma 5.12. *Let R and A be as in Proposition 5.11.*

1. *Let $A = \bigoplus_{i=0}^{c} A_i$ be graded Artinian K-algebra, and let $y \in \mathfrak{m} = \bigoplus_{i=1}^{c} A_i$, and set $e_i = \dim_K(A/y^{i+1}A) - \dim_K(A/y^i A)$ for all $i = 0, 1, \ldots$. Then $P(\times y)$ is equal to the dual partition of $e_0 \oplus e_1 \oplus \cdots \oplus e_b$, where $b = \max\{j \mid y^j \neq 0\}$.*
2. *Let I and J be two homogeneous ideals of R whose quotient rings are Artinian and have the same Hilbert function. Let y_1 and y_2 be homogeneous elements of R/I and R/J with the same degree. Suppose that the Hilbert function of $R/(I, y_1^i)$ is equal to that of $R/(J, y_2^i)$ for every $i = 1, 2, \ldots$. Then $P_{R/I}(\times y_1) = P_{R/J}(\times y_2)$.*

Proof. Noting that

$$e_i = \dim_K y^i A - \dim_K y^{i+1} A$$

$$= \dim_K(A/(0:y^i)) - \dim_K(A/(0:y^{i+1}))$$

$$= \dim_K(0:y^{i+1})/(0:y^i),$$

one sees that (1) follows from Lemma 3.60, and (2) follows from (1). $\qquad\Box$

Lemma 5.13. *With the same notation as Proposition 5.11, let y be any general linear element of A_1 and let $P(\times y) = d_1 \oplus d_2 \oplus \cdots \oplus d_r$ where $d_1 \geq d_2 \geq \cdots \geq d_r$.*

1. $y^c \neq 0$.
2. The maximum size of blocks of $P(\times y)$ is equal to $c + 1$, i.e., $d_1 = c + 1$.

Proof. Let $\mathrm{gin}(I)$ be the generic initial ideal of I with respect to the graded reverse lexicographic oder with $x_1 > x_2 > \cdots > x_n$. Then, it follows from the proof of Proposition 6.15 that $R/(I, y^k)$ and $R/(\mathrm{gin}(I), x_n^k)$ have the same Hilbert function for every $k = 1, 2, \ldots$. Hence, by Lemma 5.12 (2), we have $P_{R/I}(\times y) = P_{R/\mathrm{gin}(I)}(\times x_n)$. Since $\mathrm{gin}(I)$ is a strongly stable ideal, it follows that $x_n^c \notin \mathrm{gin}(I)$, i.e., $\overline{x_n}^c \neq 0$ in $R/\mathrm{gin}(I)$. Since $R/(I, y^c)$ and $R/(\mathrm{gin}(I), x_n^c)$ have the same Hilbert function, this implies that $y^c \neq 0$, and hence $d_1 = c + 1$. $\qquad\Box$

Proof (Proof of Proposition 5.11).
    (1) follows from Lemma 5.13 and the definition of a strong Rees element.
    (2) We may assume that $c \geq 2$. Suppose that $g \in \mathfrak{m}^2$. Then $g^{\lfloor c/2 \rfloor + 1} = 0$, but $g^c \neq 0$ by (1). This is a contradiction.
    (3) follows from Remark 5.8 (3). $\qquad\Box$

In view of Proposition 3.72 we have a natural question.

Question 5.14. Let $A = \bigoplus_{i=0}^c A_i$ be a graded Artinian K-algebra, g a strong Rees element of A and k a positive integer. Does inequality $P(\times g^k) \succeq P(\times y^k)$ hold for all linear forms $y \in A_1$?

The following is a partial answer to this question.

Proposition 5.15. *Let A be a graded Artinian K-algebra, g a strong Rees element of A and k a positive integer. Then $P(\times g^k) \succeq P(\times y^k)$ for all weak Rees elements y of A.*

Proof. We use a similar idea as in the proof of Proposition 3.72. Let $P(\times g) = m_1 \oplus m_2 \oplus \cdots \oplus m_s$ and $P(\times y) = n_1 \oplus n_2 \oplus \cdots \oplus n_r$. Since g is a strong Rees element and y is a weak Rees element, the numbers of blocks of $P(\times g)$ and $P(\times y)$ are same. Take non-negative integers q_i, r_i, q'_j and r'_j satisfying the following conditions; $m_i = q_i \times k + r_i$ ($0 \leq r_i < k$) and $n_j = q'_j \times k + r'_j$ ($0 \leq r'_j < k$). Then we have

$$P(\times g^k) = \bigoplus_{i=1}^{r} (\underbrace{(q_i + 1) \oplus (q_i + 1) \oplus \cdots \oplus (q_i + 1)}_{r_i} \oplus \underbrace{q_i \oplus q_i \oplus \cdots \oplus q_i}_{k-r_i})$$

and

$$P(\times y^k) = \bigoplus_{i=1}^{s} (\underbrace{(q_i' + 1) \oplus (q_i' + 1) \oplus \cdots \oplus (q_i' + 1)}_{r_i'} \oplus \underbrace{q_i' \oplus q_i' \oplus \cdots \oplus q_i'}_{k-r_i'}).$$

Let $P(\times g) \succ P(\times y)$. Since $s = r$, there is an integer t such that $m_1 = n_1, m_2 = n_2, \ldots, m_{t-1} = n_{t-1}$ and $m_t > n_t$. Then we have

$$\underbrace{(q_t + 1) \oplus (q_t + 1) \oplus \cdots \oplus (q_t + 1)}_{r_t} \oplus \underbrace{q_t \oplus q_t \oplus \cdots \oplus q_t}_{k-r_t}$$

$$\succ \underbrace{(q_t' + 1) \oplus (q_t' + 1) \oplus \cdots \oplus (q_t' + 1)}_{r_t'} \oplus \underbrace{q_t' \oplus q_t' \oplus \cdots \oplus q_t'}_{k-r_t'},$$

and hence this implies that $P(\times g^k) \succ P(\times y^k)$.   □

We conclude this section with a proposition on generic initial ideals (see Definition 6.12).

Proposition 5.16. *Let I be a homogeneous ideal of R whose quotient ring is Artinian. Let g be a strong Rees element of R/I. Then the equality $P_{R/I}(\times g) = P_{R/\mathrm{gin}(I)}(\times \overline{x_n})$ hold, where $\mathrm{gin}(I)$ is the generic initial ideal of I with respect to the graded reverse lexicographic order with $x_1 > x_2 > \cdots > x_n$.*

Proof. Let g be a sufficiently general linear element of A. Then, by the proof of Proposition 6.15 stated in the next chapter, $R/(I, g^k)$ and $R/(\mathrm{gin}(I), x_n^k)$ have the same Hilbert function for every $k = 1, 2, \ldots$, and hence, by Lemma 5.12 (2), we have $P_{R/I}(\times g) = P_{R/\mathrm{gin}(I)}(\times \overline{x_n})$.   □

Question 5.17. Let I be a homogeneous ideal of R whose quotient ring is Artinian. Let $\mathfrak{m}$ and $\mathfrak{n}$ be the maximal ideals of R/I and $R/\mathrm{gin}(I)$, respectively. Does the set $\{ P_{R/I}(\times y) \mid y \in \mathfrak{m} \}$ coincide with the set $\{ P_{R/\mathrm{gin}(I)}(\times z) \mid z \in \mathfrak{n} \}$?

5.4 Gorenstein Algebras with the WLP But not Having the SLP

We calculate the strongest Jordan decompositions of certain Artinian Gorenstein algebras arising from idealization, and give examples of Artinian Gorenstein algebras with the WLP but not having the SLP.

Notation and Remark 5.18 ([97, 98]). *Let* $B = \bigoplus_{i=0}^{c-1} B_i$ *be an Artinian level K-algebra with socle degree* $c - 1$, *and let* $E = \mathrm{Hom}_K(B, K)$ *be the space of K-linear maps from B to K. Then* $E = E_1 \oplus E_2 \oplus \cdots \oplus E_c$, *where* $E_i = \mathrm{Hom}_K(B_{c-i}, K)$, *giving E a graded B-module structure by* $(z\phi)(w) = \phi(zw)$ *for* $z, w \in B$ *and* $\phi \in E$. *Further we introduce a commutative ring structure on the product set* $A = B \times E$ *by* $(z, \phi) + (w, \psi) = (z + w, \phi + \psi)$ *and* $(z, \phi)(w, \psi) = (zw, z\psi + w\phi)$ *(see Definition 2.75). The gradings of B and E induce a grading of A such that* $A_i = B_i \times E_i$, *and hence the standard graded Artinian Gorenstein K-algebra* $A = \bigoplus_{i=0}^{c} A_i$ *has the Hilbert function*

$$1, H(B, 1) + H(B, c - 1), H(B, 2) + H(B, c - 2), \ldots, H(B, c - 1) + H(B, 1), 1.$$

Hence it follows that $\mathrm{Sperner}(A) \leq 2\,\mathrm{Sperner}(B)$.

Example 5.19. Let K be a field of characteristic zero and let B be the Artinian algebra $B = K[x_1, x_2, \ldots, x_n]/\mathfrak{m}^c$ and let $A = B \times \mathrm{Hom}_K(B, K)$ be the algebra obtained by idealization. Such rings are considered in Proposition 2.92. In this example we consider the algebras only for $n = 2$. For each $c = 1, 2, 3, \cdots$, denote the algebra by $A(c)$. It has the Hilbert function

$$1, \underbrace{c + 2, \ c + 2, \ \ldots, \ c + 2}_{c-1}, \ 1,$$

and $\dim_K A(c) = c^2 + c$. It is easy to compute the Sperner and the Dilworth numbers. They are $\mathrm{Sperner}(A(c)) = c + 2$ and $d(A(c)) = 2c$ if $c > 1$ and $d(A(c)) = 1$ if $c = 1$. Therefore $A(c)$ does not have the WLP except for $c = 1, 2$. Let l be a general linear form of A. Then one sees that one has the isomorphism

$$A^{(c)}/(0 : l) \cong A^{(c-1)}, \text{ for } c \geq 1.$$

(We let $A(0) = K$.) Thus it turns out that the multiplication map

$$\times l^{c-2i} : A_i^{(c)} \to A_{c-i}^{(c)}$$

does not have the full rank for all i such that $0 < i < [c/2]$. Since the ranks of the multiplication map $\times l^i : A(c) \to A(c)$ are known, it is possible to compute the Jordan block decomposition of the nilpotent element $\times l \in \mathrm{End}_K(A(c))$. In fact we have

$$P(\times l) = (c + 1) \oplus (c - 1)^3 \oplus (c - 2)^2 \oplus (c - 3)^2 \oplus \cdots \oplus 1^2.$$

For example if $c = 6$, the Jordan block decomposition $P(\times l)$ is

$$P(\times l) = 7 \oplus 5 \oplus 5 \oplus 5 \oplus 4 \oplus 4 \oplus 3 \oplus 3 \oplus 2 \oplus 2 \oplus 1 \oplus 1.$$

Example 5.20. With the same K as in Example 5.19, let B be the Artinian algebra $B = K[u, v]/(u^4, u^3v, u^2v^2, v^3)$. Then B is a level algebra with socle generators u^3, u^2v, uv^2. Let $S = B \times \mathrm{Hom}_K(B, K)$ be the ring obtained by the idealization. Then S has Hilbert function (1 5 6 5 1). A direct computation shows that S has the WLP but does not have the SLP. If l is a general element for S, then $S/(0 : l) \cong A(3)$, where $A(3)$ is the algebra defined in Example 5.19. Thus it is possible to compute the Jordan block decomposition for $\times l \in \mathrm{End}_K(S)$. It is

$$P(\times l) = 5 \oplus 3 \oplus 3 \oplus 3 \oplus 2 \oplus 2.$$

Example 5.21. With the same K as before, let B be the Artinian algebra $B = K[u, v]/(u^5, u^4v, u^3v^2, v^3)$. Then B is a level algebra with socle generators u^4, u^3v, u^2v^2. Let $T = B \times \mathrm{Hom}_K(B, K)$ be the ring obtained by the idealization. Then T has Hilbert function (1 5 6 6 5 1). As in the previous example, T has the WLP but does not have the SLP. If l is a general element for T, then $T/(0 : l) \cong S$, where S is the algebra defined in Example 5.20. Thus it is possible to compute the Jordan block decomposition for $\times l \in \mathrm{End}_K(T)$. It is given by

$$P(\times l) = 6 \oplus 4 \oplus 4 \oplus 4 \oplus 3 \oplus 3.$$

Note that $\begin{cases} \times l : T_2 \to T_3 \text{ has full rank,} \\ \times l^3 : T_1 \to T_4 \text{ has corank 1.} \end{cases}$

For more information see [41, 149, 151].

Example 5.22. Let $A = K[w, x, y, z]/\mathrm{Ann}\, F$, where $F = w^3xy + wx^3z + x^5$. Then A has Hilbert function (1 4 7 7 4 1). It turns out that A does not have the WLP and moreover if l is a general element of A, then $A/(0 : l)$ has the WLP with Hilbert function (1 4 6 4 1). Thus it is possible to compute the Jordan block decomposition for $\times l$.

$$P(\times l) = 6 \oplus 4 \oplus 4 \oplus 4 \oplus 2 \oplus 2 \oplus 1 \oplus 1.$$

Note that $\begin{cases} \times l : A_2 \to A_3 \text{ has corank 1,} \\ \times l^3 : A_1 \to A_4 \text{ has full rank.} \end{cases}$

Such an example was first constructed by Ikeda [67]. Related results can be found in Boij–Laksov [7].

Chapter 6
k-Lefschetz Properties

In this chapter we define the k-Lefschetz properties by generalizing the Lefschetz properties. The k-Lefschetz properties give us a way of computing generic initial ideals and graded Betti numbers of Artinian graded K-algebras.

6.1 k-SLP and k-WLP

6.1.1 Definitions

The following notion was first introduced by A. Iarrobino and M. Boij in a private conversation with J. Watanabe in 1995.

Definition 6.1 (k-SLP, k-WLP). Let $A = \bigoplus_{i=0}^{c} A_i$ be a graded Artinian K-algebra, and k a positive integer. We say that A has the k-**SLP** (resp. k-**WLP**) if there exist linear elements $g_1, g_2, \ldots, g_k \in A_1$ satisfying the following two conditions.

1. (A, g_1) has the SLP (resp. WLP),
2. $(A/(g_1, g_2, \ldots, g_{i-1}), g_i)$ has the SLP (resp. WLP) for all $i = 2, 3, \ldots, k$.

In this case, we say that $(A, g_1, g_2, \ldots, g_k)$ has the k-SLP (resp. k-WLP). In other words, A is said to have the k-**SLP** (resp. k-**WLP**), if A has the SLP (resp. WLP) with a Lefschetz element g_1, and $A/(g_1)$ has the $(k-1)$-SLP (resp. $(k-1)$-WLP).

Note that from the definition a graded algebra with the k-SLP (resp. k-WLP) has the $(k-1)$-SLP (resp. $(k-1)$-WLP).

Remark 6.2. When $(A, g_1, g_2, \ldots, g_k)$ has the k-SLP (resp. k-WLP), g_1 is a Lefschetz element of A from the definition. However, for some other Lefschetz element h_1 of A, there do not necessarily exist the remaining elements $h_2, h_3, \ldots, h_k \in A$ such that $(A, h_1, h_2, \ldots, h_k)$ has the k-SLP (resp. k-WLP). The following is an example. Let $A = K[x_1, x_2, x_3, x_4]/I$, where $I =$

T. Harima et al., *The Lefschetz Properties*, Lecture Notes in Mathematics 2080, DOI 10.1007/978-3-642-38206-2_6, © Springer-Verlag Berlin Heidelberg 2013

$(x_1^2, x_2x_3, x_2x_4) + (x_1, x_2, x_3, x_4)^3$. Then it is easy to see that $(A, x_1 - x_4, x_2 - x_3)$ has the 2-SLP. However, $h_1 = x_1 - x_2$ is a strong Lefschetz element of A, and $A/(h_1)$ does not have the SLP. In other words, there does not exist $h_2 \in A$ such that (A, h_1, h_2) has the 2-SLP.

In the sequel let $R = K[x_1, x_2, \ldots, x_n]$ be the polynomial ring in n variables over a field K of characteristic zero.

Remark 6.3. All graded K-algebras $K[x_1]/I$ and $K[x_1, x_2]/I$ have the SLP (see Proposition 3.15), and hence the n-SLP (resp. n-WLP) is equivalent to the $(n - 2)$-SLP (resp. $(n - 2)$-WLP) and to the $(n - 1)$-SLP (resp. $(n - 1)$-WLP) for quotient rings R/I.

6.1.2   Basic Properties

Definition 6.4. (1) (stable ideals) A monomial ideal I is said to be **stable** if I satisfies the following condition: for each monomial $u \in I$ the monomial $x_i u / x_{m(u)}$ belongs to I for every $i < m(u)$, where $m(u) = \max \{ j \mid x_j \text{ divides } u \}$.

(2) (strongly stable ideals, Borel-fixed ideals) A monomial ideal I is said to be **strongly stable** if I satisfies the following condition: for each monomial $u \in I$ and an index j satisfying $x_j | u$, the monomial $x_i u / x_j$ belongs to I for every $i < j$.

This condition is also called **Borel-fixed** when the characteristic of the field K is zero, since the ideal I is fixed under the natural action of the Borel subgroup of $GL(n; K)$. It is obvious that strongly stable ideals are stable.

(3) (almost revlex ideals, revlex ideals) A monomial ideal I is called an **almost revlex ideal**, if the following condition holds: for each monomial u in the minimal generating set of I, every monomial v with $\deg v = \deg u$ and $v >_{\text{revlex}} u$ belongs to I. Note that by Dickson's lemma (see [27] for example) there is a unique minimal generating set of the monomial ideal I consisting of monomials.

Almost revlex ideals are Borel-fixed, since each homogeneous component in which there is a monomial of the minimal generating set are Borel-fixed. In addition, if two almost revlex ideals have the same Hilbert function, then they are equal, since one can determine the minimal generators starting in low degrees using the given Hilbert function.

An ideal I of R is called a **revlex ideal** if the following condition holds: for each monomial u in I, every monomial v with $\deg v = \deg u$ and $v >_{\text{revlex}} u$ belongs to I. It is clear that revlex ideals are almost revlex.

Example 6.5. (1) Let $I = (x^2, xy, yz) + (x, y, z)^3$. It is an ideal of $K[x, y, z]$. It follows from the form of the generators that the ideal I is stable. However, I is not strongly stable, since $yz \in I$ but $xz \notin I$.

(2) Let $I = (x^2, xy) + (x, y, z)^4$ be an ideal of $K[x, y, z]$. Then I is not a revlex ideal, but an almost revlex ideal, since $x^2z \in I$ but $y^3 \notin I$.

Wiebe proved several basic results about the WLP and the SLP related to these notions.

Proposition 6.6 (Wiebe [153, Lemma 2.7]). *Let I be a stable (or Borel-fixed) ideal of R. Then R/I has the WLP (resp. SLP) if and only if $(R/I, x_n)$ has the WLP (resp. SLP).*

Proposition 6.7 (Wiebe [153, Proposition 2.9]). *Let I be a graded Artinian ideal of R, and $\mathrm{in}(I)$ the initial ideal of I with respect to the reverse lexicographic order. If $R/\mathrm{in}(I)$ has the WLP (resp. the SLP), then the same holds for R/I.*

Proposition 6.8 (Wiebe [153, Proposition 2.8]). *Let I be a graded Artinian ideal of R, and $\mathrm{gin}(I)$ denote the generic initial ideal of I with respect to the reverse lexicographic order. Then R/I has the WLP (resp. SLP) if and only if $R/\mathrm{gin}(I)$ has the WLP (resp. SLP).*

We next give generalizations of Wiebe's results. Propositions 6.9 and 6.10 are generalization of Propositions 6.6 and 6.7, respectively. A generalization of Proposition 6.8 is given by Proposition 6.15 which we take up after some preparations.

Proposition 6.9. *Let I be a homogeneous ideal of R, and $1 \le k \le n$.*

1. *When I is stable, R/I has the k-WLP (resp. k-SLP) if and only if $(R/I, x_n)$ has the WLP (resp. SLP), and $R/I + (x_n)$ has the $(k-1)$-WLP (resp. $(k-1)$-SLP).*
2. *When I is stable, R/I has the k-WLP (resp. k-SLP) if and only if $(R/I, x_n, x_{n-1}, \ldots, x_{n-k+1})$ has the k-WLP (resp. k-SLP).*

Proof. "If" part is trivial, and we prove the "only if" part next.

(1) We first remark that the group

$$H = \left\{ \begin{pmatrix} 1_{n-1} & a \\ 0 & b \end{pmatrix} \,\middle|\, a \in K^{n-1},\ b \in K^{\times} \right\} \subset GL(n; K)$$

stabilizes any stable ideal.

Let $g_1, g_2, \ldots, g_k \in R$ be elements such that $(R/I, g_1, g_2, \ldots, g_k)$ has the k-WLP (resp. k-SLP). Since such elements are generic, we may assume that the coefficient a_n of $g_1 = \sum_{j=1}^{n} a_j x_j$ is nonzero. Then there is an element $\varphi \in H$ such that $\varphi(g_1) = x_n$. By mapping $(R/I, g_1, g_2, \ldots, g_k)$ by φ, we obtain $(R/\varphi(I), \varphi(g_1), \varphi(g_2), \ldots, \varphi(g_k)) = (R/I, x_n, \varphi(g_2), \ldots, \varphi(g_k))$ has the k-WLP (resp. k-SLP).

(2) We can take x_n as the first element g_1 such that $(R/I, g_1, g_2, \ldots, g_k)$ has the k-WLP (resp. k-SLP). Set $\overline{R} = K[x_1, x_2, \ldots, x_{n-1}]$, and $\overline{I} = I \cap \overline{R}$. Then we have $\overline{R}/\overline{I} + (x_n) \simeq \overline{R}/\overline{I}$, and $\overline{I}$ is a stable ideal of $\overline{R}$. Therefore we can inductively take $g_1 = x_n, g_2 = x_{n-1}, \ldots, g_k = x_{n-k+1}$ using (1). $\square$

Proposition 6.10. *Let I be a graded Artinian ideal of R, let $\mathrm{in}(I)$ be the initial ideal of I with respect to the reverse lexicographic order, and let $1 \leq k \leq n$. If $R/\mathrm{in}(I)$ has the k-WLP (resp. the k-SLP), then the same holds for R/I.*

The following remark is needed in the proof of Proposition 6.10

Remark 6.11 ([52, Lemma 2.1, Lemma 2.2], see also [137]). The Lefschetz properties can be expressed in terms of Hilbert functions as follows. Let I be a homogeneous ideal of R, and h the Hilbert function of R/I.

1. $(R/I, L)$ has the WLP if and only if the Hilbert function of the quotient ring $R/I + (L)$ is equal to the **difference** Δh of h, where Δh is defined by

$$(\Delta h)_i = \max\{h_i - h_{i-1}, 0\} \qquad (i = 0, 1, 2, \ldots),$$

and h_{-1} is defined as zero.
2. $(R/I, L)$ has the SLP if and only if the Hilbert function of the quotient ring $R/I + (L^s)$ is equal to the sequence

$$(\max\{h_i - h_{i-s}\}, 0)_{i=0,1,2,\ldots},$$

for every positive integer s. Here $h_i = 0$ for $i < 0$.

Proof of Proposition 6.10. We can give a proof similar to Wiebe's proof of Proposition 6.7. First we consider the case of the k-WLP by induction on k. If $k = 0$ (0-WLP means the empty condition), there is nothing to prove. Let $k \geq 1$, and $R/\mathrm{in}(I)$ have the k-WLP. In particular, $R/\mathrm{in}(I)$ has the $(k-1)$-WLP, and it follows from the induction hypothesis that R/I has the $(k-1)$-WLP. Thus it suffices to show that $R/I + (g_1, g_2, \ldots, g_{k-1})$ has the WLP for generic linear forms $g_j \in R$. Let h be the $(k-1)$st difference of $h_{R/I}$, where higher differences are defined by iteration:

$$\Delta^0(h_{R/I}) = h_{R/I}, \qquad \Delta^k(h_{R/I}) = \Delta^{k-1}(\Delta(h_{R/I})) \quad (k \geq 1).$$

Then h is equal to both $h_{R/I+(g_1,g_2,\ldots,g_{k-1})}$ and $h_{R/\mathrm{in}(I)+(g_1,g_2,\ldots,g_{k-1})}$ from repeated use of Remark 6.11 (1), and moreover Δh is equal to $h_{R/\mathrm{in}(I)+(g_1,g_2,\ldots,g_k)}$ for generic linear forms $g_j \in R$. Conca proves in [23, Theorem 1.1] that

$$h(R/I + (g_1, g_2, \ldots, g_i), t) \leq h(R/\mathrm{in}(I) + (g_1, g_2, \ldots, g_i), t) \quad (i \geq 1, t \geq 0). \tag{6.1}$$

On the other hand,

$$h(R/I + (g_1, g_2, \ldots, g_k), t)$$

$$= \dim_K (R/I + (g_1, g_2, \ldots, g_{k-1}))_t - \dim_K (g_k \cdot R/I + (g_1, g_2, \ldots, g_{k-1}))_t$$

$$= \dim_K (R/I + (g_1, g_2, \ldots, g_{k-1}))_t - \dim_K g_k \cdot (R/I + (g_1, g_2, \ldots, g_{k-1}))_{t-1}$$

$$\geq h_t - h_{t-1}$$

$$= \Delta h_t.$$

Thus we have $h_{R/I+(g_1, g_2, \ldots, g_k)} = \Delta h$, and hence $R/I + (g_1, g_2, \ldots, g_{k-1})$ has the WLP. Therefore R/I has the k-WLP.

Next we consider the case of the k-SLP. From the proof of [153, Proposition 2.9], we have

$$h(R/I + (g_1, g_2, \ldots, g_{i-1}, g_i^s), t) \leq h(R/ \operatorname{in}(I) + (g_1, g_2, \ldots, g_{i-1}, g_i^s), t) \tag{6.2}$$

for generic linear forms $g_1, g_2, \ldots, g_i$, $s \geq 1$ and all $t \geq 0$. Then we can prove similarly to the case of k-WLP that the Hilbert function of $R/I + (g_1, g_2, \ldots, g_{i-1}, g_i^s)$ coincides with that of $R/ \operatorname{in}(I) + (g_1, g_2, \ldots, g_{i-1}, g_i^s)$ for every $i = 1, 2, \ldots, k$. Therefore it follows from Remark 6.11 (2) that R/I has the k-SLP. $\qquad\square$

Before proving a generalization of Wiebe's result (Proposition 6.8), we introduce the generic initial ideal and its properties.

Definition 6.12 (Generic initial ideal). Let $<$ be a term order on R such that $x_1 > x_2 > \cdots > x_n$.

1. For an ideal I of R, the **initial ideal** $\operatorname{in}_<(I)$ of I is the ideal of R generated by the initial terms $\operatorname{in}_<(f)$ of $f \in I$, where the initial term $\operatorname{in}_<(f)$ is the greatest term in f with respect to the term order $<$. We also denote the initial ideal simply by $\operatorname{in}(I)$, if there is no ambiguity.
2. We have a natural action on R of the general linear group $GL(n, K)$. Let I be a homogeneous ideal of R. Then there exist a Zariski open subset U in $GL(n, K)$ and a monomial ideal J of R such that for all $g \in U$ we have $\operatorname{in}_<(gI) = J$. We call J the **generic initial ideal** of I with respect to the term order $<$, written $J = \operatorname{gin}_<(I)$. We also write the generic initial ideal simply by $\operatorname{gin}(I)$, if there is no ambiguity.

The following is an important property of generic initial ideals. See Eisenbud [31, 15.9] for more details on generic initial ideals.

Proposition 6.13 (Galligo, Bayer–Stillman). *Generic initial ideals are Borel-fixed.*

Remark 6.14. (1) The Hilbert function of $\operatorname{gin}_<(I)$ is equal to that of I.

(2) If $I \subset R$ is a Borel-fixed ideal, then $\mathrm{gin}_<(I) = I$ for any term order $<$. Indeed, we can take the element $g \in GL(n, K)$ that achieves $\mathrm{in}_<(gI) = \mathrm{gin}_<(I)$ as an element in the Borel subgroup of $GL(n, K)$ (see [31, Theorem 15.18], e.g.).

The following proposition is a generalization of Proposition 6.8.

Proposition 6.15. *Let I be a graded Artinian ideal of R, and let $1 \le k \le n$. The following two conditions are equivalent:*

1. *R/I has the k-WLP (resp. the k-SLP),*
2. *$(R/\mathrm{gin}(I), x_n, x_{n-1}, \ldots, x_{n-k+1})$ has the k-WLP (resp. the k-SLP).*

Proof. We can give a proof using a similar idea to Wiebe's proof of Proposition 6.8. First we show that (1) and (2) are equivalent for the k-WLP. Let $1 \le i \le k$. Lemma 1.2 in [23] says that the Hilbert function of $R/I + (g_1, g_2, \ldots, g_i)$ for generic linear forms $g_1, g_2, \ldots, g_i$ is equal to the Hilbert function of $R/\mathrm{gin}(I) + (x_n, x_{n-1}, \ldots x_{n-i+1})$. In view of Proposition 6.9, it follows from Remark 6.11 that this yields the equivalence of (1) and (2).

Next we show that (1) and (2) are equivalent for the k-SLP. Let $1 \le i \le k$. By slightly generalizing the proofs of Lemma 1.2 in [23] and Proposition 2.8 in [153], it follows that the Hilbert function of $R/I + (g_1, g_2, \ldots, g_{i-1}, g_i^s)$ is equal to the Hilbert function of $R/\mathrm{gin}(I) + (x_n, x_{n-1}, \ldots, x_{n-i+2}, x_{n-i+1}^s)$ for generic linear forms $g_1, g_2, \ldots, g_i$ and $s \ge 1$. In view of Proposition 6.9, it follows that this means the equivalence of (1) and (2) for the k-SLP. $\square$

6.1.3 Almost Revlex Ideals

In the rest of this section we show that almost revlex ideals I give important examples of graded algebras R/I having the n-SLP (Proposition 6.19). The notion of x_n-chains in the following definition plays a key role in the arguments of this section and the next section.

Definition 6.16 (x_n-chain). For a monomial ideal I of R, a monomial which does not belong to I is called a **standard monomial** with respect to I. The standard monomials span the quotient ring R/I as a K-vector space (see [31, Theorem 15.3], e.g.)

Let I be an Artinian monomial ideal of R. For a standard monomial u not divisible by x_n, an x_n-**chain** beginning with u with respect to I is defined as the sequence of the standard monomials (cf. Notation 3.59)

$$u, ux_n, \ldots, ux_n^{s-1} \qquad (ux_n^{s-1} \notin I, \ ux_n^s \in I).$$

The set of the standard monomials decomposes into disjoint x_n-chains, and we call this decomposition the x_n-**chain decomposition** of the standard monomials with respect to I.

Lemma 6.17. *Let I be an Artinian stable ideal of R, and $u, ux_n, \ldots, ux_n^{s-1}$ an x_n-chain with respect to I, where u is a standard monomial not divisible by x_n, and $s \geq 1$. Then ux_n^s is a member of the minimal generators of I.*

Proof. Since $ux_n^s \in I$, there exists a minimal generator vx_n^t with $v|u$ and $t \leq s$. If $t < s$, then $vx_n^t | ux_n^{s-1}$, and this contradicts the condition that $ux_n^{s-1} \notin I$. Therefore we have $t = s$. Assume that $\deg v < \deg u$. Then there exists $i < n$ such that $x_i v | u$, and hence $x_i v x_n^{s-1} | ux_n^{s-1}$. By the definition of stable ideals, $vx_n^s \in I$ yields that $x_i v x_n^{s-1} \in I$, and we have $ux_n^{s-1} \in I$. This is a contradiction, and we have $v = u$. This means that ux_n^s is a minimal generator. □

Lemma 6.18 (Lefschetz conditions). *(cf. Proposition 3.14) Let I be an Artinian stable ideal of R.*

1. *The following two conditions are equivalent:*

   a. *R/I has the SLP.*
   b. *For any two x_n-chains $u, ux_n, \ldots, ux_n^{s-1}$ and $v, vx_n, \ldots, vx_n^{t-1}$ with respect to I, if $\deg u < \deg v$, then $\deg ux_n^{s-1} \geq \deg vx_n^{t-1}$.*

2. *Let $(h_0, h_1, \ldots, h_c)$ denote the Hilbert function of R/I. Let i be the minimum integer satisfying $h_i = \max_k\{h_k\}$, and j the maximum integer satisfying $h_j = \max_k\{h_k\}$. The following two conditions are equivalent:*

   a. *R/I has the WLP.*
   b. *Every x_n-chain starts at a degree less than or equal to i, and ends in a degree greater than or equal to j.*

*We call the conditions in (1) (1b) (resp. (2) (2b)) the **strong Lefschetz condition** (resp. the **weak Lefschetz condition**) for the x_n-chain decomposition. We call these conditions the **SL condition** (resp. the **WL condition**) for short.*

Proof. The SL condition is the condition for x_n to be a strong Lefschetz element of R/I. In addition, Proposition 6.6 says that the quotient ring R/I by a stable ideal I have the SLP if and only if x_n is a Lefschetz element. Therefore the two conditions are equivalent.

The WL condition is the condition for x_n to be a weak Lefschetz element of R/I. For the same reason for the WL condition (2b) is equivalent to (2a). □

Proposition 6.19. *For every Artinian almost revlex ideal I of $R = K[x_1, x_2, \ldots, x_n]$ where $I \subset (x_1, x_2, \ldots, x_n)^2$, the quotient ring R/I has the n-SLP.*

Proof. Let $\overline{R}$ be the subring $K[x_1, x_2, \ldots, x_{n-1}]$ of R. Since $I \cap \overline{R}$ is again an almost revlex ideal of $\overline{R}$, we have only to show that R/I has the SLP. We show that the x_n-chain decomposition with respect to I satisfies the SL condition.

Let us take two x_n-chains $u, ux_n, \ldots, ux_n^{s-1}$ and $v, vx_n, \ldots, vx_n^{t-1}$ with $\deg u < \deg v$. By Lemma 6.17, ux_n^s is a member of the minimal generators of I. Assume that $\deg v \geq \deg ux_n^s$. Since I is almost revlex, every monomial in $\overline{R}$ of the same degree as ux_n^s belongs to I, and hence v also belongs to I. This is a contradiction, and we have $\deg v < \deg u x_n^s$. Set $p = \deg u + s - \deg v$, which is positive and less than s. Then $v x_n^p$ is a monomial of the same degree as $u x_n^s$, and it is greater than $u x_n^s$ with respect to the reverse lexicographic order. Hence $v x_n^p$ is in I. Since $v x_n^{t-1} \notin I$, we have $t - 1 < p \ (< s)$, and $v x_n^{t-1}$ should be of degree less than or equal to $u x_n^{s-1}$. Thus the SL condition is satisfied.   □

6.2   Classification of Hilbert Functions

6.2.1   Hilbert Functions of k-SLP and k-WLP

Definition 6.20 (O-sequence). Let $h = (h_0, h_1, \ldots, h_j, \ldots)$ be a finite or infinite sequence of non-negative integers. If h can be the Hilbert function of some standard graded K-algebra, h is called an **O-sequence**.

Definition 6.21 (difference of sequences). For a sequence $h = (h_0, h_1, \ldots, h_c)$ of non-negative integers, we define the **difference** Δh of h by

$$(\Delta h)_j = \max\{h_j - h_{j-1}, 0\}.$$

In addition, the k-**th difference** $\Delta^k h$ is obtained by definition by applying Δ k-times to h. We use the convention that $\Delta^0 h = h$ for $k = 0$.

The following proposition is a generalization of Proposition 3.10.

Proposition 6.22. *Let k be an integer with $1 \le k \le n$, and $h = (1, n, h_2, h_3, \ldots, h_c)$ an O-sequence. The following three conditions are equivalent:*

1. *h is a Hilbert function of some graded algebra with the k-SLP,*
2. *h is a Hilbert function of some graded algebra with the k-WLP,*
3. *$\Delta^t h$ is a unimodal O-sequence for $0 \le t \le k - 1$, and $\Delta^k h$ is an O-sequence.*

See Definition 1.30 for unimodality.

Proof. It is clear that (1) implies (2). If A is a graded algebra having the WLP with a Lefschetz element g, then it follows from the proof of Proposition 3.10 (or from Remark 6.11 (1)) that the Hilbert function of the quotient $A/(g)$ is the difference of that of A. Repeated use of this fact shows that (2) implies (3). As to the implication from (3) to (1), the surjectivity in Lemma 6.24 (2) below shows the existence of an algebra with the k-SLP and with the Hilbert function h satisfying the condition in (3).   □

In addition, we have a characterization of the Hilbert functions of quotient rings R/I for Artinian almost revlex ideals I. The characterization is the same as ideals with the n-WLP.

Proposition 6.23. *Let $h = (1, n, h_2, h_3, \ldots, h_c)$ be an O-sequence. The following four conditions are equivalent:*

1. *h is a Hilbert function of R/I for some almost revlex ideal I of R,*
2. *h is a Hilbert function of some graded algebra with the n-SLP,*
3. *h is a Hilbert function of some graded algebra with the n-WLP,*
4. *$\Delta^k h$ is a unimodal O-sequence for every integer k with $0 \le k \le n$.*

Proof. It only remains to show that (2) implies (1) thanks to Proposition 6.22 above. This is proved by the following lemma.   □

Lemma 6.24. *We define the following collections of ideals in R for $0 \le k \le n$ and a given finite O-sequence h:*

$$\mathcal{B}_n^k(h) = \left\{ I \subset R \;\middle|\; \begin{array}{l} I \text{ is a Borel-fixed ideal,} \\ R/I \text{ has the Hilbert function } h, \\ \text{and } R/I \text{ has the } k\text{-SLP} \end{array} \right\},$$

$$\mathcal{A}_n(h) = \left\{ I \subset R \;\middle|\; \begin{array}{l} I \text{ is almost revlex,} \\ \text{and } R/I \text{ has the Hilbert function } h \end{array} \right\},$$

where 0-SLP means the empty condition. For $m < n$ we set $R' = K[x_1, x_2, \ldots, x_m]$, and define the map Res_m^n by

$$\mathrm{Res}_m^n : \{\text{the ideals of } R\} \to \{\text{the ideals of } R'\} \quad \text{by } (\mathrm{Res}_m^n(I) = I \cap R').$$

Then we have

1. *$\mathcal{B}_n^0(h) \supset \mathcal{B}_n^1(h) \supset \cdots \supset \mathcal{B}_n^{k-1}(h) \supset \mathcal{B}_n^k(h) \supset \cdots \supset \mathcal{B}_n^n(h) \supset \mathcal{A}_n(h)$.*
2. *$\mathrm{Res}_{n-s}^n(I) \in \mathcal{B}_{n-s}^{k-s}(\Delta^s h)$ for $I \in \mathcal{B}_n^k(h)$ and $1 \le s \le k$. Moreover, the restriction map $\mathrm{Res}_{n-s}^n \mid_{\mathcal{B}_n^k(h)}: \mathcal{B}_n^k(h) \to \mathcal{B}_{n-s}^{k-s}(\Delta^s h)$ is surjective.*
3. *$\mathcal{A}_n(h)$ is non-empty if and only if $\mathcal{B}_n^n(h)$ is non-empty. In this case $\mathcal{A}_n(h)$ consists of only one ideal I, and $\mathrm{Res}_{n-s}^n(I) \in \mathcal{A}_{n-s}(\Delta^s h)$ for $1 \le s \le k$.*

Proof. (1) The last inclusion follows from Proposition 6.19, and the other inclusions are clear.

(2) Since $\mathrm{Res}_{n-s}^n = \mathrm{Res}_{n-1}^n \circ \mathrm{Res}_{n-2}^{n-1} \circ \cdots \circ \mathrm{Res}_{n-s}^{n-s+1}$, it suffices to show the assertion when $s = 1$. Let $\overline{R} = K[x_1, x_2, \ldots, x_{n-1}]$ and $I \in \mathcal{B}_n^k(h)$, and we prove that $\mathrm{Res}_{n-1}^n(I) \in \mathcal{B}_{n-1}^{k-1}(\Delta h)$. It is clear that $\mathrm{Res}_{n-1}^n(I) = I \cap \overline{R}$ is Borel-fixed. It follows from Proposition 6.15 that $(R/I, x_n, x_{n-1}, \ldots, x_{n-k+1})$ has the k-SLP, and hence $\overline{R}/\mathrm{Res}_{n-1}^n(I) \simeq R/I + (x_n)$ has the $(k-1)$-SLP with the Lefschetz elements $x_{n-1}, x_{n-2}, \ldots, x_{n-k+1}$. The Hilbert function of $\overline{R}/\mathrm{Res}_{n-1}^n(I)$ is equal to Δh by the proof of Proposition 3.10 (or by Remark 6.11 (1)). Thus we have $\mathrm{Res}_{n-1}^n(I) \in \mathcal{B}_{n-1}^{k-1}(\Delta h)$.

For the proof of the surjectivity we define the **basic invariants** of an O-sequence h which is the Hilbert function of a graded Artinian algebra having the SLP. Let $I \in \mathcal{B}_n^1(h)$, and $v_i, v_i x_n, \ldots, v_i x_n^{t_i - 1}$ $(i = 1, 2, \ldots, l)$ be the x_n-chain decomposition with respect to I. We define the basic invariants of h as the multi-set

$$\{ (\deg v_i, t_i) \mid i = 1, 2, \ldots, l \}. \tag{6.3}$$

The basic invariants are independent of the choice of $I \in \mathscr{B}_n^1(h)$, and uniquely determined by h. It follows from the SL condition that $\deg v_i < \deg v_j$ implies $\deg v_i + t_i \geq \deg v_j + t_j$.

We next prove the surjectivity for $s = 1$. Take an ideal $\overline{I} \in \mathscr{B}_{n-1}^{k-1}(\Delta h)$, and let $v_i \in \overline{R}$ $(i = 1, 2, \ldots, l)$ be the standard monomials with respect to $\overline{I}$. We order v_i's such that $\deg v_i \leq \deg v_{i+1}$, and $v_i <_{\text{revlex}} v_{i+1}$ if $\deg v_i = \deg v_{i+1}$. Let the multi-set $\{ (d_i, t_i) \mid i = 1, 2, \ldots, l \}$ be the basic invariants of h. We order (d_i, t_i)'s such that $d_i \leq d_{i+1}$, and $t_i \geq t_{i+1}$ if $d_i = d_{i+1}$. Note that $d_i = \deg v_i$. Define the subset M of monomials of R by

$$M = \{ v_i x_n^t \mid i = 1, 2, \ldots, l, \ t = 0, 1, \ldots, t_i - 1 \},$$

and the linear subspace I of R by the linear span of the monomials of R not contained in M. If the subspace I turns out to be a Borel-fixed ideal, then the surjectivity is proved since $\mathrm{Res}_{n-1}^n(I) = \overline{I}$.

We prove that I is an ideal. It suffices to show that, for each $v_i x_n^t \in M$ $(i = 1, 2, \ldots, l, \ t = 0, 1, \ldots, t_i - 1)$, any factor $v x_n^s$ of $v_i x_n^t$ $(v | v_i$ and $s \leq t)$ is again contained in M. When $v = v_i$, it is clear that $v_i x_n^s \in M$. Suppose that $\deg v < \deg v_i$. Then v is equal to some v_j, since $\overline{I}$ is a monomial ideal of $\overline{R}$. We have $s \leq t < t_i < t_j$ from the inequality $\deg v_j + t_j \geq \deg v_i + t_i$, and hence $v x_n^s = v_j x_n^s \in M$. Thus I is an ideal of R.

We prove that I is Borel-fixed. It suffices to show that, for each $v_i x_n^t \in M$ $(i = 1, 2, \ldots, l, \ t = 0, 1, \ldots, t_i - 1)$, if x_p divides $v_i x_n^t$ $(1 \leq p \leq n - 1)$, then $v_i x_{p+1} x_n^t / x_p \in M$. If $1 \leq p \leq n - 2$, then $v_i x_{p+1}/x_p \in \overline{R}$ is a standard monomial with respect to $\overline{I}$, since $\overline{I}$ is a Borel-fixed ideal. Therefore $v_i x_{p+1}/x_p$ is equal to some v_j, and $j < i$ since $v_i x_{p+1}/x_p <_{\text{revlex}} v_i$. Hence $t_i \leq t_j$, and we have $v_i x_n^t x_{p+1}/x_p = v_j x_n^t \in M$. If $p = n - 1$, then $v_i/x_{n-1} \in \overline{R}$ is a standard monomial with respect to $\overline{I}$, and is equal to some v_j. Since $\deg v_j < \deg v_i$, we have $t_j > t_i$, and hence $t + 1 < t_j$. Thus $(v_i/x_{n-1})x_n^{t+1} = v_j x_n^{t+1} \in M$.

(3) By Proposition 6.19 it remains to show that $\mathscr{A}_n(h)$ is nonempty if $\mathscr{B}_n^n(h)$ is nonempty. When $n = 1$, this claim clearly holds. We prove that the construction of $I \in \mathscr{B}_n^k(h)$ from $\overline{I} \in \mathscr{B}_{n-1}^{k-1}(\Delta h)$ in the proof of (2) gives an almost revlex ideal I if $\overline{I}$ is almost revlex. Once this is proved, then the proof of (3) is completed by induction on n.

Take $\overline{I} \in A_{n-1}(\Delta h)$, and let v_i and $\{(d_i, t_i)\}$ be as in the proof of (2). A minimal generator $u \in R$ of I not divisible by x_n is a minimal generator of the almost revlex ideal $\overline{I}$ of $\overline{R}$, and therefore every monomial $u' \in R$ of the same degree as u satisfying $u' >_{\text{revlex}} u$ belongs to $\overline{I}$. Hence $u' \in I$. For a minimal generator $v_i x_n^{t_i} \in R$ of I divisible by x_n, consider a monomial $v_j x_n^t$ $(t \geq 0)$ satisfying $\deg v_j x_n^t = \deg v_i x_n^{t_i}$ and $v_j x_n^t >_{\text{revlex}} v_i x_n^{t_i}$. If $t = t_i$, then $\deg v_i = \deg v_j$ and $v_j >_{\text{revlex}} v_i$. By the ordering of the basic invariants, we have $t_j \leq t_i$. Hence $t \geq t_j$, and $v_j x_n^t \in I$. If $t < t_i$, then $\deg v_j > \deg v_i$, and we have $\deg v_j x_n^{t_j} \leq \deg v_i x_n^{t_i}$. Hence $t_j \leq t$, and $v_j x_n^t \in I$. We have thus proved that I is almost revlex. $\square$

6.2.2 Hilbert Functions of Artinian Complete Intersections

Hilbert functions of quotients of Artinian complete intersections are an important sub-class of Hilbert functions. Proposition 6.26 below shows that they are contained in the class of Hilbert functions for n-SLP.

Proposition 6.25. *(cf. Proposition 3.44 (2)) Let I be a homogeneous ideal of $R :=$ $K[x_1, x_2, \ldots, x_n]$ whose quotient has the k-SLP. Then the tensor product $R/I \otimes_K K[y]/(y^m)$ has the k-WLP for any m.*

Moreover, if R/I has the n-SLP, then $R/I \otimes_K K[y]/(y^m)$ has the $(n+1)$-WLP.

Proof. We prove the second statement first. By the first statement $R/I \otimes_K K[y]/(y^m)$ has the n-WLP. Then the second statement follows from Remark 6.3.

Next we prove the first statement in steps.

[Step 0] We prove a claim needed in Step 1: Let $C = \bigoplus_{i \geq 0} C_i$ be an Artinian graded algebra, and J its homogeneous ideal. If the quotient C/J satisfies the following conditions, then C/J has the k-WLP.

(i) C has the k-WLP.
(ii) $J \subset \bigoplus_{i > d} C_i$, where d is the maximal socle degree of $C/(L_C)$, and L_C is a Lefschetz element of C. In other words, the peak of the Hilbert function of C starts in degree d.

(proof of Step 0) The injectivity of $\times x_n : (C/J)_i \to (C/J)_{i+1}$ for $i < d$ is obvious. For $t \geq 1$ the multiplication map $\times x_n^t : (C/J)_d \to (C/J)_{d+t}$ is equal to the composite of the surjection $\times x_n^t : (C/J)_d = C_d \to C_{d+t}$ and the projection $C_{d+t} \to (C/J)_{d+t}$. Hence $\times x_n : (C/J)_{d+t-1} \to (C/J)_{d+t}$ is surjective, and C/J has the WLP. Since $(C/J)/(L_{C/J}) \simeq C/(L_C)$ has the $(k-1)$-WLP, C/J has the k-WLP.

[Step 1] In this step we restrict I to be a Borel-fixed ideal. We proceed by induction on k. When $k = 1$, the proposition is a consequence of Proposition 3.43 (1). Let $k \geq 2$, and assume that the assertion holds up to $k - 1$. Let $A = R/I$ have the k-SLP, and set $B = K[y]/(y^m)$. By Proposition 6.6 x_n is a Lefschetz element of A, and it follows from Proposition 3.43 (2) that $A \otimes_K B$ has the WLP with a Lefschetz element $x_n + y$. We thus have only to show that $A \otimes_K B/(x_n + y)$ has the $(k-1)$-WLP.

We have

$$A \otimes_K B/(x_n + y) \simeq K[x_1, x_2, \ldots, x_n, y]/(I) + (y^m) + (x_n + y)$$
$$\simeq R/I + (x_n^m),$$

where (I) denotes the ideal of $K[x_1, x_2, \ldots, x_n, y]$ generated by the subspace I. If we let G_1 to be the set of the minimal generators of I not divisible by x_n, and G_2 to be the set of the minimal generators of I divisible by x_n, then we have

$$
\begin{aligned}
R/I + (x_n^m) &\simeq R/(G_1) + (G_2) + (x_n^m) \\
&\simeq [R/(G_1) + (x_n^m)]/(G_2) \\
&\simeq [\overline{R}/(G_1) \otimes_K K[x_n]/(x_n^m)]/ \left(v_i x_n^{t_i} \mid i = 1, 2, \ldots, l \right) \\
&\simeq [A/(x_n) \otimes_K K[z]/(z^m)]/ \left(v_i z^{t_i} \mid i = 1, 2, \ldots, l \right),
\end{aligned}
$$

where $v_1, v_2, \ldots, v_l \in \overline{R} = K[x_1, x_2, \ldots, x_{n-1}]$ are the standard monomials with respect to $I \cap \overline{R}$, and $\{ (\deg v_i, t_i) \mid i = 1, 2, \ldots, l \}$ are the basic invariants of the Hilbert function of R/I defined in (6.3).

Then $[A/(x_n) \otimes_K K[z]/(z^m)]/ \left(v_i z^{t_i} \mid i = 1, 2, \ldots, l \right)$ satisfies the condition of Step 0. Indeed, $A/(x_n)$ has the $(k-1)$-SLP, and $A/(x_n) \otimes_K K[z]/(z^m)$ has the $(k-1)$-WLP by the assumption of induction. Let d be the maximal socle degree of $A/(x_n)$, so the peak of the Hilbert function of $A/(x_n) \otimes_K K[z]/(z^m)$ starts in degree at most d. The degrees of $v_i z^{t_i}$ are at least $d + 1$, since A has the WLP. Therefore $A \otimes_K B/(x_n + y) \simeq [A/(x_n) \otimes_K K[z]/(z^m)]/ \left(v_i z^{t_i} \mid i = 1, 2, \ldots, l \right)$ has the $(k-1)$-WLP by Step 0.

[Step 2] Let I be a homogeneous ideal such that $A = R/I$ has the k-SLP. Let $g \in GL(n; K)$ be an element for which $\mathrm{in}(gI) = \mathrm{gin}(I)$, and let $\tilde{g} \in GL(n+1, K)$ be the block diagonal matrix whose upper left block is g and $(n+1, n+1)$-entry is equal to one. Then we have

$$
\begin{aligned}
R/\mathrm{gin}(I) \otimes_K B &= R/\mathrm{in}(gI) \otimes_K B \\
&\simeq R[y]/(\mathrm{in}(gI) + (y^m)) \\
&= R[y]/\mathrm{in}(gI + (y^m)) \\
&\simeq K[x_1, x_2, \ldots, x_n, y]/\mathrm{in}(\tilde{g}(I + (y^m))),
\end{aligned}
$$

by use of [31, Proposition 15.15] for example. $R/\mathrm{gin}(I)$ has the k-SLP by Proposition 6.15. It follows from Step 1 that $R/\mathrm{gin}(I) \otimes_K B$ has the k-WLP. Hence $K[x_1, x_2, \ldots, x_n, y]/\tilde{g}(I + (y^m))$ has the k-WLP by Proposition 6.10. Thus $A \otimes_K B \simeq K[x_1, x_2, \ldots, x_n, y]/\tilde{g}(I + (y^m))$ has the k-WLP. $\square$

Proposition 6.26. *Hilbert functions of Artinian complete intersections (cf. Proposition 2.92) of embedding dimension n are in the class of Hilbert functions for the n-SLP.*

Proof. We prove the proposition by induction on n. When $n = 1$, the proposition holds trivially. Let $n \geq 1$, and assume that the proposition holds up to n. By the induction hypothesis, we can take a graded algebra A having the n-SLP with the same Hilbert function as $R/(x_1^{m_1}, x_2^{m_2}, \ldots, x_n^{m_n})$. By Proposition 6.25, the tensor product $A \otimes_K K[y]/(y^{m_n+1})$ has the $(n+1)$-WLP with the same Hilbert function as $K[x_1, x_2, \ldots, x_{n+1}]/(x_1^{m_1}, x_2^{m_2}, \ldots, x_{n+1}^{m_{n+1}})$. Since the Hilbert function of an algebras with the $(n+1)$-WLP is in the class of Hilbert functions for $(n+1)$-SLP by Proposition 6.22, we have proved the proposition. □

6.3 Generic Initial Ideals

Definition 6.27 (quasi-symmetric). A unimodal sequence $h = (h_0, h_1, \ldots, h_c)$ of positive integers is said to be **quasi-symmetric**, if the following condition holds: Let h_i be the maximum of $\{h_0, h_1, \ldots, h_c\}$. Then every integer h_j for $j > i$ is equal to one of $\{h_0, h_1, \ldots, h_i\}$.

In particular, unimodal symmetric sequences are quasi-symmetric.

Lemma 6.28. *We use the notations of Lemma 6.24.*

1. $\mathrm{Res}_2^3 \colon \mathcal{B}_3^k(h) \to \mathcal{B}_2^{k-1}(\Delta h)$ *is bijective for any O-sequence h and $1 \leq k \leq 3$.*
2. *If h is a unimodal quasi-symmetric sequence, then the mapping $\mathrm{Res}_{n-1}^n \colon \mathcal{B}_n^k(h) \to \mathcal{B}_{n-1}^{k-1}(\Delta h)$ is bijective.*

Proof. (1) The surjectivity is already shown by Lemma 6.24 (2). Take $I \in \mathcal{B}_3^k(h)$, and set $\overline{R} = K[x_1, x_2]$ and $\overline{I} = I \cap \overline{R}$. Let $v_1, v_2, \ldots, v_l \in \overline{R}$ be the standard monomials with respect to $\overline{I}$. In general, even if the standard monomials with respect to $\overline{I}$ and the basic invariants (defined in (6.3)) with respect to I are given, the minimal generators $\{v_i x_n^{t_i}\}$ of I are not uniquely determined. However, when $n = 3$, the minimal generators of I are uniquely determined as shown in what follows, and hence, the map $\mathrm{Res}_2^3 \colon \mathcal{B}_3^k(h) \to \mathcal{B}_2^{k-1}(\Delta h)$ turns out to be injective.

Let $n = 3$. If $\deg v_i = \deg v_j$, and $v_i >_{\mathrm{revlex}} v_j$, then there exists $p > 0$ such that $v_i = v_j(x_1/x_2)^p$. Therefore $v_j x_3^t \in I$ yields that $v_i x_3^t \in I$ since I is Borel-fixed. It follows from this property that $t \leq t'$ for the minimal generators $v_i x_3^t$ and $v_j x_3^{t'}$ of the Borel-fixed ideal I. Thus the minimal generators of I are uniquely determined from the standard monomials with respect to $\overline{I}$ and the basic invariants with respect to I.

(2) The surjectivity follows from Lemma 6.24 (2). Let $\{ (d_i, t_i) \mid i = 1, 2, \ldots, l \}$ be the basic invariants of h. If h is quasi-symmetric, then $t_i = t_j$ for any i and j satisfying $d_i = d_j$. Therefore the x_n-chain decomposition with respect to $I \in \mathcal{B}_n^k(h)$ can be uniquely recovered from the standard monomials with respect to $\mathrm{Res}_{n-1}^n(I)$. Thus the injectivity is proved. □

Theorem 6.29 ([19, 49]). *Let I be a graded Artinian ideal whose quotient ring R/I has the n-SLP, and h the Hilbert function of R/I. Suppose that the k-th difference $\Delta^k h$ is quasi-symmetric for every integer k with $0 \leq k \leq n-4$. Then the generic initial ideal $\mathrm{gin}(I)$ with respect to the reverse lexicographic order is almost revlex. In particular, $\mathrm{gin}(I)$ is determined by the Hilbert function alone.*

Proof. Suppose first that I is Borel-fixed. When $n \leq 2$, Borel-fixed ideals are uniquely determined by their Hilbert functions, and therefore the ideal I is almost revlex. When $n \geq 3$, the map $\mathrm{Res}_2^n \mid_{\mathscr{B}_n^n(h)} \colon \mathscr{B}_n^n(h) \to \mathscr{B}_2^2(\Delta^{n-2}h)$ is bijective by Lemma 6.28. Since $\mathscr{B}_2^2(\Delta^{n-2}h)$ consists of a single ideal, $\mathscr{B}_n^n(h) = \{I\}$. The subset $\mathscr{A}_n(h) \subset \mathscr{B}_n^n(h)$ is nonempty by Lemma 6.24 (3), and we have $\mathscr{A}_n(h) = \mathscr{B}_n^n(h) = \{I\}$. Hence I is almost revlex.

For general I, if R/I has the n-SLP, then $R/\mathrm{gin}(I)$ also has the n-SLP [49, Proposition 18], and $\mathrm{gin}(I)$ is Borel-fixed by Proposition 6.13. Hence $\mathrm{gin}(I)$ is almost revlex. $\square$

Corollary 6.30 ([1,20,49]). *Let I be a graded Artinian ideal of $K[x_1, x_2, x_3]$ (resp. $K[x_1, x_2, x_3, x_4]$) whose quotient ring has the SLP (resp. 2-SLP). Then the generic initial ideal $\mathrm{gin}(I)$ is almost revlex, and is uniquely determined by the Hilbert function.*

Proof. If $n \leq 4$, then the condition in Theorem 6.29 for the Hilbert function becomes the empty condition. Thus we have the corollary. $\square$

Example 6.31. ([56]) Let $R = K[x_1, x_2, x_3, x_4]$, and $I = (x_1^a, x_2^b, x_3^c, x_4^d)$, where at least one of a, b, c and d is equal to two. Then the generic initial ideal of I is equal to the almost revlex ideal corresponding to the same Hilbert function. Indeed, we can check that R/I has the 2-SLP, and can apply Corollary 6.30.

Example 6.32. We give four examples of complete intersection in $R = K[x_1, x_2, x_3]$ whose quotient rings have the SLP. The generic initial ideals of these ideals are the unique almost revlex ideals with corresponding Hilbert functions by Corollary 6.30.

1. Let $I = (f, g, L^r) \subset R$, where f and g are any linear forms of R, and L is any homogeneous polynomial of degree one. In this case, if I is a complete intersection, then R/I has the SLP [52, Example 6.2].
2. Let e_1, e_2 and e_3 be the elementary symmetric functions in three variables, where $\deg(e_i) = i$. Let r and s be positive integers, where r divides s. Then the quotient ring by the ideal $I = (e_1(x_1^r, x_2^r, x_3^r), e_2(x_1^r, x_2^r, x_3^r), e_3(x_1^s, x_2^s, x_3^s))$ of R has the SLP [52, Example 6.4].
3. Let p_i be the power sum symmetric function of degree i in three variables, and a be a positive integer. Then, the quotient ring of R by the ideal $I = (p_a, p_{a+1}, p_{a+2})$ of R has the SLP [53, Proposition 7.1].
4. Let $I = (e_2, e_3, f) \subset R$, where f is any homogeneous polynomial of R. In this case, if I is a complete intersection, then R/I has the SLP [53, Proposition 3.1].

To give examples of complete intersections in $K[x_1, x_2, \ldots, x_n]$ for which we can determine their generic initial ideals, we prove a preliminary proposition.

Proposition 6.33. *Let $R = K[x_1, x_2, \ldots, x_n]$, and $I = (f_1, f_2, \ldots, f_n)$ a complete intersection where f_j is a homogeneous polynomial in R. Let k be an integer satisfying $1 \leq k \leq n - 2$. Suppose that $A = R/I$ has the SLP, and that the degrees $d_j = \deg f_j$ satisfy the condition*

$$d_j \geq d_1 + d_2 + \cdots + d_{j-1} - (j-1) + 1 \qquad (6.4)$$

for all $j = n - k + 1, n - k + 2, \ldots, n$. *Then we have the following.*

1. $A = R/I$ has the k-SLP. In particular, when $k \geq n - 2$, A has the n-SLP.
2. When $k \geq n-2$, $\mathrm{gin}(I)$ coincides with the unique almost revlex ideal determined by the Hilbert function of A.
3. When $k \geq n - 2$, $\mathrm{gin}(x_1^{d_1}, x_2^{d_2}, \ldots, x_n^{d_n}) = \mathrm{gin}(I)$.

Proof. Take a Lefschetz element L of A such that $\{f_1, f_2, \ldots, f_{n-1}, L\}$ is a regular sequence. Set $S = R/LR$, and

$$B = A/\overline{L}A = S/(\overline{f_1}, \overline{f_2}, \ldots, \overline{f_{n-1}}, \overline{f_n}).$$

Noting that $d_n \geq d_1 + d_2 + \cdots + d_{n-1} - (n-1) + 1$ and $\{\overline{f_1}, \overline{f_2}, \ldots, \overline{f_{n-1}}\}$ is a regular sequence, it is easy to verify that $(\overline{f_1}, \overline{f_2}, \ldots, \overline{f_{n-1}}, \overline{f_n}) = (\overline{f_1}, \overline{f_2}, \ldots, \overline{f_{n-1}})$ and $B = S/(\overline{f_1}, \overline{f_2}, \ldots, \overline{f_{n-1}})$ is a complete intersection. Hence B also has the SLP by Example 6.2 in [52], Corollary 29 in [50] and Corollary 2.1 in [57], and A has the 2-SLP. Repeating this argument, we have the first assertion of (1). The second assertion of (1) follows from Remark 6.3. Since all k-th differences of the Hilbert function of A are symmetric by the condition on d_j, the assertion (2) follows from Theorem 6.29. The assertion (3) follows from (2).   □

Example 6.34. Let $R = K[x_1, x_2, \ldots, x_n]$. In the following two examples, I is a complete intersection and its quotient has the SLP.

1. Let f_1 and f_2 be homogeneous polynomials of degree d_i $(i = 1, 2)$, and let $g_3, g_4, \ldots, g_n$ be linear forms. Set $I = (f_1, f_2, f_3 = g_3^{d_3}, f_4 = g_4^{d_4}, \ldots, f_n = g_n^{d_n})$. Suppose that $\{f_1, f_2, g_3, g_4, \ldots, g_n\}$ is a regular sequence. Example 6.2 in [52] shows that R/I has the SLP.
2. For $i = 1, 2, \ldots, n$, let $f_i \in K[x_i, x_{i+1}, \ldots, x_n]$ be a homogeneous and monic polynomial in x_i. Set $I = (f_1, f_2, \ldots, f_n)$, and $d_i = \deg f_i$. Then R/I is always a complete intersection. Proposition 4.25 (3) shows that R/I has the SLP.

If the degrees d_j in the above examples satisfy the condition (6.4), then the generic initial ideals of these ideals are the unique almost revlex ideals with corresponding Hilbert functions by Proposition 6.33.

6.4   Graded Betti Numbers

Definition 6.35 (graded Betti numbers). Let $R = K[x_1, x_2, \ldots, x_n]$, and M a graded R-module. It is known that there is an exact sequence of graded R-modules:

$$\cdots \to F_n \to \cdots \to F_2 \to F_1 \to F_0 \to M \to 0,$$

where F_i is a free R-module. This sequence is called a **free resolution** of M. A free resolution is said to be **minimal** if each image of F_i is contained in $\mathfrak{m} F_{i-1}$, where $\mathfrak{m}$ is the homogeneous maximal ideal of R. Such a resolution is unique up to isomorphism.

The free module F_i in the minimal free resolution of M can be written as $F_i = \bigoplus_{j \geq 0} R(-j)^{\beta_{i,j}}$, where $R(-j)$ is the graded R-module whose homogeneous component of degree d is linearly isomorphic to R_{d-j}. We call the number $\beta_{i,j}(M) = \beta_{i,j}$ the **graded Betti number** of M.

Definition 6.36. For a unimodal O-sequence $h = (h_0, h_1, \ldots, h_c)$, we define a sequence ∇h by

$$(\nabla h)_j = \max\{h_{j-1} - h_j, 0\}$$

for all $j = 0, 1, \ldots, c$, where $h_{-1} = 0$. Namely $(\Delta h)_j + (\nabla h)_j = |h_j - h_{j-1}|$. In addition, we define

$$\nabla_k h = \nabla(\Delta^{k-1} h)$$

for $k \geq 1$.

Proposition 6.37. *(1) Let I be an Artinian Borel-fixed ideal of R, for which R/I has the Hilbert function $h = (h_0, h_1, \ldots, h_c)$, and $(R/I, x_n)$ has the WLP. Let $\overline{R} = K[x_1, x_2, \ldots, x_{n-1}]$ and $\overline{I} = I \cap \overline{R}$. Then the graded Betti numbers $\beta_{i,i+j}(R/I)$ of R/I are given as follows:*

$$\beta_{i,i+j}(R/I) = \beta_{i,i+j}(\overline{R}/\overline{I}) + \binom{n-1}{i-1} \times (\nabla h)_{j+1} \qquad (i, j \geq 0),$$

where we use the convention that $h_{-1} = 0$.

(2) Let I be an Artinian Borel-fixed ideal of R, and suppose that R/I has the k-WLP. Set $R' = K[x_1, x_2, \ldots, x_{n-k}]$ and $I' = I \cap R'$. Then we have

$$\beta_{i,i+j}(R/I) = \beta_{i,i+j}(R'/I') + \binom{n-k}{i-1} \cdot (\nabla_k h)_{j+1} + \cdots$$

$$\cdots + \binom{n-2}{i-1} \cdot (\nabla_2 h)_{j+1} + \binom{n-1}{i-1} \cdot (\nabla_1 h)_{j+1}$$

When $k = n$, we regard $\beta_{i,i+j}(R'/I')$ as zero, and, in particular, $\beta_{i,i+j}(R/I)$ is determined only by the Hilbert function.

Proof. (1) It follows from the WL condition and Lemma 6.17 that the minimal generators divisible by x_n occur only in degrees d satisfying $h_{d-1} > h_d$, and the number of such generators of degree d is equal to $h_{d-1} - h_d$. By the formula of Eliahou and Kervaire [32], we have

$$\beta_{i,i+j}(R/I) = \sum_{m=0}^{n} \sum_{\substack{u: \text{ minimal generator} \\ \deg(u)=j \\ \max(u)=m}} \binom{m-1}{i-1}$$

$$= \sum_{m=0}^{n-1} \sum_{\substack{u: \text{ minimal generator} \\ \deg(u)=j \\ \max(u)=m}} \binom{m-1}{i-1}$$

$$+ \binom{n-1}{i-1} \times \left| \left\{ u \; \middle| \; \begin{array}{l} u: \text{ a minimal generator} \\ \deg(u) = j, \quad x_n | u \end{array} \right\} \right|$$

$$= \beta_{i,i+j}(\overline{R}/\overline{I}) + \binom{n-1}{i-1} \times (\nabla h)_{j+1},$$

where $\max(u)$ denotes the maximum index i satisfying $x_i | u$.

(2) By using (1) repeatedly we have the formula of (2). □

Let h be the Hilbert function of a graded Artinian K-algebra R/I. Then there is the uniquely determined lex-segment ideal $J \subset R$ such that R/J has h as its Hilbert function. We define

$$\beta_{i,i+j}(h, R) = \beta_{i,i+j}(R/J).$$

The numbers $\beta_{i,i+j}(h, R)$ can be computed without considering lex-segment ideals. Explicit formulas can be found in [32].

The following theorem gives a sharp upper bound on the Betti numbers among graded Artinian K-algebras having the k-WLP. Moreover the upper bound is achieved by a graded Artinian K-algebra with the k-SLP. Note that Constantinescu [25] and Cho–Park [19] also obtained the same result independently.

Theorem 6.38 ([1, 20, 49]). *Let h be the Hilbert function of some graded Artinian K-algebra with the k-WLP. Then there is a Borel-fixed ideal I of R such that R/I has the k-SLP, the Hilbert function of R/I is h, and $\beta_{i,i+j}(A) \leq \beta_{i,i+j}(R/I)$ for any i and j for all graded Artinian K-algebras A having the k-WLP and h as Hilbert function.*

In particular, when $k = n$, the unique almost revlex ideal for the Hilbert function h attains the upper bound.

These upper bounds are explicitly computed in [49], and they are uniquely determined by the Hilbert functions h.

Proof. Let h be the Hilbert function of a graded Artinian algebra having the k-WLP. Let $R' = K[x_1, x_2, \ldots, x_{n-k}]$, and I' the lex-segment ideal of R', where the Hilbert function of R'/I' is equal to $\Delta^k h$. We can take a Borel-fixed ideal I of R, which is an inverse image of $\text{Res}^n_{n-k} : \mathcal{B}^k_n(h) \rightarrow \mathcal{B}^0_{n-k}(\Delta^k h)$ (see Lemma 6.24 (2) for the surjectivity of Res^n_{n-k}). Then, by construction, R/I has the k-SLP, and has the Hilbert function h. Furthermore it follows from Proposition 6.37 that R/I has the maximal Betti numbers among graded algebras having the k-WLP and the Hilbert function h.

In particular, when $k = n$, any Borel-fixed ideal $I \in \mathscr{B}^n_n(h)$ of R attains the maximal Betti numbers. Hence the unique almost revlex ideal attains them.

By the construction of I and the formula of Proposition 6.37, the maximal Betti numbers are explicitly given by

$$\beta_{i,i+j}(R/I) = \beta_{i,i+j}(\Delta^k h, R') + \binom{n-k}{i-1} \cdot (\nabla_k h)_{j+1} + \cdots$$
$$+ \binom{n-2}{i-1} \cdot (\nabla_2 h)_{j+1} + \binom{n-1}{i-1} \cdot (\nabla_1 h)_{j+1}, \tag{6.5}$$

where $\beta_{i,i+j}(\Delta^k h, R')$ is zero when $k = n$. In particular, the maximal Betti numbers are determined by the Hilbert function h alone. □

Chapter 7
Cohomology Rings and the Strong Lefschetz Property

The Lefschetz property originates in the Hard Lefschetz Theorem for compact Kähler manifolds, so it is natural that some results discussed in the former chapters have geometric backgrounds. For example, Corollary 4.17 on the flat extension can be understood from the cohomology ring of projective space bundles in a geometric setting.

At the same time, the cohomology rings of compact Kähler manifolds provide a lot of examples of Artinian Gorenstein algebras with the Lefschetz property. We can utilize powerful geometric tools such as ampleness criteria for analyzing the ring structure of the cohomology of projective manifolds.

Basic facts on complex manifolds and their cohomology rings including the proof of the Hard Lefschetz Theorem can be found in [43].

7.1 Hard Lefschetz Theorem

Let (X, ω) be a compact Kähler manifold of $\dim_{\mathbb{C}} X = d$ with a Kähler form ω. Let $\mathscr{A}^{p,q}$ the space of (p, q)-forms on X. The space of k-forms $\mathscr{A}^k$ decomposes into the direct sum of the spaces of (p, q)-forms with $p + q = k$, i.e., $\mathscr{A}^k = \bigoplus_{p+q=k} \mathscr{A}^{p,q}$. Put $\mathscr{A}^* = \bigoplus_{k=0}^{2d} \mathscr{A}^k$. Define the operator $\Lambda \colon \mathscr{A}^* \to \mathscr{A}^*$ as an adjoint of the multiplication by ω:

$$L := \times \omega \colon \mathscr{A}^* \to \mathscr{A}^*$$

with respect to the hermitian inner product

$$(\alpha, \beta) = \int_X (\alpha(z), \beta(z)) \, d\mathrm{Vol},$$

where $d\mathrm{Vol}$ is the volume form of (X, ω). It is known that $\{L, \Lambda, H := [L, \Lambda]\}$ defines a representation of $\mathfrak{sl}_2$ on the cohomology ring $H^*(X, \mathbb{C})$.

T. Harima et al., *The Lefschetz Properties*, Lecture Notes in Mathematics 2080, DOI 10.1007/978-3-642-38206-2_7, © Springer-Verlag Berlin Heidelberg 2013

Theorem 7.1 (Hard Lefschetz Theorem [79], see also [43]). *The map L induces linear isomorphisms*

$$\times L^{d-k}: H^k(X, \mathbb{C}) \xrightarrow{\sim} H^{2d-k}(X, \mathbb{C})$$

for $k = 1, 2, \ldots, d$.

The subspace

$$H^k_{\text{prim}}(X) := \text{Ker}(L^{d-k+1}: H^k(X, \mathbb{C}) \to H^{2d-k+2}(X, \mathbb{C}))$$

is called the **primitive cohomology**, and we have the **Lefschetz decomposition**

$$H^m(X, \mathbb{C}) = \bigoplus_k L^k H^{m-2k}_{\text{prim}}(X).$$

The Lefschetz decomposition is a consequence of the irreducible decomposition of $H^*(X, \mathbb{C})$ as an $\mathfrak{sl}_2$-module.

Example 7.2 (Cohomology of projective spaces). The projective space $\mathbb{P}^n = \text{Proj}\, \mathbb{C}[X_0, X_1, \ldots, X_n]$ has a standard Kähler metric called **Fubini–Study metric**. The $(1, 1)$-form on the affine open set $U_0 = \{X_0 \neq 0\} \subset \mathbb{P}^n$ defined by

$$\omega_{\mathbb{P}^n} = \frac{\sqrt{-1}}{2\pi} \partial \bar{\partial} \log(1 + \sum_{i=1}^{n}(X_i/X_0)^2)$$

extends to a globally defined differential form on $\mathbb{P}^n$. The Fubini–Study metric is the hermitian metric on $\mathbb{P}^n$ given by $\omega_{\mathbb{P}^n}$. The complex submanifolds M of $\mathbb{P}^n$ also have a Kähler form obtained by restricting $\omega_{\mathbb{P}^n}$ to M.

The cohomology ring of $\mathbb{P}^n$ is given by $H^*(\mathbb{P}^n) \cong \mathbb{C}[x]/(x^{n+1})$, where x is identified with the multiplication by $[\omega_{\mathbb{P}^n}] \in H^2(\mathbb{P}^n)$. More generally, since $H^*(\mathbb{P}^{n_1} \times \mathbb{P}^{n_2} \times \cdots \times \mathbb{P}^{n_k})$ is isomorphic to the monomial complete intersection

$$\mathbb{C}[x_1, x_2, \ldots, x_k]/(x_1^{n_1+1}, x_2^{n_2+1}, \ldots, x_k^{n_k+1}),$$

we get a geometric proof of Corollary 3.35. If we choose $n_1 = n_2 = \cdots = n_k = 1$, the monomial basis of the algebra $H^*((\mathbb{P}^1)^k) \cong \mathbb{C}[x_1, x_2, \ldots, x_k]/(x_1^2, x_2^2, \ldots, x_k^2)$ can be identified with the Boolean lattice $2^{\{1,2,\ldots,k\}}$ via the correspondence between a square-free monomial $x_{i_1} \cdots x_{i_a}$ and the subset $\{i_1, \ldots, i_a\}$ of $\{1, \ldots, k\}$. So it can be shown that the Boolean lattice has the Sperner property from the Hard Lefschetz Theorem for the product of the projective lines.

In case $n_1 = n_2 = \cdots = n_k = n$, the symmetric group S_k acts on $H^*((\mathbb{P}^n)^k)$ via permutation of the components of $(\mathbb{P}^n)^k$, so $H^*((\mathbb{P}^n)^k)$ has the structure of $(\mathfrak{sl}_2, S_k)$-bimodule. This is the module occuring in Schur–Weyl duality.

7.2 Numerical Criterion for Ampleness

Let L be a line bundle over a complex manifold X. Fix a linear basis $s_1, \ldots, s_N$ of $H^0(X, L)$. If L is generated by its global sections, we have a natural morphism

$$\Phi_{|L|}: X \to \mathbb{P}H^0(X, L)$$
$$x \mapsto (s_1(x) : \cdots : s_N(x)).$$

When L is generated by global sections and $\Phi_{|L|}$ is a closed embedding, L is called **very ample**. In this case, we have $L = \Phi_{|L|}^*(\mathscr{O}_{\mathbb{P}H^0(X,L)}(1))$. If $L^{\otimes m}$ is very ample for some positive integer m, L is called an **ample** line bundle.

For a line bundle $L \in \mathrm{Pic}(X)$ and a curve C on X, the pairing (L, C) is defined by $(L, C) := \deg L|_C$. Let $L, L' \in \mathrm{Pic}(X)$. If $(L, C) = (L', C)$ for all curves C on X, the line bundles L and L' are said to be **numerically equivalent**. Let C, C' be curves on X. Similarly, if $(L, C) = (L, C')$ for all line bundles L on X, the curves C and C' are **numerically equivalent**. For a projective complex manifold X, we define

$$N^1(X) := \mathrm{Pic}(X)_{\mathbb{R}}/(\text{numerical equivalence}).$$

The **ample cone** $\mathrm{Amp}(X) \subset N^1(X)$ is defined to be the convex cone generated by the classes of ample line bundles. Let $Z_1(X)_{\mathbb{R}}$ be the $\mathbb{R}$-linear space of the formal linear combinations of curves on X. We also define

$$N_1(X) := Z_1(X)_{\mathbb{R}}/(\text{numerical equivalence}).$$

Denote by $\overline{NE}(X)$ the closed convex cone spanned by the images of curves in $Z_1(X)$.

Theorem 7.3 (Kleiman [72]). *An element $L \in N^1(X)$ is contained in the ample cone $\mathrm{Amp}(X)$ if and only if $(L, C) > 0$ for all $C \in \overline{NE}(X) \setminus \{0\}$.*

The class $[\mathscr{L}] \in H^2(X, \mathbb{R})$ of an ample line bundle $\mathscr{L}$ gives a Lefschetz element of the algebra $H^*(X)$. The cone in $H^2(X, \mathbb{R})$ of a Kähler manifold X consisting of the Kähler forms is called the Kähler cone. Since both the ample cone and the Kähler cone are convex, any element in $\mathrm{Amp}(X)$ is a Lefschetz element.

The ampleness criterion plays a crucial role in the study of algebraic geometry. Various kinds of numerical criteria can be found in [78].

7.3 Cohomology Rings

In the following, we are interested in Artinian commutative rings obtained as the even part $H^{\mathrm{even}}(X)$ of a compact complex manifold X. In this section we define the grading of $H^{\mathrm{even}}(X)$ by $\deg H^{2d}(X) = d$.

7.3.1 Projective Space Bundle

Let us consider a vector bundle $\mathscr{E}$ of rank $r + 1$ over a compact manifold M and the associated $\mathbb{P}^r$-bundle

$$\mathbb{P}(\mathscr{E}) \to M.$$

The cohomology ring of $\mathbb{P}(\mathscr{E})$ is given by

$$H^*(\mathbb{P}(\mathscr{E})) \cong H^*(M)[t]/(t^r + c_1(\mathscr{E})t^{r-1} + c_2(\mathscr{E})t^{r-2} + \cdots + c_r(\mathscr{E})),$$

where $c_1(\mathscr{E}), c_2(\mathscr{E}), \ldots, c_r(\mathscr{E}) \in H^*(M)$ are the Chern classes of $\mathscr{E}$, and $t \in H^2(\mathbb{P}(\mathscr{E}))$ is the class of $\mathscr{O}(1)$. If M is a Kähler manifold equipped with a Kähler form ω, then the cohomology ring $H^*(\mathbb{P}(\mathscr{E}))$ also has the strong Lefschetz property with the Lefschetz element $[\omega] + t$. Hence the projective space bundles give an explanation for the geometric background of Corollary 4.17.

7.3.2 Homogeneous Spaces

Let G be a connected simply-connected semisimple Lie group over $\mathbb{C}$. Fix $B \subset G$ a Borel subgroup. Denote by $\mathfrak{h}_{\mathbb{Z}}^*$ the weight lattice of the maximal torus $T \subset B$, i.e., $\mathfrak{h}_{\mathbb{Z}}^* = \mathrm{Hom}(T, \mathbb{C}^\times)$, and assume that rank $\mathfrak{h}_{\mathbb{Z}}^* = r$. Let us describe the cohomology ring $H^*(G/B, \mathbb{R})$. The cohomology ring $H^*(G/B, \mathbb{R})$ has the structure of the coinvariant algebra of the corresponding Weyl group W.

$$H^*(G/B, \mathbb{R}) \cong \mathrm{Sym}(\mathfrak{h}_{\mathbb{R}}^*)/(f_1, f_2, \ldots, f_r),$$

where $\mathfrak{h}_{\mathbb{R}}^* := \mathfrak{h}_{\mathbb{Z}}^* \otimes \mathbb{R}$, $\mathrm{Sym}(\mathfrak{h}_{\mathbb{R}}^*)$ the symmetric algebra on $\mathfrak{h}_{\mathbb{R}}^*$, and $f_1, f_2, \ldots, f_r$ are a set of fundamental W-invariants (see Theorem 8.10).

For an element $w \in W$, the Schubert subvariety $\Omega_w \subset G/B$ is defined as the closure of the orbit $\overline{Bw_0wB/B} \subset G/B$, where $w_0 \in W$ is the element of maximal length. The codimension of Ω_w in G/B is the length $l(w)$ (see Definition 8.5). The corresponding dual classes $[\Omega_w]$ to the subvarieties Ω_w, $w \in W$, form a linear basis on $H^*(G/B)$. Let $\{\alpha_1, \alpha_2, \ldots, \alpha_r\}$ be the set of simple roots.

Theorem 7.4 (Chevalley formula [18]). *In the cohomology ring $H^*(G/B)$, we have*

$$\Omega_{s_{\alpha_i}} \Omega_w = \sum_{\substack{\gamma \in \Phi^+, \\ l(ws_\gamma) = l(w)+1}} \langle \omega_i, \gamma^\vee \rangle \Omega_{ws_\gamma},$$

where $\omega_i = \omega_{\alpha_i}$ is the fundamental weight corresponding to α_i (see Definition 8.11).

Let $P \subset G$ be a parabolic subgroup. The cohomology ring $H^*(G/P)$ of the homogeneous space G/P is a subalgebra of $H^*(G/B)$ generated by parabolic invariants corresponding to P. Since the homogeneous spaces G/P are projective, the cohomology ring $H^*(G/P)$ has the strong Lefschetz property. The anti-canonical class $-K_{G/B} = \sum_{i=1}^{r} \Omega_{s_{\alpha_i}}$ is ample, so it gives a Lefschetz element. In Sect. 8.2, we will again discuss the Lefschetz property of the cohomology ring $H^*(G/P)$ from the viewpoint of the coinvariant algebras of reflection groups.

Proposition 7.5 (Chevalley [17]). *The cohomology group $H^2(G/B, \mathbb{Z})$ is identified with the weight lattice $\mathfrak{h}_{\mathbb{Z}}^*$. The ample cone in $H^2(G/B, \mathbb{R})$ coincides with the fundamental chamber under the identification $H^2(G/B, \mathbb{R}) = \mathfrak{h}_{\mathbb{R}}^*$.*

7.3.3 Toric Variety

It is remarkable that the cohomology ring of a toric variety has a nice combinatorial description. We get many examples of Artinian Gorenstein algebras with the strong Lefschetz property as cohomology rings of toric manifolds. The reader can consult [110] for a more detailed exposition on the theory of toric varieties. In the following, let $N = \mathbb{Z}^r$ and $M = \mathrm{Hom}(N, \mathbb{Z})$.

Definition 7.6. The set Σ of strongly convex rational polyhedral cones in $N_{\mathbb{R}}$ is called a **fan** if the following conditions are satisfied:

1. if $\sigma \in \Sigma$, then every face of σ is also in Σ,
2. if $\sigma, \tau \in \Sigma$, then $\sigma \cap \tau$ is a face of σ and of τ.

In the following, we denote by $\Sigma(i)$ the set of i-dimensional cones of Σ. Let Σ be a strongly convex rational polyhedral fan in $N_{\mathbb{R}}$. For a cone σ in $N_{\mathbb{R}}$, define the dual cone $\sigma^{\vee}$ in $M_{\mathbb{R}}$ by

$$\sigma^{\vee} := \{ x \in M_{\mathbb{R}} \mid \langle x, y \rangle \geq 0, \ \forall y \in \sigma \}.$$

The toric variety X_{Σ} is defined by

$$X_{\Sigma} := \bigcup_{\sigma \in \Sigma} U_{\sigma}, \quad U_{\sigma} := \mathrm{Spec}\, \mathbb{C}[\sigma^{\vee} \cap M],$$

where U_{σ} and U_{τ} are patched along $U_{\sigma \cap \tau}$ (i.e. $U_{\sigma} \cap U_{\tau} = U_{\sigma \cap \tau}$).

Proposition 7.7 (See e.g. [110, Chapter 1]).

1. *The toric variety X_{Σ} is nonsingular if and only if each cone $\sigma \in \Sigma$ is nonsingular, i.e. there exists a $\mathbb{Z}$-basis $\{n_1, n_2, \ldots, n_r\}$ of N and $s \leq r$ such that $\sigma = \mathbb{R}_+ n_1 + \mathbb{R}_+ n_2 + \cdots + \mathbb{R}_+ n_s$.*
2. *The toric variety X_{Σ} is compact if and only if Σ is finite and complete, i.e., $N_{\mathbb{R}} = \mathrm{Supp}(\Sigma)$.*

We assume that X_Σ is compact and nonsingular in the following. Denote by $n(\rho) \in N$ the primitive generator of a ray $\rho \in \Sigma(1)$. Let $Q_\mathbb{Z} = \mathbb{Z}[X_\rho | \rho \in \Sigma(1)]$ be a polynomial ring generated by the symbols X_ρ. We define the ideals $I, J \subset Q_\mathbb{Z}$ by

$$I := \left(\sum_{\rho \in \Sigma(1)} \langle m, n(\rho) \rangle X_\rho \,\middle|\, m \in M \right),$$

and

$$J := \left(X_{\rho_1} X_{\rho_2} \cdots X_{\rho_s} \,\middle|\, \rho_1, \rho_2, \ldots, \rho_s \text{ are distinct and do not generate a cone in } \Sigma \right).$$

The ideal J is called the **Stanley–Reisner ideal**. The cohomology ring $H^*(X_\Sigma, \mathbb{Z})$ of X_Σ is generated by the torus invariant divisors

$$V(\rho) := \text{closure}(U_\rho \setminus \text{Spec }\mathbb{C}[M])$$

corresponding to $\rho \in \Sigma(1)$. More precisely, the following is known.

Theorem 7.8 (Danilov [28], Jurkiewicz [69]). *For a nonsingular compact toric variety X_Σ, we have the isomorphism of algebras $Q_\mathbb{Z}/(I + J) \xrightarrow{\sim} H^*(X_\Sigma, \mathbb{Z})$ given by sending X_ρ to $V(\rho)$.*

Definition 7.9 (Σ-linear support function). Let Σ be a finite and complete fan of N.

1. A continuous function $\varphi \colon N_\mathbb{R} \to \mathbb{R}$ is called Σ**-linear support function** if the restriction $\varphi|_\sigma$ to each cone $\sigma \in \Sigma$ is a linear form and φ takes values in $\mathbb{Z}$ on N. We denote by $SF(\Sigma)$ the module of Σ-linear support functions.
2. Let φ be a Σ-linear support function. For $\sigma \in \Sigma(r)$, take the linear form $l_\sigma \in M$ such that $\langle l_\sigma, n \rangle = \varphi(n)$ for $n \in \sigma$. We call φ **strictly upper convex** if the following conditions (a) and (b) are satisfied:

   a. $\langle l_\sigma, n \rangle \geq \varphi(n)$ for all $\sigma \in \Sigma(r)$ and $n \in N_\mathbb{R}$,
   b. $\langle l_\sigma, n \rangle = \varphi(n)$ if and only if $n \in \sigma$.

For a Σ-linear support function φ, define the divisor D_φ on X_Σ by

$$D_\varphi := - \sum_{\rho \in \Sigma(1)} \varphi(n(\rho)) V(\rho).$$

Then the following is well-known.

Proposition 7.10 (See e.g. [110, Chapter 2].). *Let X_Σ be compact.*

1. The sequence

$$0 \to M \to SF(\Sigma) \to \text{Pic}(X_\Sigma) \to 0$$

is exact, where the map $SF(\Sigma) \to \mathrm{Pic}(X_\Sigma)$ sends $\varphi \in SF(\Sigma)$ to $\mathcal{O}(D_\varphi) \in \mathrm{Pic}(X_\Sigma)$.

2. The line bundle $\mathcal{O}(D_\varphi)$ is ample if and only if φ is strictly upper convex. In particular, X_Σ is projective if and only if there exists a strictly upper convex Σ-linear support function.

Proposition 7.11. 1. If X_Σ is compact and nonsingular, then $H^*(X_\Sigma, \mathbb{Q})$ is Gorenstein.

2. If X_Σ is projective and nonsingular, then $H^*(X_\Sigma, \mathbb{Q})$ has the strong Lefschetz property.

Proof. The first statement follows from the Poincaré duality theorem and Theorem 2.79. The second one is a consequence of the Hard Lefschetz Theorem.

□

Remark 7.12. It is known that Poincaré duality and the Hard Lefschetz Theorem hold for a wider class of toric varieties under the assumption that Σ is simplicial. If Σ is complete and simplicial, then the cohomology ring of X_Σ has the presentation

$$H^*(X_\Sigma, \mathbb{Q}) \cong Q/(I + J), \quad Q = Q_\mathbb{Z} \otimes \mathbb{Q},$$

as in Theorem 7.8 (but over $\mathbb{Q}$) and satisfies Poincaré duality. If we assume that Σ is simplicial and X_Σ is projective, then the Hard Lefschetz Theorem also holds.

If X_Σ is a toric Fano manifold, the anti-canonical divisor

$$-K_{X_\Sigma} = \sum_{\rho \in \Sigma(1)} V(\rho)$$

is ample. So it is a Lefschetz element of the cohomology ring $H^*(X_\Sigma, \mathbb{Q})$.

Let $\Sigma(1) = \{\rho_1, \rho_2, \ldots, \rho_k\}$. Define the polynomial $F \in \mathbb{Z}[x_1, x_2, \ldots, x_k]$ by the intersection number

$$F := \left(\sum_{i=1}^{k} x_i V(\rho_i) \right)^r.$$

When X_Σ is projective, the polynomial F is essentially the volume function. Then, by Lemma 3.74, we have a presentation of the cohomology ring as follows:

$$H^*(X_\Sigma) = Q/\mathrm{Ann}_Q(F).$$

The above presentation of the cohomology ring of toric varieties was found by Khovanskii and Pukhlikov [118]. More generally, Kaveh [71] has described the defining ideal of the cohomology ring of the spherical varieties as the annihilator of the volume polynomial.

It is known that

$$\mathrm{Hilb}(H^*(X_\Sigma), t) = \sum_{i=0}^{r} |\Sigma(i)| (t-1)^{r-i}.$$

For $\sigma \in \Sigma$, the monomial $\mu_\sigma := \prod_{\tau \in \Sigma(1), \tau \prec \sigma} X_\tau$ is called the **face monomial** of σ. Face monomials μ_σ, $\sigma \in \Sigma$ linearly span $H^*(X_\Sigma)$.

Example 7.13 (Hirzebruch surface). Let us consider the complete fan of dimension two determined by the rays

$$\rho_1 := \mathbb{R}_+ n_1, \quad \rho_2 := \mathbb{R}_+ n_2, \quad \rho_3 := \mathbb{R}_+(-n_1 + a n_2), \quad \rho_4 := \mathbb{R}_+(-n_2).$$

The corresponding toric surface is the Hirzebruch surface $\mathscr{F}_a$, which has the structure of a $\mathbb{P}^1$-bundle over $\mathbb{P}^1$:

$$\mathscr{F}_a = \mathbb{P}(\mathscr{O} \oplus \mathscr{O}(a)) \to \mathbb{P}^1.$$

The cohomology ring $H^*(\mathscr{F}_a)$ is generated by the classes of the fiber $f := V(\rho_1)$ and the section $s := V(\rho_2)$. It is easy to see that the intersection form on $H^2(\mathscr{F}_a)$ is given by

$$f^2 = 0, \qquad\qquad fs = 1, \qquad\qquad s^2 = -a.$$

Then it is immediate to see the following hold:

1. The element $\alpha f + \beta s \in H^2(\mathscr{F}_a)$ is ample $\iff \alpha > a\beta$ and $\beta > 0$.
2. The element $\alpha f + \beta s \in H^2(\mathscr{F}_a)$ is Lefschetz $\iff \beta \neq 0$ and $2\alpha \neq a\beta$.

Though the classes of form $c(af + s)$ with $c > 0$ belong to the boundary of the ample cone, they are Lefschetz.

Example 7.14 (Non-projective compact toric manifold, Miyake–Oda [109]). It is known that there exist examples of non-projective compact toric manifolds in dim ≥ 3. In such a case, we can not decide from a geometric argument whether $H^*(X_\Sigma)$ has the SLP or not (Fig. 7.1). One of the simplest example of non-projective compact toric manifolds is due to Miyake–Oda. Let $N = \mathbb{Z}n_1 + \mathbb{Z}n_2 + \mathbb{Z}n_3$ be the lattice of rank 3. We define a complete fan Σ of N consisting of the following three-dimensional cones and their faces:

$$\mathbb{R}_+ n_1 + \mathbb{R}_+ n_2 + \mathbb{R}_+ n_3, \quad \mathbb{R}_+ n_1 + \mathbb{R}_+ n_2 + \mathbb{R}_+ n_1', \quad \mathbb{R}_+ n_2 + \mathbb{R}_+ n_3 + \mathbb{R}_+ n_2',$$

$$\mathbb{R}_+ n_1 + \mathbb{R}_+ n_3 + \mathbb{R}_+ n_3', \quad \mathbb{R}_+ n_1 + \mathbb{R}_+ n_1' + \mathbb{R}_+ n_3', \quad \mathbb{R}_+ n_2 + \mathbb{R}_+ n_2' + \mathbb{R}_+ n_1',$$

$$\mathbb{R}_+ n_3 + \mathbb{R}_+ n_2' + \mathbb{R}_+ n_3', \quad \mathbb{R}_+ n_0 + \mathbb{R}_+ n_1' + \mathbb{R}_+ n_2', \quad \mathbb{R}_+ n_0 + \mathbb{R}_+ n_2' + \mathbb{R}_+ n_3',$$

$$\mathbb{R}_+ n_0 + \mathbb{R}_+ n_1' + \mathbb{R}_+ n_3',$$

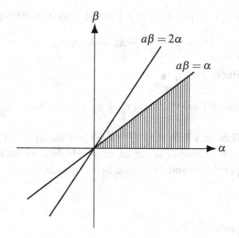

Fig. 7.1 Ample cone of $\mathscr{F}_a$

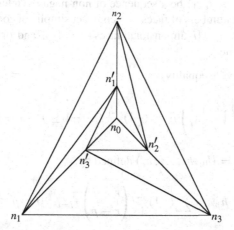

Fig. 7.2 Non-projective complete fan

where

$$n_0 := -n_1 - n_2 - n_3, \quad n_1' := n_0 + n_1, \quad n_2' := n_0 + n_2, \quad n_3' := n_0 + n_3.$$

There exist no strictly upper convex Σ-linear support functions, so the fan Σ gives an example of compact non-projective toric manifold X_Σ (Fig. 7.2). The cohomology ring of X_Σ has the presentation

$$H^*(X_\Sigma, \mathbb{C}) = \mathbb{C}[X_0, X_1, X_2, X_3]/\operatorname{Ann}(F),$$

where

$$F = (x_0 V(\rho_0) + x_1 V(\rho_1) + x_2 V(\rho_2) + x_3 V(\rho_3))^3$$
$$= -4x_0^3 - x_1^3 - x_2^3 - x_3^3 + 3x_1 x_2 x_3.$$

See Lemma 3.74. Since

$$\text{Hess}(F) = -1296x_0(x_1^3 + x_2^3 + x_3^3 - 3x_1 x_2 x_3),$$

we can see from Theorem 3.76 that $H^*(X_\Sigma)$ has the strong Lefschetz property, though X_Σ is non-projective. The set of non-Lefschetz elements in $H^2(X_\Sigma, \mathbb{C})$ consists of four hyperplanes and one cubic hypersurface.

7.3.4   O-Sequences

Let $f = (f_0, f_1, \ldots, f_{r-1})$ be a sequence of non-negative integers. The sequence f is realized as the numbers of faces of a compact simplicial convex r-dimensional polytope P (i.e. $f_i = |\{i\text{-dimensional faces of } P\}|$) if and only if the following conditions are satisfied.

1. (Dehn–Sommerville equality)

$$\sum_{k=j}^{r-1}(-1)^k \binom{k+1}{j+1} f_k = (-1)^{r-1} f_j, \quad -1 \le j \le r-1, \, f_{-1} := 1,$$

2. The sequence $h = (h_0, h_1, \ldots, h_r)$ defined by

$$h_p := \sum_{j=0}^{p}(-1)^{p-j} \binom{r-j}{r-p} f_{j-1}, \quad 0 \le p \le r$$

satisfies $h_p = h_{r-p}$ for $0 \le p \le r$ and Δh is an O-sequence.

These conditions were conjectured by McMullen [96]. The necessity of the second condition was proved by Stanley [135] by using the Hard Lefschetz Theorem for the cohomology ring of projective toric varieties. Here we sketch the outline of Stanley's argument.

We may assume that the polytope $P \subset N_\mathbb{R}$ contains the origin and the vertices $v_1, v_2, \ldots, v_{f_0}$ of P are in $N_\mathbb{Q}$. Let $\rho_1, \rho_2, \ldots, \rho_{f_0}$ be the rays generated by the vertices of $v_1, v_2, \ldots, v_{f_0}$. Since P is simplicial, $\rho_1, \rho_2, \ldots, \rho_{f_0}$ define a rational simplicial fan Σ. In this case, the algebra Q/J is Gorenstein. Fix a linear basis $m_1, m_2, \ldots, m_r$ of M. The elements

$$\sum_{i=1}^{f_0} \langle m_j, n(\rho_i) \rangle X_{\rho_i}, \quad 1 \le j \le r$$

form a regular sequence of Q/J. Hence the cohomology ring $H^*(X_\Sigma, \mathbb{Q}) \cong Q/(I + J)$ is a Cohen–Macaulay ring with the Hilbert function $\sum_{i=0}^r h_i t^i$. We obtain $h_p = h_{r-p}$ for $0 \le p \le r$ from Poincaré duality. Since the Hard Lefschetz Theorem holds for the projective variety X_Σ, we can pick a Lefschetz element $L \in H^2(X_\Sigma)$. Let us consider the ideal $\mathscr{I}$ of $H^*(X_\Sigma)$ generated by L. Then the quotient algebra of $H^*(X_\Sigma)$ by $\mathscr{I}$ has the Hilbert function Δh. So Δh is an O-sequence.

Remark 7.15. In the argument above, since $\dim H^2(X_\Sigma) = f_0 - r$, one has the inequality

$$h_p \le \binom{f_0 - r + p - 1}{p}.$$

This inequality is known as Motzkin's upper bound conjecture [104], which was proved by McMullen [95].

Chapter 8
Invariant Theory and Lefschetz Properties

In this chapter we discuss topics of invariant theory such as coinvariant algebras of reflection groups. In particular the coinvariant algebras of real reflection groups have the SLP, and the set of Lefschetz elements is explicitly determined in most cases.

8.1 Reflection Groups

We briefly review some facts concerning reflection group. See the books of Humphreys [61] and Hiller [59] for more details.

Definition 8.1 (Finite reflection group). Let V be a finite-dimensional real vector space endowed with a positive definite symmetric bilinear form (λ, μ). A linear map $s \in GL(V)$ is called a **reflection** if there exist a nonzero vector $\alpha \in V$ such that s is the identity on the hyperplane H_α orthogonal to α, and s sends α to $-\alpha$. In other words, s act on $\lambda \in V$ by

$$s\lambda = \lambda - \frac{2(\lambda, \alpha)}{(\alpha, \alpha)}\alpha.$$

We denote the reflection with respect to α by s_α, and call H_α the **reflecting hyperplane** of s_α.

A finite subgroup of the orthogonal group $O(V)$ generated by a set of reflections is called a **finite reflection group**.

Remark 8.2. The class of the finite reflection groups contains the class of **Weyl groups**. The class of the finite reflection groups is contained in the class of the **complex reflection groups**. (See Theorem 8.9.)

Let W be a finite reflection group. Since $t s_\alpha t^{-1} = s_{t\alpha}$ for any $t \in O(V)$, W permutes the lines $\mathbb{R}\alpha$ ($s_\alpha \in W$), that is, for any $w \in W$ and a reflection s_u, $w(\mathbb{R}\alpha) = \mathbb{R}\beta$ for some reflection s_β. This leads to the notion of root systems as we explain next.

T. Harima et al., *The Lefschetz Properties*, Lecture Notes in Mathematics 2080,
DOI 10.1007/978-3-642-38206-2_8, © Springer-Verlag Berlin Heidelberg 2013

Definition 8.3 (Root system, positive system, simple system). Let V be a finite-dimensional real vector space. A finite subset Φ of $V \setminus \{0\}$ is called a **root system**, if Φ satisfies the following two axioms.

1. $\Phi \cap \mathbb{R}\alpha = \{\alpha, -\alpha\}$ for all $\alpha \in \Phi$,
2. $s_\alpha \Phi = \Phi$.

Each vector in the root system is called a **root**. If $\{ s_\alpha \mid \alpha \in \Phi \}$ is equal to the set of all the reflections of a reflection group W, then Φ is called a **root system** of W.

For a root α in the root system Φ of W, we define $\alpha^\vee = 2\alpha/(\alpha, \alpha)$, and $\alpha^\vee$ is called the **coroot** of α. Using coroots we can write the action of the reflection s_α as $s_\alpha \lambda = \lambda - (\lambda, \alpha^\vee)\alpha$.

It is known that there exists a subset Δ of the root system Φ such that Δ forms a linear basis of the linear span of Φ, and any $\alpha \in \Phi$ can be expressed as a linear combination of Δ with coefficients of equal sign (all positive or all negative). In this case we call Δ a **simple system** of Φ, $\alpha \in \Delta$ a **simple root**, and s_α ($\alpha \in \Delta$) a **simple reflection**. The number of the simple roots is called a **rank** of Φ. The subset Φ^+ of Φ consisting of the roots expressed as a linear combination of Δ with positive coefficients is called a **positive system**. Clearly $\Phi = \Phi^+ \cup (-\Phi^+)$.

Root systems in Lie theory play a central role. However, note that the above definition of the root system of reflection groups is weaker than the definition of the root system of a Lie algebra.

Theorem 8.4 (see [61, §1, e.g.]).

(1) If Δ is a simple system of a root system Φ, then there is a unique positive system Φ^+ of Φ containing Δ.
(2) Every positive system contains a unique simple system.
(3) For a fixed simple system Δ of a reflection group W, the simple reflections s_α ($\alpha \in \Delta$) generate W.

Definition 8.5 (length). By the above theorem we can consider the length of a presentation of an element $w \in W$ using simple reflections. We call the minimum length simply the **length** of w, and denote by $l(w)$.

Definition 8.6. Let W be a finite reflection group. The Bruhat ordering on W is defined by the cover relation

$$u < v \iff \exists \gamma \in \Phi^+, v = us_\gamma, l(v) = l(u) + 1.$$

Theorem 8.7 (generators and relations). *Let Φ be a root system. Fix a simple system Δ in Φ. Then the generating set $\{ s_\alpha \mid \alpha \in \Delta \}$ is subject only to the relations*

$$(s_\alpha s_\beta)^{m(\alpha,\beta)} = 1,$$

where $m(\alpha, \beta)$ denotes the order of the product $s_\alpha s_\beta$.

Table 8.1 Connected Coxeter graphs

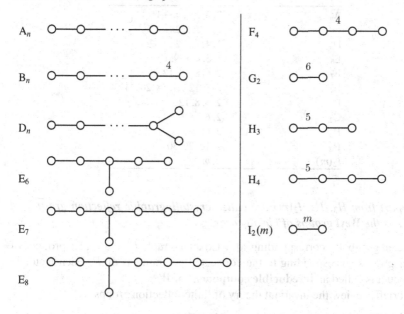

Definition 8.8 (Coxeter graph, Coxeter group). Let W be a finite reflection group, Φ the corresponding root system, and Δ a simple system for Φ. The **Coxeter graph** of W is the graph with nodes corresponding to the simple roots and the following labeled edges: If two simple roots α and β satisfy $m(\alpha, \beta) = 2$, namely they are orthogonal, then there is no edge connecting them. If $m(\alpha, \beta) \geq 3$ then, there is an edge labeled with $m(\alpha, \beta)$. We usually omit the label if the number is equal to three.

Any group having such a presentation by generators and relations is called a **Coxeter group**. Namely, a Coxeter group is a group that has generators $\{ s_\alpha \mid \alpha \in \Phi \}$ which are subject only to the relation $(s_\alpha s_\beta)^{m(\alpha, \beta)} = 1$ and $s_\alpha^2 = 1$.

It is known that the finite Coxeter groups are precisely the finite real reflection groups. A Coxeter group W has a representation as a real reflection group of dimension equal to the number of generators $\{s_\alpha\}$, and this representation is unique up to isomorphism.

A basic fact in the theory of reflection groups is that the Coxeter graph of a finite reflection group is independent of the choice of a root system or choice of a simple system. If two finite reflection groups have the same Coxeter graph, it is clear that the two groups are isomorphic. The Coxeter graphs are classified in the following theorem.

Theorem 8.9 (classification). *Connected Coxeter graphs are classified as in Table 8.1, where a Coxeter graph X_n has n vertices. Every Coxeter graph is a union of connected Coxeter graphs. We remark that the finite Coxeter groups of*

Table 8.2 Fundamental degrees

A_n	$2, 3, \ldots, n-1$
B_n	$2, 4, \ldots, 2n$
D_n	$2, 4, \ldots, 2n-2; n$
E_6	$2, 5, 6, 8, 9, 12$
E_7	$2, 6, 8, 10, 12, 14, 18$
E_8	$2, 8, 12, 14, 18, 20, 24, 30$
F_4	$2, 6, 8, 12$
G_2	$2, 6$
H_2	$2, 6, 10$
H_3	$2, 12, 20, 30$
$I_2(m)$	$2, m$

*types apart from H_3, H_4, $I_2(m)$ are called **crystallographic reflection group** and realized as the **Weyl groups** of Lie algebras.*

A Coxeter group W corresponding to a Coxeter graph Γ is a direct product of Coxeter groups corresponding to the connected components of Γ. Each factor in this product is called an **irreducible component** of W.

We briefly review the invariant theory of finite reflection groups.

Theorem 8.10 (Chevalley). *Let $W \subset GL(n, \mathbb{R})$ be a reflection group of rank n, and $R = \mathbb{R}[x_1, x_2, \ldots, x_n]$. Then the subalgebra R^W of W-invariant polynomials is generated as an $\mathbb{R}$-algebra by n homogeneous, algebraically independent elements of positive degrees.*

*Moreover, the degrees of generators do not depend on the choice of generators. These degrees are called the **fundamental degrees** of W, and such generators are called a set of **fundamental invariants** of W.*

Table 8.2 is the list of fundamental degrees.

Definition 8.11 (fundamental weight). The **fundamental weights** ω_α, $\alpha \in \Delta$, are linear forms on V defined by the condition

$$\langle \omega_\alpha, \beta^\vee \rangle = \begin{cases} 1, & \text{if } \alpha = \beta, \\ 0, & \text{if } \alpha \neq \beta, \end{cases}$$

for $\alpha, \beta \in \Delta$.

8.2 Coinvariant Algebras

Let W be a finite Coxeter group and V its reflection representation. Then W acts on the ring of polynomial functions $R = \mathrm{Sym}\, V^*$. The W-invariant subalgebra $(\mathrm{Sym}\, V^*)^W$ is generated by fundamental W-invariants $f_1, f_2, \ldots, f_r$. The **coinvariant algebra** R_W is defined by

$$R_W := \mathrm{Sym}(V^*)/(f_1, f_2, \ldots, f_r).$$

The coinvariant algebra R_W is isomorphic to the regular representation $\mathbb{R}\langle W \rangle$ as a left W-module. Note that $R \cong R^W \otimes_{\mathbb{R}} R_W$ as a W-module. If W is crystallographic R_W is isomorphic to the cohomology ring $H^*(G/B)$ of the flag variety G/B for an algebraic group G with Weyl group W. In particular, the coinvariant algebra R_W for the crystallographic Coxeter groups has the SLP.

8.2.1   BGGH Polynomial

We define a family of operators acting on R by

$$\partial_\alpha(f) = (s_\alpha(f) - f)/\alpha,$$

for the simple root α. Then the operators $\{\partial_\alpha\}_{\alpha \in \Delta}$ satisfy $\partial_\alpha^2 = 0$ and the Coxeter relations:

$$(\partial_\alpha \partial_\beta)^{[m(\alpha,\beta)/2]} \partial_\alpha^{\nu(\alpha,\beta)} = (\partial_\beta \partial_\alpha)^{[m(\alpha,\beta)/2]} \partial_\beta^{\nu(\alpha,\beta)},$$

where $\alpha \neq \beta$, $[m(\alpha, \beta)/2]$ stands for the integer part of $m(\alpha, \beta)/2$ and $\nu(\alpha, \beta) := m(\alpha, \beta) - 2[m(\alpha, \beta)/2]$. For a reduced expression (i.e., an expression of minimal length) $w = s_{\alpha_1} s_{\alpha_2} \cdots s_{\alpha_l}$ of an element of W, we define the operator $\partial_w := \partial_{\alpha_1} \partial_{\alpha_2} \cdots \partial_{\alpha_l}$. The definition of the operator ∂_w is independent of the choice of reduced expression thanks to the Coxeter relations.

Next we introduce a family of polynomials $\{p_w\}_{w \in W}$ which gives a standard linear basis of R_W. We define $p_{w_0} := |W|^{-1} \prod_{\gamma \in \Phi^+} \gamma$ for the maximal length element $w_0 \in W$, and

$$p_w := \partial_{w^{-1} w_0} p_{w_0}$$

for general $w \in W$. We call the polynomials $\{p_w\}_{w \in W}$ the **Bernstein–Gelfand–Gelfand–Hiller polynomials** (or BGGH polynomials for short). The following formula determines the multiplicative structure of R_W. See also Theorem 7.4.

Proposition 8.12 (Bernstein–Gelfand–Gelfand [6], Hiller [58]). *Let α be a simple root. wE have the following formula in R_W.*

$$p_{s_\alpha} p_w = \sum_{\gamma \in \Phi^+, l(w s_\gamma) = l(w) + 1} \langle \omega_\alpha, \gamma^\vee \rangle p_{w s_\gamma}$$

8.2.2   Coinvariant Algebra

When W is a Weyl group of a semisimple Lie group G, the coinvariant algebra R_W is isomorphic to the cohomology ring $H^*(G/B)$ of the flag variety G/B.

The polynomials p_w represent the Schubert subvarieties Ω_w under the identification of R_W with $H^*(G/B)$.

The bilinear pairing

$$(\, , \,): R_W \times R_W \to \mathbb{R}$$

given by $(f, g) := \partial_{w_0}(fg)$ is non-degenerate. Hence, R_W is Gorenstein.

Theorem 8.13 (Stanley [136], Numata–Wachi [108] for H_4). *The coinvariant algebra R_W of finite Coxeter group has the strong Lefschetz property.*

Theorem 8.14 (MNW [90]). *Assume that W is a finite Coxeter group that does not contain the Coxeter group of type H_4 as an irreducible component. Then the set of the Lefschetz elements of R_W is the complement of the union of the reflection hyperplanes of W in $R_W^1 = V^*$.*

Proof. The crystallographic case is easily shown by Proposition 7.5 and the Weyl group action. The noncrystallographic case is done by direct computation. See [90] for the details.   □

Stanley [136] has pointed out that the Hard Lefschetz Theorem for $H^*(G/B)$ implies that the Bruhat ordering on the Weyl group has the Sperner property. The Lefschetz property for the coinvariant algebra of H_4 has been shown more recently by Numata–Wachi [108].

Remark 8.15. 1. In the H_4 case, the determination of the set of Lefschetz elements is still open.
2. Theorem 8.14 is true only over $\mathbb{R}$. For example, we can use the Hessian of the difference product $D = \prod_{i<j}(x_i - x_j)$ to see this. Let us consider the S_3-case. In this case, we have $D = (x - y)(x - z)(y - z)$ and

$$\text{Hess}_{X,Y}^{(1)} D = -4(x^2 + y^2 + z^2 - xy - yz - zx).$$

If we extend the field of coefficients to $\mathbb{C}$, the locus of non-Lefschetz elements has an extra component in addition to the reflection hyperplanes.

Since $\text{Hilb}(R^W, t) = \prod_{i=1}^r (1 - t^{m_i+1})^{-1}$ and $R = R^W \otimes R_W$, we have the Hilbert function of R_W as follows:

$$\text{Hilb}(R_W, t) = \sum_{w \in W} t^{l(w)} = \frac{\prod_{i=1}^r (1 - t^{m_i+1})}{(1 - t)^r},$$

where $m_1, m_2, \ldots, m_r$ are the fundamental degrees of W.

Definition 8.16. For a given subset I of the simple system Δ, the **parabolic subgroup** W_I is defined to be the subgroup of W generated by the simple reflections of I. We also define $W^I := \{ w \in W \mid l(ws_\alpha) > l(w), \forall \alpha \in I \}$. For $w \in W$, we have a unique decomposition $w = uv$, $u \in W^I$, $v \in W_I$. In this decomposition, u is the unique element of minimal length in the coset wW_I. The set W^I is called the **minimal coset representatives** of W/W_I.

For $I \subset \Delta$, the family of polynomials $\{p_w\}_{w \in W^I}$ gives a linear basis of the invariant subalgebra $R_W^{W_I}$. A closed subgroup $P \subset G$ is called a **parabolic subgroup** if G/P is compact. A parabolic subgroup P containing the fixed Borel subgroup B determines the subset $I_P \subset \Delta$ such that $P = BW_{I_P}B$. The cohomology ring $H^*(G/P)$ is identified with the invariant subalgebra $R_W^{W_{I_P}}$. Note that the cohomology ring $H^*(G/P) = \bigoplus_{w \in W_{I_P}} \mathbb{R}[\Omega_w]$ has a nonstandard grading for $P \neq B$, but it also has the SLP because G/P is a projective manifold.

Example 8.17 (Grassmannian $G(k, N)$). The Grassmannian $G(k, N)$ of k-dimensional subspaces in a fixed N-dimensional complex vector space has the structure of projective complex manifold of dimension $N(N - k)/2$. The Grassmannian $G(k, N)$ is a homogeneous space of $SL(N, \mathbb{C})$. The corresponding parabolic subgroup $W_I \subset W = S_N$ is the product of two symmetric groups $S_k \times S_{N-k}$, where the first component S_k is the permutation group of the indices $\{1, 2, \ldots, k\}$ and the second component S_{N-k} is that of $\{k + 1, k + 2, \ldots, N\}$. The minimal coset representatives for $S_N/S_k \times S_{N-k}$ are given by so-called **Grassmannian permutation**s, which are defined as permutations with a unique descent. For an arbitrary permutation $w \in S_N$, an index $i \in \{1, 2, \ldots, N - 1\}$ such that $w(i) > w(i + 1)$ is called a **descent** of w. Let r be the unique descent of a Grassmannian permutation w. Then we define a partition $\lambda(w)$ by

$$\lambda(w) = (w(r) - r, w(r - 1) - (r - 1), \ldots, w(1) - 1).$$

The cohomology ring of the Grassmannian $G(k, N)$ has a linear basis parameterized by the Young diagrams contained in the rectangular diagram of shape $((N - k)^k)$. At the same time, $H^*(G(k, N))$ is isomorphic to the subalgebra of R_{S_N} linearly spanned by the classes of the Schur polynomials $\{s_\lambda\}_{\lambda \subset ((N-k)^k)}$. More precisely, we have the following isomorphism:

$$H^*(G(k, N)) \to R_{S_N}^{S_k \times S_{N-k}} \subset R_{S_N}$$
$$\Omega_w \mapsto s_{\lambda(w)}$$

By applying the Hard Lefschetz Theorem to the cohomology ring $H^*(G(k, N-k))$, we get the Sperner property for the Young lattice.

Remark 8.18. In [136], Stanley showed the poset obtained by restricting the Bruhat ordering on the Weyl group W to a set of minimal coset representatives W^I has the Sperner property by using the cohomology ring of the corresponding homogeneous space G/P. By taking $G = SO(2n + 1, \mathbb{C})$ and a certain maximal parabolic subgroup P, he applied this result to prove the Erdös–Moser conjecture [35] on the number of subsets X of a given finite subset of $\mathbb{R}$ such that the elements of X sum to a given real number. See also [117].

In the following, we regard $\mathrm{Sym}(V^*)$ as an algebra of differential operators acting on $\mathrm{Sym}(V)$.

Proposition 8.19. *Let $D \in \mathrm{Sym}(V)$ be the product of all positive roots of W. Then the coinvariant algebra R_W is presented as $\mathrm{Sym}(V^*)/\mathrm{Ann}(D)$.*

Proof. It is enough to show that $\mathrm{Ann}(D)$ is generated by the homogeneous W-invariants of positive degree. When $\phi(X) \in (\mathrm{Sym}\, V^*)^W$ is homogeneous of positive degree, we have $w(\phi(X)D(x)) = (-1)^{l(w)}\phi(X)D(x)$ for $w \in W$ and $\deg(\phi(X)D(x)) < |\Phi^+|$. This means $\phi(X)D(x) = 0$. Hence $\mathrm{Ann}(D)$ contains all the homogeneous invariants of positive degree. On the other hand, R_W has the same socle degree as $\mathrm{Sym}(V^*)/\mathrm{Ann}(D)$. Thus we have $R_W \cong \mathrm{Sym}(V^*)/\mathrm{Ann}(D)$.   □

The above proposition shows that the coinvariant algebra R_W is isomorphic to the space of harmonic polynomials as remarked in Remark 4.19 for the root system of type A. We can see from the above proposition that the condition of an element $L \in (R_W)_1$ to be Lefschetz is given by the (higher) Hessians of D. However it seems difficult in general to compute the Hessians of D explicitly.

Problem 8.20 (Algebraic proof of the Lefschetz property for $H^*(G/B)$). When G is of exceptional type, prove that the cohomology ring $H^*(G/B) \cong R_W$ has the strong Lefschetz property without using the Hard Lefschetz Theorem. Is it possible to prove the Lefschetz property for $H^*(G/B)$ from the non-vanishing of the Hessians $\mathrm{Hess}^{(d)}\, D$ of the product D of the positive roots?

Remark 8.21. McDaniel [94] has proved the Lefschetz property for the coinvariant algebra of the finite Coxeter groups and some subalgebras generated by parabolic invariants algebraically based on Theorem 4.10.

8.2.3   Complex Reflection Groups

It seems that the theory of the Lefschetz property for the coinvariant algebras might be formulated for complex reflection groups.

Definition 8.22. Let V be a finite-dimensional complex vector space. If a non-identity element $g \in GL(V)$ has a finite order and there exists a subspace $U \subset V$ of codimension one such that $g|_U = \mathrm{id}_U$, then g is called a **pseudo-reflection**.

Theorem 8.23 (Shephard–Todd [124], Chevalley [16]). *Let V be a finite dimensional complex vector space and $G \subset GL(V)$ a finite subgroup. Then $\mathbb{C}[V]^G$ is a polynomial algebra if and only if G is generated by pseudo-reflections.*

Remark 8.24. Watanabe [145] also proved essentially the same result in a more general setup.

Finite groups generated by pseudo-reflections were classified by Shephard–Todd in [124] into the infinite series $G(m,q,n)$ and 34 exceptional types $G_4, G_5, \ldots, G_{37}$. Let $G \subset GL(V)$ be a finite subgroup generated by pseudo-reflections. The coinvariant algebra R_G is the quotient of $\mathbb{C}[V]$ by the ideal generated by a set of fundamental invariants $f_1, f_2, \ldots, f_r$, $r = \dim V$. Since R_G is a complete intersection, it gives further examples of Artinian Gorenstein algebras with finite group actions.

Let m, q, n be positive integers with $q \mid m$. The group $G(m,q,n) \subset GL(n, \mathbb{C})$ is generated by S_n (permutation matrices) and the diagonal matrices

$$\begin{pmatrix} \lambda_1 & 0 & \cdots & 0 \\ 0 & \lambda_2 & \cdots & 0 \\ \vdots & \vdots & \ddots & \vdots \\ 0 & 0 & \cdots & \lambda_n \end{pmatrix},$$

where λ_i's are m-th roots of unity satisfying

$$(\lambda_1 \lambda_2 \cdots \lambda_n)^{m/q} = 1.$$

When $G = G(m, 1, n)$, then the coinvariant algebra R_G is given as follows:

$$R_G = \mathbb{C}[x_1, x_2, \ldots, x_n]/(f_1, f_2, \ldots, f_n),$$

where

$$f_i := e_i(x_1^m, x_2^m, \ldots, x_n^m), \quad 1 \le i \le n.$$

In this case, the coinvariant algebra R_G has the strong Lefschetz property from Proposition 4.26.

Conjecture 8.25. The coinvariant algebra R_G has the strong Lefschetz property for a general finite complex reflection group G.

Chapter 9
The Strong Lefschetz Property and the Schur–Weyl Duality

The purpose of this chapter is to illustrate a role played by the SLP in connection with the theory of Artinian rings and the Schur–Weyl duality. We assume that the reader is familiar with commutative algebra but perhaps without knowledge of representation theory, but we are hopeful that the expert in representation theory may also find the following sections of interest.

9.1 The Schur–Weyl Duality

Let $R = K[x_1, x_2, \ldots, x_n]$ be the polynomial algebra in n variables $x_1, x_2, \ldots, x_n$ over a field K of characteristic zero, and let I be the ideal $(x_1^d, x_2^d, \ldots, x_n^d)$ of R. Denote by S_n the symmetric group on n letters. We let S_n act on R by

$$(\sigma f)(x_1, x_2, \ldots, x_n) = f(x_{\sigma(1)}, x_{\sigma(2)}, \ldots, x_{\sigma(n)})$$

for $\sigma \in S_n$ and $f = f(x_1, x_2, \ldots, x_n) \in R$. Since the ideal I is invariant under the action of the group S_n, the algebra

$$A = R/I = K[x_1, x_2, \ldots, x_n]/(x_1^d, x_2^d, \ldots, x_n^d) \tag{9.1}$$

has a graded S_n-module structure. We are interested in decomposing this S_n-module A into irreducible modules.

We remark that the algebra is isomorphic to the n-fold tensor product $V^{\otimes n}$ of a d-dimensional vector space $V = K[x]/(x^d)$ over K:

$$A \cong K[x_1]/(x_1^d) \otimes K[x_2]/(x_2^d) \otimes \cdots \otimes K[x_n]/(x_n^d) \cong V^{\otimes n}.$$

The second isomorphism is a linear isomorphism given by

$$x_i^k \mapsto (e_k \text{ at the } i\text{-th position}),$$

T. Harima et al., *The Lefschetz Properties*, Lecture Notes in Mathematics 2080, DOI 10.1007/978-3-642-38206-2_9, © Springer-Verlag Berlin Heidelberg 2013

for $k = 0, 1, \ldots, d - 1$, $i = 1, 2, \ldots, n$, where $e_0, e_1, \ldots, e_{d-1}$ is a basis of V. The tensor space $V^{\otimes n}$ is the underlying space for the **Schur–Weyl duality**. The reader can consult [40, 152] for a comprehensive description. Let $GL(V)$ be the general linear group of the space V. The group $GL(V)$ acts on the space $V^{\otimes n}$ **diagonally**, i.e.,

$$g(v_1 \otimes v_2 \otimes \cdots \otimes v_n) = g v_1 \otimes g v_2 \otimes \cdots \otimes g v_n,$$

for $v_1 \otimes v_2 \otimes \cdots \otimes v_n \in V^{\otimes n}$. The action of S_n on A is translated into an action on $V^{\otimes n}$ by

$$\sigma(v_1 \otimes v_2 \otimes \cdots \otimes v_n) = v_{\sigma^{-1}(1)} \otimes v_{\sigma^{-1}(2)} \otimes \cdots \otimes v_{\sigma^{-1}(n)},$$

for $v_1 \otimes v_2 \otimes \cdots \otimes v_n \in V^{\otimes n}$. Evidently, these two actions **commute** with each other, i.e., $g \circ \sigma = \sigma \circ g$. Thus the space $V^{\otimes n}$ is given a structure of a module for the product group $GL(V) \times S_n$. It is known that the irreducible modules for the product group $GL(V) \times S_n$ are precisely the tensor products $W \otimes U$ where W and U are irreducible modules for $GL(V)$ and S_n respectively (see e.g., [40]). The Schur–Weyl duality gives an irreducible decomposition of $V^{\otimes n}$ as a $GL(V) \times S_n$-module. Recall that a **partition** λ of n is a finite non-increasing sequence $\lambda = (\lambda_1, \lambda_2, \ldots, \lambda_l)$ with the size $|\lambda| = \sum_{i=1}^{l} \lambda_i$ of n. The equality $\lambda_l = 0$ is in general allowed, but if we add the assumption $\lambda_l > 0$ to the definition here, then the number l is called the **length** of λ, denoted by $l(\lambda)$. It is well-known that the (isomorphism classes of) irreducible S_n-modules U are in one-to-one correspondence with the partitions of n [37, 121], and we denote such a U by U_λ if the corresponding partition is $\lambda \vdash n$. Let M be an irreducible $GL(V) \times S_n$-submodule of $V^{\otimes n}$. Under this parameterization, there exists an irreducible $GL(V)$-module W_λ satisfying $M \cong_{GL(V) \times S_n} W_\lambda \otimes U_\lambda$, where the partition λ is uniquely determined from M. The Schur–Weyl duality Theorem states the following.

Proposition 9.1. *The tensor space $V^{\otimes n}$ is isomorphic to $\bigoplus_{\substack{\lambda \vdash n \\ l(\lambda) \leq d}} W_\lambda \otimes U_\lambda$ as $GL(V) \times$*

S_n-modules.

We would like to make some clarifying remarks for better understanding. The isomorphism shows that the tensor space $V^{\otimes n}$ as an S_n-module decomposes into a direct sum $\bigoplus_{\substack{\lambda \vdash n \\ l(\lambda) \leq d}} U_\lambda^{\oplus \dim W_\lambda}$ of irreducible S_n-modules, where each irreducible component of $V^{\otimes n}$ isomorphic to U_λ appears as many times as the dimension of its counterpart W_λ. Likewise as a $GL(V)$-module, the number of irreducible components isomorphic to W_λ is equal to the dimension of U_λ. Since $V^{\otimes n}$ is finite-dimensional, it is obviously expected that the two modules W_λ and U_λ should be finite-dimensional. In fact, it is known (see e.g., [40]) that the dimension of the module W_λ is given by the formula

$$\dim W_\lambda = \frac{\Delta(\lambda + \delta)}{\Delta(\delta)},$$

where $\delta = (\delta_1, \delta_2, \ldots, \delta_d) = (d-1, d-2, \ldots, 0)$ and $\Delta(\delta)$ is the **difference product** $\Delta(\delta) = \prod_{1 \le i < j \le d} (\delta_i - \delta_j)$, and $\Delta(\lambda + \delta)$ is similarly defined for $\lambda + \delta$. On the other hand, the dimension of U_λ equals the number of standard Young tableaux of shape λ, which is finite (see [37, 121]).

Example 9.2. We will examine the example in the case where $d = 2$ and $n = 3$. The dimension of $V^{\otimes n}$ is $2^3 = 8$. There exist three partitions (3), $(2, 1)$ and $(1, 1, 1)$ whose total sum is $n = 3$, but the third one does not contribute to the decomposition of $V^{\otimes 3}$ because of the condition $l(\lambda) \le 2$. Let $\lambda = (3)$ and $\mu = (2, 1)$. Since $d = 2$ we have $\delta = (1, 0)$ and the dimensions of W_λ and W_μ are 4 and 2 respectively. On the other hand, there exist only one standard Young tableau of shape λ and two for μ. Thus the dimension of $\bigoplus_{\substack{\lambda \vdash n \\ l(\lambda) \le d}} W_\lambda \otimes U_\lambda$ is $4 \times 1 + 2 \times 2 = 8$, which coincides with that of $V^{\otimes n}$ just as it should.

Proposition 9.1 and all known results about the decomposition of $V^{\otimes n}$ can be applied directly to the monomial complete intersection A as defined in (9.1). In the following sections we will discuss how the strong Lefschetz property of A can be used to decompose A into irreducible $GL(V) \times S_n$-modules. The following is a restatement of Theorem 3.35.

Theorem 9.3. *The algebra* $A = K[x_1, x_2, \ldots, x_n]/(x_1^d, x_2^d, \ldots, x_n^d)$ *has the strong Lefschetz property, with a Lefschetz element* $l = x_1 + x_2 + \cdots + x_n$.

Proof. It is obvious that the algebra $K[x_i]/(x_i^d)$ has the SLP with a Lefschetz element $l = x_i$. Since the algebra A is isomorphic to the tensor product of those algebras, A itself has the SLP with a Lefschetz element $l = x_1 + x_2 + \cdots + x_n$ by Theorem 3.34. $\square$

We would like to emphasize that the Lefschetz element $l = x_1 + x_2 + \cdots + x_n$ is invariant under the action of S_n, and this fact makes it possible to construct the irreducible decomposition in a more explicit way; we could expect to obtain an explicit basis for each irreducible component systematically by virtue of the SLP. (Maeno's construction in [87] works in certain cases in this way.)

Since each homogeneous component A_k is S_n-fixed, it suffices to consider the irreducible decomposition for each A_k to decompose A into irreducible S_n-modules. The fact that l is an S_n-invariant ensures that the map

$$E : A \to A, \ x \mapsto lx$$

is an S_n-module homomorphism, i.e., $E(\sigma x) = \sigma E(x)$ for each $x \in A$. This guarantees that an irreducible component U in A_k is mapped by E either isomorphically to a subspace of A_{i+1} or onto $\{0\}$, since the kernel of $E|_U : U \to A_{k+1}$ is either $\{0\}$ or U. Let F be the operator constructed in Sect. 3.3.5 so that $\{E, F, [E, F]\}$ is an $\mathfrak{sl}_2$-triple. In addition to E, the operator F also commutes with the action of S_n; we prove this in the next proposition.

Proposition 9.4. *The operator* $F : A \to A$ *is an* S_n*-module homomorphism.*

Proof. Let $U \subset A_k$ be an irreducible S_n-module. Recall that A_k consists of weight vectors of the same weight. Then, since E is S_n-equivariant, the vector space $EU \subset A_{k+1}$ is isomorphic to U as an S_n-module unless it is $\{0\}$. Thus, by inductive reasoning, there is $r \geq 0$ such that $E^r U$ consists of highest weight vectors. Then by the strong Lefschetz property of A, there exists a vector space $U' \subset A_{k'}$ spanned by lowest weight vectors such that $E^s U' = E^r U$ for some $s \geq 0$. It follows that U' is an irreducible S_n-module isomorphic to U, since E^s is given by multiplication with the element l^s. Moreover we have $F^j E^r U = E^{s-j} U'$ for $0 \leq j \leq s$. In particular, $F^r E^r U = U = E^{s-r} U'$ and $FU = E^{s-r-1} U'$. This shows that $F|_U : U \to A_{k-1}$ commutes with the action of S_n. Since A_k is a direct sum of irreducible S_n-modules, the proof is complete.   □

We have shown that the operator F is an S_n-module homomorphism $F : A \to A$, hence $\ker F$ is an S_n-submodule. The successive images $E^k U$ of an irreducible component U of $\ker F$ are isomorphic to U as long as $E^k U \neq 0$. Since all the lowest weight vectors of A belong to $\ker F$, one can readily see that these nonzero successive images $\{ E^k U \}$ are the irreducible components of A:

$$A = \bigoplus_{U, k \geq 0} E^k U,$$

where the sum runs over all the irreducible components U of $\ker F$, and non-negative integers k satisfying $E^k U \neq 0$.

9.2 An Example

Proposition 9.5. *In the case where $d = 2$, the map $F : A \to A$ is given explicitly by $F = \partial_1 + \partial_2 + \cdots + \partial_n$, where ∂_k denotes the partial derivation with respect to the variable x_k $(k = 1, 2, \ldots, n)$.*

Proof. Note that A is isomorphic to the n-fold tensor product of the two-dimensional algebra $K[x]/(x^2)$. The vector space $K[x]/(x^2)$ is made into an $\mathfrak{sl}_2$-module with the three operators

$$\left\{ \times x, \; \frac{d}{dx}, \; x\frac{d}{dx} - \frac{d}{dx}x \right\}.$$

(Note that the third operator is the commutator of the first two.) Hence the assertion follows by Theorem 3.34.   □

Let $d = 2$ and $n = 3$. In this case, irreducible components U_λ of the S_3-module A correspond to $\lambda = (3)$ or $\lambda = (2, 1)$, since the corresponding partitions should be of length strictly less than three. The multiplicity of $U_{(3)}$ is $\Delta(4, 0)/\Delta(1, 0) = 4$, and $\Delta(3, 1)/\Delta(1, 0) = 2$ for $U_{(2,1)}$. The Hilbert series of A

is $(1+t)^3 = 1 + 3t + 3t^2 + t^3$. The corresponding partition μ depicting the sizes of Jordan blocks of E is given by $\mu = (4, 2, 2) = (2^2 4^1)$. As an $\mathfrak{sl}_2$-module, A has one irreducible submodule isomorphic to V_4, and two isomorphic to V_2, where V_i denotes the irreducible $\mathfrak{sl}_2$-module of dimension $i + 1$ as in Sect. 3.3.4. We employ again the symbol $v_k^{(i)}$ to describe a lowest weight vector of the $\mathfrak{sl}_2$-module A which generates an $(i + 1)$-dimensional irreducible component of A. The homogeneous component $A_0 = K$ is the trivial S_3-module with the basis $\{1\}$. Since l, l^2, l^3 do not vanish in A, the SLP guarantees that A_1, A_2, A_3 include a subspace isomorphic to the trivial module; $K(x_1 + x_2 + x_3)$, $K(x_1 x_2 + x_1 x_3 + x_2 x_3)$ and $Kx_1 x_2 x_3$, respectively. These are the consecutive non-zero images of $A_0 = K$ under the map E. The explicit image of $v_1^{(4)} = 1$ by E is $Ev_1^{(4)} = lv_1^{(4)} = x_1 + x_2 + x_3$. Similarly, $E^2 v_1^{(4)} = 2(x_1 x_2 + x_1 x_3 + x_2 x_3)$, $E^3 v_1^{(4)} = 6x_1 x_2 x_3$. The following figure describes the irreducible decomposition of A under the $\mathfrak{sl}_2$-action. The last row of length four depicts the four-dimensional irreducible $\mathfrak{sl}_2$-submodule with the basis consisting of the successive images of $v_1^{(4)} = 1 \in A_0$ by E.

$$v_1^{(2)} \overset{E}{\longmapsto} Ev_1^{(2)}$$
$$v_2^{(2)} \overset{E}{\longmapsto} Ev_2^{(2)}$$
$$v_1^{(4)} \overset{E}{\longmapsto} Ev_1^{(4)} \overset{E}{\longmapsto} E^2 v_1^{(4)} \overset{E}{\longmapsto} E^3 v_1^{(4)}$$
$$A_0 \qquad A_1 \qquad A_2 \qquad A_3$$

Obviously, $E^k v_1^{(4)} = l^k v_1^{(4)}$ belongs to A_k for each $k = 0, 1, 2, 3$. Regarding the $\mathfrak{sl}_2$-action, the lowest weight vectors, i.e., a basis of $\ker F$, are $v_1^{(4)}$, $v_1^{(2)}$ and $v_2^{(2)}$, and the last two vectors $v_1^{(2)}, v_2^{(2)}$ belong to A_1. These two vectors are expected to form a basis of an irreducible component isomorphic to $U_{(2,1)}$, and this will be confirmed soon after we introduce Specht polynomials. The action of $H = [E, F]$ is given by the diagonal matrices $\mathrm{diag}(-3, -1, 1, 3)$ and $\mathrm{diag}(-1, 1)$ with respect to the basis here, and the action of the operator F is given by

$$F(E^k v_j^{(i)}) = (i - k)k E^{k-1} v_j^{(i)}.$$

For example, we have $H(Ev_j^{(2)}) = E(v_j^{(2)})$ and $H(v_j^{(2)}) = -v_j^{(2)}$ for $j = 1, 2$; and $F(E^3 v_1^{(4)}) = 3E^2 v_1^{(4)}$, $F(E^2 v_1^{(4)}) = 4Ev_1^{(4)}$ and $F(Ev_1^{(4)}) = 3v_1^{(4)}$, and $F(Ev_j^{(2)}) = v_j^{(2)}$, $F(v_j^{(2)}) = 0$, etc. We can confirm these equalities coincide with a direct calculation: $F(E^3 v_1^{(4)}) = (\partial_1 + \partial_2 + \partial_3)6x_1 x_2 x_3 = 6(x_1 x_2 + x_1 x_3 + x_2 x_3) = 3E^2 v_1^{(4)}$, etc.

The homogeneous component A_1 contains a trivial S_3-module $Klv_1^{(4)} = K(x_1 + x_2 + x_3)$, hence the complementary submodule W in the irreducible decomposition of A_1 has dimension two. There exist two possibilities: either the complementary submodule is itself irreducible, or the direct sum of two one-dimensional irreducible submodules. In any case, the complementary submodule is mapped isomorphically by E, and the image EW is a two-dimensional submodule of A_2, irreducible or not according as W is irreducible or not. We can see that it is possible to obtain an irreducible decomposition of A together with a basis for each irreducible component, once we construct an irreducible decomposition of $\ker F$. In this example, we can actually construct these basis elements $v_1^{(2)}$ and $v_2^{(2)}$ as follows:

$$v_1^{(2)} = x_1 - x_3, \quad v_2^{(2)} = x_1 - x_2.$$

Obviously, $v_1^{(2)}$ and $v_2^{(2)}$ are linearly independent and contained in A_1. It is also easy to see that they belong to $\ker F$ by Proposition 9.5, and one can check directly that $E^2 v_1^{(2)} = E^2 v_2^{(2)} = 0$ in A. The polynomials $v_1^{(2)}$ and $v_2^{(2)}$ are known as Specht polynomials (see Sect. 9.3); they give an explicit basis for each irreducible S_n-module. In fact, the subspace of A_1 spanned by these two vectors affords the irreducible S_3-module corresponding to the partition $(2, 1)$.

9.3 Specht Polynomials

Let n be a positive integer, λ a partition of n. We have already remarked that the irreducible S_n-modules U are in one-to-one correspondence with the partitions λ of n, and the dimension of $U = U_\lambda$ coincides with the number of standard Young tableaux of shape λ. (See the paragraph preceding Proposition 9.1 and Remark 1.74.) The problem of concern to us here is a method of construction for basis elements for U_λ in the polynomial ring $R = K[x_1, x_2, \ldots, x_n]$, each of which can be constructed directly from a standard Young tableau T. Let t_{ij} be the integer assigned to the box (i, j) of T. The set of positive integers assigned to the j-th column of T is then given by $\{t_{1j}, t_{2j}, \ldots, t_{\lambda'_j j}\}$, where λ'_j is the size of the j-th column. Let $\Delta_T(j)$ denote the difference product of variables whose subscripts belong to the j-th column of T, and Δ_T the product of all $\Delta_T(j)$ $(j = 1, 2, \ldots, \lambda_1)$: Namely

$$\Delta_T = \prod_{j=1}^{\lambda_1} \Delta_T(j), \quad \Delta_T(j) = \Delta(x_{t_{1j}}, x_{t_{2j}}, \ldots, x_{t_{\lambda'_j j}}).$$

The polynomial Δ_T is called the **Specht polynomial** corresponding to T (Cf. [131]).

Proposition 9.6. *Let λ be a partition of n, $\mathrm{STab}(\lambda)$ the set of standard Young tableaux of shape λ. Then we have:*

- *The Specht polynomials $\{\Delta_T \mid T \in \mathrm{STab}(\lambda)\}$ are linearly independent.*
- *The linear subspace $\bigoplus_{T \in \mathrm{STab}(\lambda)} K\Delta_T$ affords the irreducible representation U_λ of S_n corresponding to λ.*

Let $n = 3$ and $\lambda = (2, 1)$. There exist two standard Young tableaux of shape λ:

$$T_1 = \begin{array}{|c|c|}\hline 1 & 2 \\\hline 3 \\\cline{1-1}\end{array}, \qquad T_2 = \begin{array}{|c|c|}\hline 1 & 3 \\\hline 2 \\\cline{1-1}\end{array}.$$

The corresponding Specht polynomials are $\Delta_{T_1} = x_1 - x_3$, $\Delta_{T_2} = x_1 - x_2$, which are obviously linearly independent, and the irreducible representation of S_3 corresponding to λ is realized on the subspace of $R = K[x_1, x_2, x_3]$ spanned by these two polynomials. A transposition $\sigma = (1, 2)$, for example, acts on the Specht polynomials by $\sigma\Delta_{T_1} = x_2 - x_3 = \Delta_{T_1} - \Delta_{T_2}$ and $\sigma\Delta_{T_2} = x_2 - x_1 = -\Delta_{T_2}$. The other Specht polynomials are 1 and $\Delta(x_1, x_2, x_3) = (x_1 - x_2)(x_1 - x_3)(x_2 - x_3)$. The polynomial 1 is a basis of the trivial representation that corresponds to the partition $\lambda = (3)$. The polynomial $\Delta(x_1, x_2, x_3)$ is a basis of the sign representation that corresponds to $\lambda = (1, 1, 1)$. It clearly follows from the definition that, if the length of the partition λ is not larger than d, then the degree of a Specht polynomial Δ_T ($T \in \mathrm{STab}(\lambda)$) with respect to each single variable is smaller than d, as one can see in the examples presented above. This shows, in general, that such Δ_T's do not vanish in the quotient ring $A = R/(x_1^d, x_2^d, \ldots, x_n^d)$ as long as $l(\lambda) \leq d$ for T. Hence, the space $\bigoplus_{T \in \mathrm{STab}(\lambda)} K\Delta_T$ affords the irreducible representation U_λ in the algebra A for any partition λ with length $l(\lambda) \leq d$.

In the case where $n = d = 3$, the Hilbert series of the algebra $A = \bigoplus_i A_i$ is given by $1 + 3t + 6t^2 + 7t^3 + 6t^4 + 3t^5 + t^6$. Since $d = 3$ is the maximum length of the partitions $\lambda \vdash 3$, every partition $\lambda \vdash 3$ can occur as a parameter of an irreducible component $U = U_\lambda$ of A. The multiplicity of U_λ equals $\dim W_\lambda$: $\dim W_{(3)} = 10, \dim W_{(2,1)} = 8, \dim W_{(1,1,1)} = 1$. This shows that there exist 10 irreducible components isomorphic to the trivial representation, and the same holds for the other irreducible components. Since the non-zero consecutive images $E^k U$ of an irreducible component U by $E \in \mathrm{End}(A)$ are mutually equivalent, the algebra A contains 10 copies of the trivial S_3-module. The following figure describes the $\mathfrak{sl}_2$-action on A, where each row depicts an irreducible component, and each column depicts a homogeneous component of A. The circles $\circ$ are the homogeneous basis elements of those components obtained by successive images of the lowest vectors $v_j^{(i)}$:

$$
\begin{array}{c}
v_1^{(1)} \\
v_1^{(3)} \overset{E}{\mapsto} \circ \overset{E}{\mapsto} \circ \\
v_2^{(3)} \overset{E}{\mapsto} \circ \overset{E}{\mapsto} \circ \\
v_3^{(3)} \overset{E}{\mapsto} \circ \overset{E}{\mapsto} \circ \\
v_1^{(5)} \overset{E}{\mapsto} \circ \overset{E}{\mapsto} \circ \overset{E}{\mapsto} \circ \overset{E}{\mapsto} \circ \\
v_2^{(5)} \overset{E}{\mapsto} \circ \overset{E}{\mapsto} \circ \overset{E}{\mapsto} \circ \overset{E}{\mapsto} \circ \\
v_1^{(7)} \overset{E}{\mapsto} \circ \overset{E}{\mapsto} \circ \overset{E}{\mapsto} \circ \overset{E}{\mapsto} \circ \overset{E}{\mapsto} \circ \overset{E}{\mapsto} \circ \\
A_0 \quad A_1 \quad A_2 \quad A_3 \quad A_4 \quad A_5 \quad A_6
\end{array}
\tag{9.2}
$$

The preceding observation shows that, among the lowest vectors, $v_1^{(7)}$ and one of the $v_k^{(3)}$'s are S_3-invariant, that is, they afford the trivial representation $U_{(3)}$. Let $v_1^{(3)}$ be a unique vector among $v_k^{(3)}$'s affording the trivial representation. One can also see that $\{\, v_1^{(1)} \,\}$ is a basis of the sign representation $U_{(1^3)}$. On the other hand, $\{\, v_2^{(3)}, v_3^{(3)} \,\}$ and $\{\, v_1^{(5)}, v_2^{(5)} \,\}$ span two-dimensional irreducible submodules isomorphic to $U_{(2,1)}$. These basis vectors of degree 0, 1 and 3 can be realized by Specht polynomials, but those of degree 2 can not. However, in the case where $d = 2$, we can show that all the lowest vectors can be realized by Specht polynomials using the following proposition. Consult [103] for a proof of this proposition.

Proposition 9.7. *In the case* $d = 2$ *for* $F = \partial_1 + \partial_2 + \cdots + \partial_n$, *one has that* $\ker F = K[x_1 - x_i | i = 2, 3, \ldots, n]$.

Since any Specht polynomial is a product of bilinear forms $x_i - x_j$, it is clear that it belongs to $\ker F$. Since the dimension of homogeneous components A_k is given by the binomial coefficient $\binom{n}{k}$, the following lemma shows that in the case where $d = 2$, a basis of $\ker F$ consists of Specht polynomials.

Lemma 9.8. *Let* $\lambda = (n - k, k)$ *be a partition with the length not larger than two, where* $0 \leq k \leq [n/2]$. *Then the number of standard Young tableaux is given by the difference* $\binom{n}{k} - \binom{n}{k-1}$ *of binomial coefficients.*

Proof. Use induction on n.   □

Theorem 9.9 ([103]). *The subspace* $\ker F$ *of the* $\mathfrak{sl}_2$-*module*

$$A = K[x_1, x_2, \ldots, x_n]/(x_1^2, x_2^2, \ldots, x_n^2)$$

is spanned by the Specht polynomials corresponding to the partitions λ *with* $l(\lambda) \leq 2$.

Proof. Specht polynomials defined for partitions of length less than or equal to two belong to $\ker F$ by Proposition 9.7, and are linearly independent by Proposition 9.6. Thus they form a basis of $\ker F$ by Lemma 9.8.   □

Therefore, the non-zero consecutive images of the Specht polynomials with respect to the map E form a basis of the whole algebra A thanks to the SLP. As we can see from the example for $d = n = 3$, the Specht polynomials contained in $\mathrm{Ker}(F)$ do not exhaust basis elements of $\mathrm{Ker}(F)$ for $d \geq 3$.

We can construct highest weight vectors explicitly in the case where the number of variables is two (the power d may be arbitrary). In this case, one can easily construct the highest weight vectors, and the remaining basis vectors are obtained by applying $F \in \mathrm{End}(A)$ that represents $f \in \mathfrak{sl}_2$. Let x and y be indeterminates, and $A = K[x, y]/(x^d, y^d)$ the quotient of the polynomial algebra by the ideal (x^d, y^d). It follows from Theorem 9.3 that the algebra A has the SLP with the Lefschetz element $l = x + y$. The Hilbert series of A is $(1 + t + \cdots + t^{d-1})^2$. Thanks to the SLP, it obviously follows that there exist d highest weight vectors that belong to the corresponding homogeneous spaces of degree $d - 1, d, d + 1, \ldots, 2(d - 1)$. For each positive integer $k = d - 1, d, d + 1, \ldots, 2(d - 1)$, let φ_k be the image of the polynomial $\sum_{i=0}^{k} (-1)^i x^i y^{d-i}$ in the algebra A. If $d = 3$ for example, the algebra A has three (up to scalar multiple) highest weight vectors $\varphi_2 = x^2 - xy + y^2$, $\varphi_3 = -x^2 y + xy^2$, $\varphi_4 = x^2 y^2$. Then we have the following theorem.

Theorem 9.10 ([103]). *The elements φ_k $(k = d - 1, d, d + 1, \ldots, 2(d - 1))$ are the highest weight vectors of $A = K[x, y]/(x^d, y^d)$.*

9.4 Irreducible Decomposition of $\bigwedge \Omega$

Let $R = K[x_1, x_2, \ldots, x_n]$ be the polynomial ring over a field K, Char $K = 0$. Let Ω be the module of differentials:

$$\Omega := R dx_1 \oplus R dx_2 \oplus \cdots \oplus R dx_n.$$

Let $G := S_n$ be the symmetric group in n letters, so G acts on R by permutation of the variables. Extend the action to $\bigwedge^j \Omega$ for $j = 0, 1, 2, \ldots, n$. We consider the following problems:

Problem 9.11. Decompose $\bigwedge^j \Omega$ into irreducible S_n-modules.

Problem 9.12. Determine the Hilbert series of the isotypic components of $\bigwedge^j \Omega$.

Recall that the irreducible modules of S_n are parameterized by the partitions of n. Thus we write U_λ for the (isomorphism type of) irreducible S_n-module corresponding to $\lambda \vdash n$. Let $Y^\lambda(-)$ be the functor from the category of S_n-modules to itself that extracts the isotypic component belonging to λ. Schur's Lemma tells us that Y^λ can be taken to be the functor

$$Y^\lambda(-) = U_\lambda \otimes_{\text{End}_{S_n}(U_\lambda)} \text{Hom}_{S_n}(U_\lambda, -).$$

Note that it is an exact functor. For a graded vector space M, we write

$$\text{Hilb}(M, q)$$

for the Hilbert series of M. This is a power series in q with positive integers as coefficients.

Before we proceed to a detail analysis of $\bigwedge \Omega$ as an S_n-module we examine two easy cases, where (1) $n = 2$ and $j = 0, 1, 2$, and (2) an arbitrary n and $j = 0$.

Table 9.1 Hilbert series for
$Y^\lambda(\bigwedge^j \Omega)$

	$j = 0$	$j = 1$	$j = 2$
$\lambda = (2)$	$\frac{1}{(1-q)(1-q^2)}$	$\frac{1}{(1-q)^2}$	$\frac{q}{(1-q)(1-q^2)}$
$\lambda = (11)$	$\frac{q}{(1-q)(1-q^2)}$	$\frac{1}{(1-q)^2}$	$\frac{1}{(1-q)(1-q^2)}$

9.4.1 Examples

Example 9.13. Assume $n = 2$. Then $R = K[x, y]$ and $G = S_2$. There are only two partitions: $\lambda = (2)$ and $\lambda = (11)$. The irreducible representation corresponding to $\lambda = (2)$ is the trivial representation and $\lambda = (11)$ the sign representation. We want to determine

$$\mathrm{Hilb}(Y^\lambda(\overset{j}{\bigwedge} \Omega), q)$$

for $j = 0, 1, 2$. For $j = 0$, it is easy to determine the Hilbert series since

$$R^G = K[x + y, xy]$$

and R is the direct sum of symmetric polynomials and the alternating polynomials. The module of alternating polynomials being written as

$$(x - y)R^G.$$

For $j = 2$, we have

$$\overset{2}{\bigwedge} \Omega \cong R(dx \wedge dy).$$

Thus we have

$$\begin{cases} \mathrm{Hilb}(Y^{(2)}(\bigwedge^2 \Omega), q) = \mathrm{Hilb}(Y^{(11)}(R), q), \\ \mathrm{Hilb}(Y^{(11)}(\bigwedge^2 \Omega), q) = \mathrm{Hilb}(Y^{(2)}(R), q). \end{cases}$$

For $j = 1$, we have to decompose

$$\Omega \cong Rdx \oplus Rdy.$$

into irreducible S_2-modules. As is easily seen, symmetric 1-forms are of the form either $sdx + sdy$ with $s \in R^G$ or $adx - ady$ with $a \in (x - y)R^G$, and alternating 1-forms are either $sdx - sdy$ or $adx + ady$.

Thus we have Table 9.1 for $\mathrm{Hilb}(Y^\lambda(\bigwedge^j \Omega), q)$.

Example 9.14. Fix $j = 0$. So $\bigwedge^j \Omega = R$. If $\lambda = (n)$, the trivial partition, then $Y^\lambda(R)$ is R^G, the ring of invariants. So we have

$$\mathrm{Hilb}(R^G, q) = \frac{1}{(1 - q)(1 - q^2) \cdots (1 - q^n)},$$

since the algebra R^G is generated by the elementary symmetric polynomials.

Put $R_G = R/(R_+^G)$, where (R_+^G) is the ideal generated by the elementary symmetric polynomials. Then it is easy to see that

$$\mathrm{Hilb}(Y^\lambda(R), q) = \mathrm{Hilb}(R^G, q) \, \mathrm{Hilb}(Y^\lambda(R_G), q)$$

$$= \frac{\mathrm{Hilb}(Y^\lambda(R_G), q)}{(1 - q)(1 - q^2) \cdots (1 - q^n)}.$$

The numerator $\mathrm{Hilb}(Y^\lambda(R_G), q)$, which is a polynomial, is known to be the q-analog of the hook length formula multiplied by the dimension of U_λ with a certain shift of degree determined by λ. For details see Sect. 9.4.4.

9.4.2 General Case

We write

$$\Omega(-d) = R dx_1 \oplus R dx_2 \oplus \cdots \oplus R dx_n,$$

where we give dx_i degree d. The module Ω we have been considering up to this point is denoted $\Omega = \Omega(0)$ in this notation. Similarly $R(-d)$ denotes the free module of rank one generated by an element of degree d. Thus

$$\bigwedge^j (\Omega(-d)) \cong R^{\binom{n}{j}}(-jd)$$

$$\cong \left(\bigwedge^j \Omega \right)(-jd).$$

Also put

$$A(d) = R/(x_1^d, x_2^d, \ldots, x_n^d).$$

For simplicity put $F = \Omega(-d)$. We would like to construct a minimal free resolution of $A(d)$ as an R-module such that the boundary maps are compatible with the action of S_n. For this the usual minimal free resolution that is the Koszul resolution suffices:

$$0 \to \overset{n}{\bigwedge} F \to \overset{n-1}{\bigwedge} F \to \cdots \to \overset{2}{\bigwedge} F \to \overset{1}{\bigwedge} F \to \overset{0}{\bigwedge} F \to A(d) \to 0 \qquad (9.3)$$

We want to compute

$$\xi_j := \mathrm{Hilb}(Y^\lambda(\overset{j}{\bigwedge} \Omega), q).$$

Put

$$h_d := \mathrm{Hilb}(Y^\lambda(A(d)), q),$$

for $d = 0, 1, 2, \ldots$. Fix $\lambda \vdash n$ and apply the functor $Y^\lambda(-)$ to the sequence (9.3) above. Then we have

$$0 \to Y^\lambda(\overset{n}{\bigwedge} F) \to Y^\lambda(\overset{n-1}{\bigwedge} F) \to \cdots \to Y^\lambda(\overset{2}{\bigwedge} F) \to Y^\lambda(\overset{1}{\bigwedge} F) \to Y^\lambda(\overset{0}{\bigwedge} F)$$
$$\to Y^\lambda(A(d)) \to 0.$$

Since the sequence is exact it gives us:

$$h_d = \sum_{j=0}^{n}(-1)^j q^{jd}\xi_j \qquad (9.4)$$

Thus we have an infinite set of linear equations relating $\{\xi_i\}$ and $\{h_i\}$. For example if $n = 3$, we have

$$\begin{pmatrix} 1 & -1 & 1 & -1 \\ 1 & -q & q^2 & -q^3 \\ 1 & -q^2 & q^4 & -q^6 \\ 1 & -q^3 & q^6 & -q^9 \\ 1 & -q^4 & q^8 & -q^{12} \\ \vdots & \vdots & \vdots & \vdots \end{pmatrix} \begin{pmatrix} \xi_0 \\ \xi_1 \\ \xi_2 \\ \xi_3 \end{pmatrix} = \begin{pmatrix} h_0 \\ h_1 \\ h_2 \\ h_3 \\ h_4 \\ \vdots \end{pmatrix}$$

Theorem 9.15. *With $\lambda \vdash n$ fixed, any $(n + 1)$ of*

$$h_0, h_1, h_2, \ldots$$

determine

$$\xi_0, \xi_1, \xi_2, \ldots, \xi_n.$$

Proof. No $(n + 1) \times (n + 1)$ minor of the matrix

$$\left((-1)^j q^{jd}\right)_{j=0,1,2,\ldots,n;\, d=0,1,2,\ldots}$$

vanishes. Hence the assertion follows from (9.4).   □

Corollary 9.16. *The infinite sequence*

$$h_0, h_1, h_2, \ldots.$$

is determined by any $n+1$ terms.

Remark 9.17. A result of Morita–Wachi–Watanabe (see [103] which we discuss in the next section) says that

$$\mathrm{Hilb}(Y^\lambda(A(d)), q)$$

is the q-analog of the Weyl dimension formula. This means that we have determined

$$\mathrm{Hilb}(Y^\lambda(\overset{j}{\bigwedge} \Omega), q)$$

for all $\lambda \vdash n$, $j = 0, 1, 2, \ldots, n$.

9.4.3 The q-Analog of the Weyl Dimension Formula

We keep the notation of the previous section, so

$$A(d) = K[x_1, x_2, \ldots, x_n]/(x_1^d, x_2^d, \ldots, x_n^d).$$

Note that the set of monomials of degrees in the individual variables at most $(d-1)$:

$$\left\{ x_1^{i_1} x_2^{i_2} \cdots x_n^{i_n} \mid 0 \le i_1, i_2, \ldots, i_n \le d-1 \right\}$$

is a basis of $A(d)$.

The Hilbert series

$$\mathrm{Hilb}(Y^\lambda(A(d)), q)$$

of the module $Y^\lambda(A(d))$ is in fact a polynomial, since $A(d)$ is Artinian. It is obtained as the q-analog of the Weyl dimension formula multiplied by $\dim U_\lambda$ with a certain shift of degrees.

First we explain the Weyl dimension formula. (In the next subsection we explain the hook length formula.)

Let $(K^d)^{\otimes n}$ be the n-fold tensor space of the d-dimensional vector space K^d. Let $\{ v_0, v_1, \ldots, v_{d-1} \}$ be a basis for K^d. Then $(K^d)^{\otimes n}$ is spanned by the set

$$\{ v_{i_1} \otimes v_{i_2} \otimes \cdots \otimes v_{i_n} \mid 0 \le i_1, i_2, \ldots, i_n \le d - 1 \}.$$

The general linear group $GL(d, K)$ acts on the tensor space $(K^d)^{\otimes n}$ by

$$v_{i_1} \otimes v_{i_2} \otimes \cdots \otimes v_{i_n} \mapsto (v_{i_1} \otimes v_{i_2} \otimes \cdots \otimes v_{i_n})^g := v_{i_1}^g \otimes v_{i_2}^g \otimes \cdots \otimes v_{i_n}^g$$

for $g \in GL(d, K)$. Since we are identifying the spaces

$$(K^d)^{\otimes n} \xrightarrow{\sim} A(d)$$

via the correspondence of the basis elements

$$v_{i_1} \otimes v_{i_2} \otimes \cdots \otimes v_{i_n} \mapsto x_1^{i_1} x_2^{i_2} \cdots x_n^{i_n},$$

the above action of $g \in GL(d, K)$ gives us the representation

$$\Phi : GL(d, K) \to GL(A(d))$$

$$\Phi(g)(x_1^{i_1} x_2^{i_2} \cdots x_n^{i_n}) = (\sum_{\beta=0}^{d-1} g_{i_1 \beta} x_1^\beta)(\sum_{\beta=0}^{d-1} g_{i_2 \beta} x_2^\beta) \cdots (\sum_{\beta=0}^{d-1} g_{i_n \beta} x_n^\beta)$$

for $g = (g_{\alpha\beta}) \in GL(d, K)$. Here the indices α, β of the matrix entries for $g = (g_{\alpha\beta}) \in GL(d, K)$ range over $0, 1, 2, \ldots, n - 1$. In addition $A(d)$ is a $GL(d, K)$-module and the symmetric group S_n acts on $A(d)$ as the permutation of the variables. We are interested in the decomposition of $A(d)$ into irreducible S_n-modules. According to the Schur–Weyl duality this will give us an irreducible decomposition of $A(d)$ as $(GL(d, K) \times S_n)$-modules.

We take it for granted that the tensor space $(K^d)^{\otimes n}$ decomposes into irreducible $GL(d, K)$-modules, and these modules are parameterized by the partitions of n with at most d terms.

Let V_λ be the irreducible $GL(d, K)$-module corresponding to $\lambda \vdash n$ with at most d terms. Then the Weyl dimension formula says that

$$\dim V_\lambda = \prod_{1 \le i < j \le d} \frac{\lambda_i - \lambda_j + j - i}{j - i}, \tag{9.5}$$

where

$$\lambda = (\lambda_1, \lambda_2, \ldots, \lambda_d) \vdash n.$$

(If λ has more than d parts, we let $\dim V_\lambda = 0$.)

In Morita–Wachi–Watanabe [103], the Hilbert function of $Y^\lambda(A(d))$ is obtained in terms of the q-analog of the Weyl dimension formula (e.g., [40] Theorem 7.1.7). For a non-negative integer a, the q-integer $[a]_q$ is defined by

$$[a]_q = \frac{1 - q^a}{1 - q}$$

$$= 1 + q + \cdots + q^{a-1}.$$

If we replace every integer by $[a]_q$ on the right-hand side of (9.5), the resulting formula is called the q-analog of the Weyl dimension formula. Since there are the same number of integers in the denominator and enumerator of (9.5), it is equivalent if we replace a by $1 - q^a$.

With these preliminaries we may describe the Hilbert function for $Y^\lambda(A(d))$.

Proposition 9.18 (Morita–Wachi–Watanabe [103]). *For $\lambda = (\lambda_1, \lambda_2, \ldots, \lambda_d) \vdash n$ let V be an irreducible $GL(d, K)$-submodule of $A(d)$ isomorphic to the Weyl module V_λ. Then its Hilbert series is independent of V, and is given by*

$$\mathrm{Hilb}(V, q) = q^{\lambda_2 + 2\lambda_3 + \cdots + (d-1)\lambda_d} \prod_{1 \le i < j \le d} \frac{[\lambda_i - \lambda_j + j - i]_q}{[j - i]_q}.$$

Here $[a]_q$ denotes $\dfrac{1 - q^a}{1 - q}$. In particular all V isomorphic to V_λ occur in $A(d)$ with the same shift of degrees: $\lambda_2 + 2\lambda_3 + \cdots + (d - 1)\lambda_d$.

Remark 9.19. In the notation of Proposition 9.18, it is remarkable that all V isomorphic to V_λ occur in $A(d)$ with the same shift of degrees: $\lambda_2 + 2\lambda_3 + \cdots + (d - 1)\lambda_d$, which is not a priori obvious. It is forced by the fact that $Y^\lambda(A(d))$ has the SLP. (Cf. Proposition 9.27).

To describe the Hilbert series of $Y^\lambda(A(d))$, we only need to know that

$$Y^\lambda(A(d)) \cong V_\lambda \otimes U_\lambda.$$

Here U_λ is the irreducible S_n-submodule corresponding to λ. An explicit basis for U_λ is well known and with this basis U_λ is usually called the **Specht module**. We give more detail for it in Sect. 9.4.4.

Theorem 9.20. *For $\lambda = (\lambda_1, \lambda_2, \ldots, \lambda_r) \vdash n$, the Hilbert series of $Y^\lambda(A(d))$ is*

$$\mathrm{Hilb}(Y^\lambda(A(d)), q) = q^{\lambda_2 + 2\lambda_3 + \cdots + (r-1)\lambda_r} \prod_{1 \le i < j \le d} \frac{[\lambda_i - \lambda_j + j - i]_q}{[j - i]_q} \dim U_\lambda,$$

where U_λ is the Specht module for λ.

Proof. By the Schur–Weyl duality Theorem, the isotypic component $Y^\lambda(A(d))$ in $A(d)$ corresponding to the Specht module U_λ is also the isotypic component in $A(d)$ corresponding to the Weyl module V_λ. Namely $Y^\lambda(A(d)) \cong V_\lambda \otimes U_\lambda$, as indicated above. Hence the multiplicity of V_λ in this isotypic component is equal to dim U_λ. One last thing to be noticed is that U_λ occuring in $Y^\lambda(A(d))$ is concentrated in one graded piece. Thus the assertion follows. $\square$

Example 9.21. Let $n = 3$. In this case $A(0) = 0$, $A(1) \cong K$, $A(2) \cong K[x,y,z]/(x^2,y^2,z^2)$ and $A(3) \cong K[x,y,z]/(x^3,y^3,z^3)$. There are three irreducible S_3-modules, $U_{(3)}$, $U_{(2,1)}$, and $U_{(1,1,1)}$. $U_{(3)}$ is the trivial module and dim $U_{(3)} = 1$, and $U_{(1,1,1)}$ is the module for the sign representation and dim $U_{(1,1,1)} = 1$. For $\lambda = (2,1)$ we have dim $U_\lambda = 2$. Without using the q-analog of the Weyl dimension formula, it is easy to compute $\mathrm{Hilb}(Y^\lambda(A(d)), q)$ for $d = 0,1,2,3$. (Cf. (9.2) in Sect. 9.3) Note that $U_{(1,1,1)}$ occurs only once in $A(3)$ at degree three, with a single basis element $(x-y)(x-z)(y-z)$. Thus it is possible to determine the series $\xi_0, \xi_1, \xi_2, \xi_3$ for the modules $Y^\lambda(\bigwedge^j \Omega)$ over $R = K[x,y,z]$ for each $\lambda \vdash 3$. We show the results of the computations below.

$\lambda = (3), d = 0,1,2,3$

$$\mathrm{Hilb}(Y^\lambda(A(0)), q) = 0$$

$$\mathrm{Hilb}(Y^\lambda(A(1)), q) = 1$$

$$\mathrm{Hilb}(Y^\lambda(A(2)), q) = 1 + q + q^2 + q^3$$

$$\mathrm{Hilb}(Y^\lambda(A(3)), q) = 1 + q + 2q^2 + 2q^3 + 2q^4 + q^5 + q^6.$$

$\lambda = (2,1), d = 0,1,2,3$

$$\mathrm{Hilb}(Y^\lambda(A(0)), q) = 0$$

$$\mathrm{Hilb}(Y^\lambda(A(1)), q) = 0$$

$$\mathrm{Hilb}(Y^\lambda(A(2)), q) = 2(q + q^2)$$

$$\mathrm{Hilb}(Y^\lambda(A(3)), q) = 2(q + 2q^2 + 2q^3 + 2q^4 + q^5).$$

$\lambda = (1,1,1), d = 0,1,2,3$

$$\mathrm{Hilb}(Y^\lambda(A(0)), q) = 0$$

$$\mathrm{Hilb}(Y^\lambda(A(1)), q) = 0$$

$$\mathrm{Hilb}(Y^\lambda(A(2)), q) = 0$$

$$\mathrm{Hilb}(Y^\lambda(A(3)), q) = q^3.$$

Put

$$M = \begin{pmatrix} 1 & -q^0 & q^0 & -q^0 \\ 1 & -q^1 & q^2 & -q^3 \\ 1 & -q^2 & q^4 & -q^6 \\ 1 & -q^3 & q^6 & -q^9 \end{pmatrix}.$$

For $\lambda = (3)$, we compute:

$$M^{-1} \begin{pmatrix} \mathrm{Hilb}(Y^\lambda(A(0)), q) \\ \mathrm{Hilb}(Y^\lambda(A(1)), q) \\ \mathrm{Hilb}(Y^\lambda(A(2)), q) \\ \mathrm{Hilb}(Y^\lambda(A(3)), q) \end{pmatrix} = M^{-1} \begin{pmatrix} 0 \\ 1 \\ 1 + q + q^2 + q^3 \\ 1 + q + 2q^2 + 2q^3 + 2q^3 + q^5 + q^6 \end{pmatrix}.$$

This gives us

$$\xi_0 = \frac{1}{(1-q)^3(1 + 2q + 2q^2 + q^3)},$$

$$\xi_1 = \frac{1}{(1-q)^3(1+q)},$$

$$\xi_2 = \frac{q}{(1-q)^3(1+q)},$$

$$\xi_3 = \frac{q^3}{(1-q)^3(1 + 2q + 2q^2 + q^3)}.$$

For $\lambda = (2, 1)$, we compute:

$$M^{-1} \begin{pmatrix} \mathrm{Hilb}(Y^\lambda(A(0)), q) \\ \mathrm{Hilb}(Y^\lambda(A(1)), q) \\ \mathrm{Hilb}(Y^\lambda(A(2)), q) \\ \mathrm{Hilb}(Y^\lambda(A(3)), q) \end{pmatrix} = M^{-1} \begin{pmatrix} 0 \\ 0 \\ 2q + 2q^2 \\ 2q + 4q^2 + 4q^3 + 4q^4 + 2q^5 \end{pmatrix}.$$

This gives us

$$\xi_0 = \frac{2q}{(1-q)^3(1 + q + q^2)},$$

$$\xi_1 = \frac{2}{(1-q)^3},$$

$$\xi_2 = \frac{2}{(1-q)^3},$$

$$\xi_2 = \frac{2q}{(1-q)^3(1 + q + q^2)}.$$

For $\lambda = (1, 1, 1)$, we compute:

$$M^{-1} \begin{pmatrix} \text{Hilb}(Y^\lambda(A(0)), q) \\ \text{Hilb}(Y^\lambda(A(1)), q) \\ \text{Hilb}(Y^\lambda(A(2)), q) \\ \text{Hilb}(Y^\lambda(A(3)), q) \end{pmatrix} = M^{-1} \begin{pmatrix} 0 \\ 0 \\ 0 \\ q^3 \end{pmatrix}.$$

This gives us

$$\xi_0 = \frac{q^3}{(1-q)^3(1+2q+2q^2+q^3)},$$

$$\xi_1 = \frac{q}{(1-q)^3(1+q)},$$

$$\xi_2 = \frac{1}{(1-q)^3(1+q)},$$

$$\xi_3 = \frac{1}{(1-q)^3(1+2q+2q^2+q^3)}.$$

In the computation we have used $m_{(3)} = m_{(1,1,1)} = 1$ and $m_{(2,1)} = 2$. These weights are the multiplicities of the irreducible components of U_λ in $R/(R_+^G)$. Generally these weights can be computed by the hook length formula.

One may wish to verify the equality

$$\sum_\lambda \text{Hilb}(Y^\lambda(A(d)), q) = (1 + q + q^2 + \cdots + q^{d-1})^3,$$

for $d = 2, 3$.

Example 9.22. We show how Proposition 9.18 is used to get $\text{Hilb}(Y^\lambda(A(4)), q)$ for $n = 3$.

First recall the Weyl dimension formula (9.5)

$$\dim V_\lambda = \prod_{1 \le i < j \le d} \frac{\lambda_i - \lambda_j + j - i}{j - i},$$

where

$$\lambda = (\lambda_1, \lambda_2, \ldots, \lambda_d) \vdash n.$$

We apply it with $d = 4$. For all λ the denominator is the difference product of the vector

$$(4, 3, 2, 1).$$

The numerator is the difference product of the vector

$$(4, 3, 2, 1) + \lambda \text{ (appended with } 0).$$

Replacing every integer a by $[a]_q$ and making an appropriate shift of degrees we obtain the following:

$$\text{Hilb}(Y^{(3)}(A(4)), q) = \frac{[6]_q \, [5]_q \, [4]_q}{[3]_q \, [2]_q \, [1]_q} = (1 + q^2)(1 + q^3)(1 + q + q^2 + q^3 + q^4)$$

$$\text{Hilb}(Y^{(2,1)}(A(4)), q) = 2q \frac{[5]_q \, [4]_q \, [2]_q}{[1]_q \, [2]_q \, [1]_q} = 2q(1 + q + q^2 + q^3)(1 + q + q^2 + q^3 + q^4)$$

$$\text{Hilb}(Y^{(1,1,1)}(A(4)), q) = q^3 \frac{[4]_q \, [2]_q}{[2]_q \, [1]_q} = q^3(1 + q + q^2 + q^3)$$

9.4.4 The q-Analog of the Hook Length Formula

As is often customary the partitions of n and the Young diagrams of n boxes are identified. For example the partition

$$10 = 5 + 4 + 1$$

is regarded as the same as the Young diagram below.

The (i, j)-hook of a Young diagram is the set of boxes consisting of the box at the (i, j)-th position together with the boxes on the same row to the right of it and with the boxes in the same column that are below it. For example the $(1, 2)$-hook of the above diagram is the set of the boxes containing a $\bullet$.

The hook length is the number of boxes in a hook. Let h_{ij} be the hook length of the (i, j)-hook. We can show these numbers as part of the Young diagram by filling in the boxes of the diagram with these numbers. The following shows the hook lengths of the Young diagram $10 = 5 + 4 + 1$.

7	5	4	3	1
5	3	2	1	
1				

Let S_n be the group of permutations of n letters. It is well-known that there exists a one-to-one correspondence between the set of irreducible S_n-modules and the set of partitions of the integer n. We denote by U_λ the irreducible S_n-module corresponding to the partition $\lambda \vdash n$. What follows is the **hook length formula** which describes the dimension of the module U_λ.

Proposition 9.23. *Denote by h_{ij} the length of the (i, j)-hook of the Young diagram corresponding to λ. Then*

$$\dim(U_\lambda) = \frac{n!}{\prod h_{ij}}.$$

Here the indices i, j run through all boxes in the Young diagram.

Example 9.24. $\dim(U_{(541)}) = \dfrac{10!}{7 \cdot 5 \cdot 4 \cdot 3 \cdot 1 \cdot 5 \cdot 3 \cdot 2 \cdot 1 \cdot 1} = 72$

To construct an explicit representation space, one uses Specht polynomials. Then the dimension formula can be obtained by counting the number of Specht polynomials.

Let $R = K[x_1, x_2, \ldots, x_n]$ be the polynomial ring over a field K of characteristic zero and let $(R_+^G) = (e_1, e_2, \ldots, e_n)$ be the ideal of R generated by the elementary symmetric polynomials. We define

$$R_G := R/(R_+^G).$$

The permutation group S_n acts on R_G by permutation of the variables. Thus R_G is an S_n-module. The decomposition of R_G into irreducible S_n-modules is well-known. Surprisingly enough it turns out to be the q-analog of the hook length formula for λ, which we state in the next theorem. One may picture the decomposition of R_G as similar to the decomposition of the group ring $K[S_n]$. Maschke's theorem says that every irreducible module U_λ occurs in $K[S_n]$ as many times as its dimension. It is a consequence of the fact that a semi-simple ring is a direct product of the matrix rings (provided that K is algebraically closed and Char K equals 0 or does not divide $n!$).

Remark 9.25. The decomposition of R_G into irreducible S_n-modules has a long and twisted history. See Humphreys [61, §3.6].

Theorem 9.26.

$$\mathrm{Hilb}(Y^\lambda(R_G), q) = q^{\lambda_2 + 2\lambda_3 + \cdots + (r-1)\lambda_r} \frac{[n]_q!}{\prod_{i,j} [h_{ij}]_q} \dim U_\lambda$$

$$= q^{\lambda_2 + 2\lambda_3 + \cdots + (r-1)\lambda_r} \frac{[n]_q!}{\prod_{i,j} [h_{ij}]_q} \frac{n!}{\prod_{i,j} h_{ij}}.$$

Proof. The multiplicity $m_{\lambda,i}$ of the irreducible S_n-module U_λ in the homogeneous component $(R_G)_i$ of degree i is given in Terasoma–Yamada [142]. It is determined by the formula

$$\sum_i m_{\lambda,i} q^i = q^{\lambda_2 + 2\lambda_3 + \cdots + (r-1)\lambda_r} \frac{[n]_q!}{\prod_{i,j} [h_{ij}]_q}$$

for a partition $\lambda = (\lambda_1, \lambda_2, \ldots, \lambda_r)$ of n. Thus the claimed equality follows. □

Remark 9.27. 1. $\mathrm{Hilb}(Y^\lambda(A(d)), q) = q^{\lambda_2 + 2\lambda_3 + \cdots + (r-1)\lambda_r} \prod_{1 \le i < j \le d} \frac{[\lambda_i - \lambda_j + j - i]_q}{[j-i]_q}$
is symmetric unimodal and the graded vector space $Y^\lambda(A(d))$ has the SLP for every λ.

2. $\mathrm{Hilb}(Y^\lambda(R_G), q) = q^{\lambda_2 + 2\lambda_3 + \cdots + (r-1)\lambda_r} \frac{[n]_q!}{\prod_{i,j} [h_{ij}]_q}$ is symmetric, but not necessarily unimodal. The graded vector space $Y^\lambda(R_G)$ does not necessarily have the SLP.

Proof. 1. Recall that we have set $A(d) = K[x_1, x_2, \ldots, x_n]/(x_1^d, x_2^d, \ldots, x_n^d)$, which has the SLP with a strong Lefschetz element $x_1 + x_2 + \cdots + x_n$. It is easy to notice that $Y^\lambda(A(d))$ has a symmetric Hilbert function whose reflecting degree is the same as that of $A(d)$. Moreover the multiplication map

$$\times (x_1 + x_2 + \cdots + x_n): A(d) \to A(d)$$

restricts to give a map of the graded vector subspace

$$\times (x_1 + x_2 + \cdots + x_n): Y^\lambda(A(d)) \to Y^\lambda(A(d)).$$

This shows that $Y^\lambda(A(d))$ has the strong Lefschetz property.
2. It is easy to notice that $Y^\lambda(R_G)$ has a symmetric Hilbert function with the same reflecting degree as that of $K[x_1, x_2, \ldots, x_n]/(e_1, e_2, \ldots, e_n)$. Consider the special case for R_G, where $R_G = K[x_1, x_2, x_3, x_4]/(e_1, e_2, e_3, e_4)$. This gives us a non-unimodal Hilbert function:

$$\mathrm{Hilb}(Y^{(22)}(R_G), q) = 2q^2 + 2q^4. \qquad \qquad □$$

9.5   The Homomorphism $\Phi : GL(V) \to GL(V^{\otimes n})$

In this section we give another application of the strong Lefschetz property of monomial complete intersections to representation theory. We assume that K is an algebraically closed field of characteristic zero.

By $J(a,d)$ we denote the $d \times d$ matrix

$$
J(a,d) = \begin{bmatrix} a & & & 0 \\ 1 & a & & \\ & \ddots & \ddots & \\ 0 & & 1 & a \end{bmatrix}.
$$

It is an elementary fact that if a matrix M has a single eigenvalue a, then, by conjugation, M decomposes into a direct sum of blocks as follows:

$$
P^{-1}MP = J(a,d_1) \dotplus J(a,d_2) \dotplus \cdots \dotplus J(a,d_r)
$$

$$
:= \begin{bmatrix} J(a,d_1) & & & 0 \\ & J(a,d_2) & & \\ & & \ddots & \\ 0 & & & J(a,d_r) \end{bmatrix},
$$

where $d_1 + d_2 + \cdots + d_r = d$. The decomposition is unique up to permutation of blocks. Thus the Jordan canonical form of M is represented by a partition of d the size of the matrix M. In what follows, we regard a partition of an integer as an expression of the integer as a sum of positive integers.

Theorem 9.28. *Let $\Phi : GL(d, K) \to GL((K^d)^{\otimes n})$ be the tensor representation of the general linear group $GL(d, K)$. Let $M = J(a,d)$. Then a^n is the unique eigenvalue of $\Phi(M)$ and the Jordan canonical form of $\Phi(M)$ is given by*

$$
J(a^n, u_1) \dotplus \cdots \dotplus J(a^n, u_r),
$$

where $d^n = u_1 + \cdots + u_r$ is the dual partition to the partition of d^n as the sum of the coefficients of the Hilbert polynomial $\mathrm{Hilb}(A,t) = (1 + t + \cdots + t^{d-1})^n$.

Proof. Let V be a d-dimensional vector space and let

$$
\Phi : GL(V) \to GL(V^{\otimes n})
$$

be the tensor representation. Put $A = K[x_1, x_2, \ldots, x_n]/(x_1^d, x_2^d, \ldots, x_n^d)$ and $B = K[x]/(x^d)$. As a vector space the algebra B is isomorphic to V and the multiplicative group B^* is represented by the matrices

$$\left\{ \left(\begin{array}{cccccc} a_0 & & & & & \\ a_1 & a_0 & & & \mathbf{0} & \\ a_2 & a_1 & a_0 & & & \\ \vdots & \vdots & \vdots & \ddots & & \\ a_{d-2} & \vdots & \vdots & \ddots & \ddots & \\ a_{d-1} & a_{d-2} & a_{d-3} & \cdots & \cdots & a_0 \end{array} \right) \right\} , \quad a_0 \neq 0$$

as an algebraic subgroup of $GL(V)$. Similarly the multiplicative group A^* is an algebraic subgroup group of $GL(V^{\otimes n})$.

Define the map

$$\Phi' : B^* \to A^*$$

by

$$f(x) \mapsto f(x_1) f(x_2) \cdots f(x_n).$$

Then it is easy to see that Φ' is a group homomorphism. Note that Φ' is nothing but the restriction of Φ. In other words we have the commutative diagram:

$$\begin{array}{ccc} GL(V) & \xrightarrow{\Phi} & GL(A) \\ \uparrow & & \uparrow \\ B^* & \xrightarrow{\Phi'} & A^* \end{array} , \tag{9.6}$$

where the vertical maps are natural inclusions. The element $a + x \in B$ is a unit if $a \neq 0$ and the image of $a + x$ by the vertical map in the diagram is $J(a, d)$. It follows that a^n is a unique eigenvalue of $\Phi'(a + x)$, since $\Phi'(a + x) - a^n$ is a nilpotent element. Put $l := x_1 + x_2 + \cdots + x_n$ and $l' = \Phi'(a + x) - a^n$. The degree one part of l' is a constant multiple of l, and moreover l is a strong Rees element in the sense of Definition 5.7. So l and l' have the same Jordan canonical form. Since A has the SLP and l is the strong Lefschetz element, the Jordan block decomposition is obtained as the dual partition of $\mathrm{Hilb}(A, t)$. Thus the assertion follows. □

Remark 9.29. Some related results can be found in the papers of Louis Solomon listed below, but the matters are dealt with in very different ways than with the SLP.

- *Invariants of Finite Reflection Groups*, Nagoya Math. J. 22, 57–64 (1963).
- *Invariants of Euclidian Reflection groups*, Trans. Am. Math. Soc. 113, 274–286 (1964).
- *Partition Identities and Invariants of Finite Groups*, Nagoya Math. J. 22, 57–64 (1963).

Remark 9.30. In §9.4, the Koszul complex plays an important role. Though the construction goes back to D. Hilbert in his 1980 paper on invariants and syzygies, the reference for the reintroduction in modern times should be J. L. Koszul, *Homologie et Cohomologie des Algébre de Lie*, Bull. Soc. Math. de France 78, 65–127 (1950).

Appendix A
The WLP of Ternary Monomial Complete Intersections in Positive Characteristic

So far in this monograph we have considered the Lefschetz properties mostly in characteristic zero. In this appendix we indicate some amazing results on the WLP in positive characteristic recently discovered by J. Li and F. Zanello [80] and also by C. Chen et al. [15].

The following is due to Migliore et al. [102]. (Here we state it in a weaker form than the original.)

Proposition A.1. *Suppose that* $A = \oplus A^e_{i=0} A_i$, $A_e \neq 0$ *is a standard Artinian algebra over* A_0, *an arbitrary field. Suppose that* $l \in A$ *is a linear form and* $\phi_d :$ $A_d \to A_{d+1}$ *is the homomorphism defined by the multiplication by* l *for* $d \geq 0$.

1. *If* ϕ_{d_0} *is surjective, then* ϕ_d *is surjective for all* $d \geq d_0$.
2. *If* ϕ_{d_0} *is injective, then* ϕ_d *is injective for all* $d \leq d_0$.
3. *If* A *is Gorenstein and if* e *is odd, then* A *has the WLP if and only if* $\phi_{(e-1)/2}$ *is bijective.*

The numerical function $M(a, b, c)$ defined below plays a crucial role in determining the WLP for a monomial complete intersection in three variables over a field of characteristic $p > 0$.

Definition A.2. For any integers a, b, c, define the numerical function $M(a, b, c)$ by

$$M(a, b, c) = \begin{cases} \prod_{i=1}^{a} \prod_{j=1}^{b} \prod_{k=1}^{c} \dfrac{i + j + k - 1}{i + j + k - 2} & \text{if } a, b, c \text{ are positive,} \\ 1 & \text{otherwise.} \end{cases}$$

Theorem A.3. *1. Let* α, β, γ *be positive integers such that* $\alpha + \beta + \gamma$ *is even, and let*

$$A = \mathbb{Q}[x, y, z]/(x^\alpha, y^\beta, z^\gamma).$$

T. Harima et al., *The Lefschetz Properties*, Lecture Notes in Mathematics 2080,
DOI 10.1007/978-3-642-38206-2, © Springer-Verlag Berlin Heidelberg 2013

Let U be the square matrix representing the multiplication map,

$$\times(x + y + z) : R_m \to R_{m+1}, \text{ where } m = (\alpha + \beta + \gamma - 4)/2,$$

in the monomial bases of R_m and R_{m+1}. Then

$$|\det U| = M \left(\frac{-\alpha + \beta + \gamma}{2}, \frac{\alpha - \beta + \gamma}{2}, \frac{\alpha + \beta - \gamma}{2} \right).$$

2. *Let K be a field of characteristic $p > 0$, α, β, γ positive integers such that $\alpha + \beta + \gamma$ is even, and let $A = K[x, y, z]/(x^\alpha, y^\beta, z^\gamma)$. Then A fails to have the WLP if and only if*

$$p \left| M \left(\frac{-\alpha + \beta + \gamma}{2}, \frac{\alpha - \beta + \gamma}{2}, \frac{\alpha + \beta - \gamma}{2} \right) \right.$$

(If $a + b + c$ is odd, there is an analogous criterion but it is a little more complicated. We omit the description.)

Remark A.4. In the above Theorem we may assume that $\alpha \le \beta \le \gamma$. We note that the socle degree e of A is odd; $e = \alpha + \beta + \gamma - 3$. So U is a square matrix. Thanks to Theorem A.1, the WLP of A is determined by the injectivity of U. Li and Zanello proved that $|\det U| = \det \left(\binom{c}{\beta - c + i - j} \right)_{1 \le i, j \le c}$, where $c = \frac{\alpha + \beta - \gamma}{2}$. It turns out that this determinant is equal to $M(\frac{-\alpha + \beta + \gamma}{2}, \frac{\alpha - \beta + \gamma}{2}, \frac{\alpha + \beta - \gamma}{2})$. This was proved in Krattenthaler [74, 2.17]. Visibly (1) implies (2).

Definition A.5. A plane partition of a positive integer n is a two-dimensional array $N = (n_{i,j})$ such that $n_{i,j} \ge n_{i,j+1} \ge 1$, $n_{i,j} \ge n_{i+1,j} \ge 1$ for all i, j and $\sum_{i,j} n_{i,j} = n$. We say that a plane partition $A = (a_{i,j})$ is contained in a box of size $a \times b \times c$ if $1 \le i \le a, 1 \le j \le b, 1 \le n_{i,j} \le c$.

The set of Young diagrams contained in a rectangle of size $a \times b$ may be considered as a lattice of ideals of two chains. In Chap. 1 this was denoted by $\mathscr{I}(a^b)$. It was shown that it is isomorphic to the lattice of ideals of two chains $\mathscr{I}(a^b) \cong \mathscr{I}(b^a) \cong J(C(a-1) \times C(b-1))$. Likewise, with a little contemplation, one sees that the set of plane partitions contained in a box of size $a \times b \times c$ is, with the containment order, isomorphic to the lattice of ideals of three chains

$$J(C(a-1) \times C(b-1) \times C(c-1)).$$

Theorem A.6. *The number of plane partitions contained in a box of size $a \times b \times c$ is equal to $M(a, b, c)$.*

This was discovered by MacMahon [86]. For a combinatorial proof see Krattenthaler [75].

C. Chen et al. [15] gave a bijective proof for this fact: $|\det U|$ counts the number of plane partitions in a box of size $a \times b \times c$. This is stated in the next theorem.

Theorem A.7 ([15]). *Let* $R = \mathbb{Z}[x, y, z]/(x^{a+b}, y^{b+c}, z^{c+a})$ *with* a, b, c *positive integers and set* $m = a + b + c - 2$. *Let* U *be the square matrix defining the multiplication map* $\times(x + y + z) : R_m \to R_{m+1}$ *in the monomial* $\mathbb{Z}$*-bases. Then the determinant of* U *is equal, up to sign, to the permanent of* U *and each nonzero term in its permanent corresponds to a plane partition in a box of size* $a \times b \times c$. *Therefore* $|\det U| = M(a, b, c)$.

As a corollary we have

Corollary A.8. *Let* K *be a field of characteristic* $p > 0$, a, b, c *positive integers and* $A = K[x, y, z]/(x^{a+b}, y^{b+c}, z^{c+a})$. *Then* A *fails to have the WLP if and only if* $p | M(a, b, c)$.

If $a = 1$, it gives us the following

Corollary A.9. *With the same* K *as above, let* b, c *be positive integers. Then* p *divides* $M(1, b, c) = \binom{b+c}{b}$ *if and only if* $K[x, y, z]/(x^{b+1}, y^{b+c}, z^{c+1})$ *fails to have the WLP.*

References

1. Ahn, J., Cho, Y.H., Park, J.P.: Generic initial ideals of Artinian ideals having Lefschetz properties or the strong Stanley property. J. Algebra **318**(2), 589–606 (2007). doi:10.1016/j.jalgebra.2007.09.016. http://dx.doi.org/10.1016/j.jalgebra.2007.09.016
2. Aigner, M.: Combinatorial theory. In: Classics in Mathematics. Springer, Berlin (1997). Reprint of the 1979 original
3. Anderson, I.: Combinatorics of finite sets. Dover Publications Inc., Mineola (2002). Corrected reprint of the 1989 edition
4. Anderson, D.D., Winders, M.: Idealization of a module. J. Commut. Algebra **1**(1), 3–56 (2009). doi:10.1216/JCA-2009-1-1-3. http://dx.doi.org/10.1216/JCA-2009-1-1-3
5. Bass, H.: On the ubiquity of Gorenstein rings. Math. Z. **82**, 8–28 (1963)
6. Bernšteĭn, I.N., Gel'fand, I.M., Gel'fand, S.I.: Schubert cells, and the cohomology of the spaces G/P. Uspehi Mat. Nauk **28**(3(171)), 3–26 (1973)
7. Boij, M., Laksov, D.: Nonunimodality of graded Gorenstein Artin algebras. Proc. Am. Math. Soc. **120**(4), 1083–1092 (1994). doi:10.2307/2160222. http://dx.doi.org/10.2307/2160222
8. Boij, M., Migliore, J.C., Miró-Roig, R.M., Nagel, U., Zanello, F.: On the shape of a pure O-sequence. Mem. Am. Math. Soc. **218**(1024), viii+78 (2012)
9. Bollobás, B.: Combinatorics: Set Systems, Hypergraphs, Families of Vectors and Combinatorial Probability. Cambridge University Press, Cambridge (1986).
10. Brenner, H., Kaid, A.: Syzygy bundles on $\mathbb{P}^2$ and the weak Lefschetz property. Ill. J. Math. **51**(4), 1299–1308 (2007). http://projecteuclid.org/getRecord?id=euclid.ijm/1258138545
11. de Bruijn, N.G., van Ebbenhorst Tengbergen, C., Kruyswijk, D.: On the set of divisors of a number. Nieuw Arch. Wiskunde (2) **23**, 191–193 (1951)
12. Bruns, W., Herzog, J.: Cohen-Macaulay rings. In: Cambridge Studies in Advanced Mathematics, vol. 39. Cambridge University Press, Cambridge (1993)
13. Buchsbaum, D.A., Eisenbud, D.: Algebra structures for finite free resolutions, and some structure theorems for ideals of codimension 3. Am. J. Math. **99**(3), 447–485 (1977)
14. Canfield, E.R.: On a problem of Rota. Adv. Math. **29**(1), 1–10 (1978)
15. Chen, C.P., Guo, A., Jin, X., Liu, G.: Trivariate monomial complete intersections and plane partitions. J. Commut. Algebra **3**(4), 459–489 (2011). doi:10.1216/JCA-2011-3-4-459. http://dx.doi.org/10.1216/JCA-2011-3-4-459
16. Chevalley, C.: Invariants of finite groups generated by reflections. Am. J. Math. **77**, 778–782 (1955)
17. Chevalley, C.: Classification des groups de Lie algébriques. Séminaire C. Chevalley, 1956–1958, 2 vols. Secrétariat mathématique, Paris (1958)

18. Chevalley, C.: Sur les décompositions cellulaires des espaces G/B. In: Algebraic Groups and Their Generalizations: Classical Methods, University Park, PA, 1991. Proceedings of the Symposium on Pure Mathematics, vol. 56, pp. 1–23. American Mathematical Society, Providence (1994). With a foreword by Armand Borel

19. Cho, Y.H., Park, J.P.: Conditions for generic initial ideals to be almost reverse lexicographic. J. Algebra **319**(7), 2761–2771 (2008). doi:10.1016/j.jalgebra.2008.01.014. http://dx.doi.org/ 10.1016/j.jalgebra.2008.01.014

20. Cimpoeaş, M.: Generic initial ideal for complete intersections of embedding dimension three with strong Lefschetz property. Bull. Math. Soc. Sci. Math. Roum. (N.S.) **50(98)**(1), 33–66 (2007)

21. Cook II, D., Nagel, U.: The weak lefschetz property, monomial ideals, and lozenges. Illinois J. Math. **55**(1), 377–395 (2012). MR3006693

22. Cook II, D., Nagel, U.: Enumerations deciding the weak lefschetz property (2011). Preprint, arXiv:1105.6062v2 [math.AC]

23. Conca, A.: Reduction numbers and initial ideals. Proc. Am. Math. Soc. **131**(4), 1015–1020 (electronic) (2003). doi:10.1090/S0002-9939-02-06607-8. http://dx.doi.org/10.1090/S0002-9939-02-06607-8

24. Conca, A., Krattenthaler, C., Watanabe, J.: Regular sequences of symmetric polynomials. Rend. Semin. Mat. Univ. Padova **121**, 179–199 (2009)

25. Constantinescu, A.: Hilbert function and Betti numbers of algebras with Lefschetz property of order m. Commun. Algebra **36**(12), 4704–4720 (2008). doi:10.1080/00927870802174074. http://dx.doi.org/10.1080/00927870802174074

26. Cook II, D., Nagel, U.: Hyperplane sections and the subtlety of the Lefschetz properties. J. Pure Appl. Algebra **216**(1), 108–114 (2012). doi:10.1016/j.jpaa.2011.05.007. http://dx. doi.org/10.1016/j.jpaa.2011.05.007

27. Cox, D., Little, J., O'Shea, D.: Ideals, varieties, and algorithms: An introduction to computational algebraic geometry and commutative algebra. In: Undergraduate Texts in Mathematics, 3rd edn. Springer, New York (2007). doi:10.1007/978-0-387-35651-8. http://dx.doi.org/10. 1007/978-0-387-35651-8.

28. Danilov, V.I.: The geometry of toric varieties. Uspekhi Mat. Nauk **33**(2(200)), 85–134, 247 (1978)

29. Dilworth, R.P.: A decomposition theorem for partially ordered sets. Ann. Math. (2) **51**, 161–166 (1950)

30. Dilworth, R.P.: Some combinatorial problems on partially ordered sets. In: Proceedings of Symposia in Applied Mathematics, vol. 10, pp. 85–90. American Mathematical Society, Providence (1960)

31. Eisenbud, D.: Commutative algebra. In: Graduate Texts in Mathematics, vol. 150. Springer, New York (1995). With a view toward algebraic geometry

32. Eliahou, S., Kervaire, M.: Minimal resolutions of some monomial ideals. J. Algebra **129**(1), 1–25 (1990). doi:10.1016/0021-8693(90)90237-I. http://dx.doi.org/10.1016/0021-8693(90)90237-I

33. Engel, K.: Sperner theory. In: Encyclopedia of Mathematics and Its Applications, vol. 65. Cambridge University Press, Cambridge (1997). doi:10.1017/CBO9780511574719. http:// dx.doi.org/10.1017/CBO9780511574719

34. Engel, K., Gronau, H.D.O.F.: Sperner theory in partially ordered sets. In: Teubner-Texte zur Mathematik [Teubner Texts in Mathematics], vol. 78. BSB B.G. Teubner Verlagsgesellschaft, Leipzig (1985). With German, French and Russian summaries

35. Erdős, P.: Extremal problems in number theory. In: Proceedings of Symposia in Pure Mathematics, vol. VIII, pp. 181–189. American Mathematical Society, Providence (1965)

36. Freese, R.: An application of Dilworth's lattice of maximal antichains. Discrete Math. **7**, 107–109 (1974)

37. Fulton, W.: Young tableaux. In: London Mathematical Society Student Texts, vol. 35. Cambridge University Press, Cambridge (1997). With applications to representation theory and geometry

38. Geramita, A.V.: Inverse systems of fat points: Waring's problem, secant varieties of Veronese varieties and parameter spaces for Gorenstein ideals. In: The Curves Seminar at Queen's, vol. X (Kingston, ON, 1995). Queen's Papers in Pure and Applied Mathematics, vol. 102, pp. 2–114. Queen's University, Kingston (1996)
39. Geramita, A.V., Harima, T., Migliore, J.C., Shin, Y.S.: The Hilbert function of a level algebra. Mem. Am. Math. Soc. **186**(872), vi+139 (2007)
40. Goodman, R., Wallach, N.R.: Representations and invariants of the classical groups. In: Encyclopedia of Mathematics and its Applications, vol. 68. Cambridge University Press, Cambridge (1998)
41. Gordan, P., Nöther, M.: Ueber die algebraischen Formen, deren Hesse'sche Determinante identisch verschwindet. Math. Ann. **10**(4), 547–568 (1876). doi:10.1007/BF01442264. http://dx.doi.org/10.1007/BF01442264
42. Greene, C., Kleitman, D.J.: Proof techniques in the theory of finite sets. In: Studies in Combinatorics. MAA Studies in Mathematics, vol. 17, pp. 22–79. Mathematical Association of America, Washington (1978)
43. Griffiths, P., Harris, J.: Principles of Algebraic Geometry. Wiley Classics Library. Wiley, New York (1994). Reprint of the 1978 original
44. Gröbner, W.: Über irreduzible Ideale in kommutativen Ringen. Math. Ann. **110**(1), 197–222 (1935). doi:10.1007/BF01448025. http://dx.doi.org/10.1007/BF01448025
45. Gunston, T.K.: Cohomological degrees, Dilworth numbers and linear resolution. Thesis (Ph.D.)–Rutgers The State University of New Jersey - New Brunswick. ProQuest LLC, Ann Arbor (1998). http://gateway.proquest.com/openurl?url_ver=Z39.88-2004&rft_val_fmt=info: ofi/fmt:kev:mtx:dissertation&res_dat=xri:pqdiss&rft_dat=xri:pqdiss:9915442
46. Hall, P.: On representatives of subsets. J. Lond. Math. Soc. **S1-10**(1), 26–30 (1935). doi:10.1112/jlms/s1-10.37.26
47. Hall, B.C.: Lie groups, Lie algebras, and representations. In: Graduate Texts in Mathematics, vol. 222. Springer, New York (2003). An elementary introduction
48. Hara, M., Watanabe, J.: The determinants of certain matrices arising from the Boolean lattice. Discrete Math. **308**(23), 5815–5822 (2008). doi:10.1016/j.disc.2007.09.055. http://dx.doi.org/10.1016/j.disc.2007.09.055
49. Harima, T., Wachi, A.: Generic initial ideals, graded Betti numbers, and k-Lefschetz properties. Commun. Algebra **37**(11), 4012–4025 (2009). doi:10.1080/00927870802502753. http://dx.doi.org/10.1080/00927870802502753
50. Harima, T., Watanabe, J.: The finite free extension of Artinian K-algebras with the strong Lefschetz property. Rend. Sem. Mat. Univ. Padova **110**, 119–146 (2003). See errata in [51].
51. Harima, T., Watanabe, J.: Erratum to: "The finite free extension of Artinian K-algebras with the strong Lefschetz property" [Rend. Sem. Mat. Univ. Padova **110**, 119–146 (2003); mr2033004]. Rend. Sem. Mat. Univ. Padova **112**, 237–238 (2004)
52. Harima, T., Watanabe, J.: The central simple modules of Artinian Gorenstein algebras. J. Pure Appl. Algebra **210**(2), 447–463 (2007). doi:10.1016/j.jpaa.2006.10.016. http://dx.doi.org/10.1016/j.jpaa.2006.10.016
53. Harima, T., Watanabe, J.: The strong Lefschetz property for Artinian algebras with non-standard grading. J. Algebra **311**(2), 511–537 (2007). doi:10.1016/j.jalgebra.2007.01.019. http://dx.doi.org/10.1016/j.jalgebra.2007.01.019
54. Harima, T., Watanabe, J.: The commutator algebra of a nilpotent matrix and an application to the theory of commutative Artinian algebras. J. Algebra **319**(6), 2545–2570 (2008). doi:10.1016/j.jalgebra.2007.09.011. http://dx.doi.org/10.1016/j.jalgebra.2007.09.011
55. Harima, T., Migliore, J.C., Nagel, U., Watanabe, J.: The weak and strong Lefschetz properties for Artinian K-algebras. J. Algebra **262**(1), 99–126 (2003). doi:10.1016/S0021-8693(03)00038-3. http://dx.doi.org/10.1016/S0021-8693(03)00038-3
56. Harima, T., Sakaki, S., Wachi, A.: Generic initial ideals of some monomial complete intersections in four variables. Arch. Math. (Basel) **94**(2), 129–137 (2010). doi:10.1007/s00013-009-0088-2. http://dx.doi.org/10.1007/s00013-009-0088-2

57. Herzog, J., Popescu, D.: The strong lefschetz property and simple extensions (2005). Preprint, arXiv:math/0506537
58. Hiller, H.L.: Schubert calculus of a Coxeter group. Enseign. Math. (2) **27**(1, 2), 57–84 (1981)
59. Hiller, H.: Geometry of Coxeter groups. In: Research Notes in Mathematics, vol. 54. Pitman (Advanced Publishing Program), Boston (1982)
60. Humphreys, J.E.: Introduction to Lie algebras and representation theory. In: Graduate Texts in Mathematics, vol. 9. Springer, New York (1978). Second printing, revised
61. Humphreys, J.E.: Reflection groups and Coxeter groups. In: Cambridge Studies in Advanced Mathematics, vol. 29. Cambridge University Press, Cambridge (1990)
62. Huneke, C., Ulrich, B.: General hyperplane sections of algebraic varieties. J. Algebr. Geom. **2**(3), 487–505 (1993)
63. Huybrechts, D., Lehn, M.: The Geometry of Moduli Spaces of Sheaves, 2nd edn. Cambridge Mathematical Library. Cambridge University Press, Cambridge (2010). doi:10.1017/CBO9780511711985. http://dx.doi.org/10.1017/CBO9780511711985
64. Iarrobino, A.: Associated graded algebra of a Gorenstein Artin algebra. Mem. Am. Math. Soc. **107**(514), viii+115 (1994)
65. Iarrobino, A., Kanev, V.: Power sums, Gorenstein algebras, and determinantal loci. In: Lecture Notes in Mathematics, vol. 1721. Springer, Berlin (1999). Appendix C by Iarrobino and Steven L. Kleiman
66. Ikeda, H.: Results on Dilworth and Rees numbers of Artinian local rings. Jpn. J. Math. (N.S.) **22**(1), 147–158 (1996)
67. Ikeda, H., Watanabe, J.: The Dilworth lattice of Artinian rings. J. Commut. Algebra **1**(2), 315–326 (2009)
68. Jacobson, N.: Lie Algebras. Dover Publications, New York (1979). Republication of the 1962 original
69. Jurkiewicz, J.: Chow ring of projective nonsingular torus embedding. Colloq. Math. **43**(2), 261–270 (1980/1981)
70. Kantor, W.M.: On incidence matrices of finite projective and affine spaces. Math. Z. **124**, 315–318 (1972)
71. Kaveh, K.: Note on cohomology rings of spherical varieties and volume polynomial. J. Lie Theory **21**(2), 263–283 (2011)
72. Kleiman, S.L.: Toward a numerical theory of ampleness. Ann. Math. (2) **84**, 293–344 (1966)
73. Koszul, J.L.: Homologie et Cohomologie des Algébre de Lie. Bull. Soc. Math. de France **78**, 65–127 (1950)
74. Krattenthaler, C.: Advanced determinant calculus. Sém. Lothar. Combin. **42**, Art. B42q, 67 pp. (electronic) (1999). The Andrews Festschrift (Maratea, 1998)
75. Krattenthaler, C.: Another involution principle-free bijective proof of Stanley's hook-content formula. J. Comb. Theory Ser. A **88**(1), 66–92 (1999). doi:10.1006/jcta.1999.2979. http://dx.doi.org/10.1006/jcta.1999.2979
76. Krull, W.: Idealtheorie. Zweite, ergänzte Auflage. Ergebnisse der Mathematik und ihrer Grenzgebiete, Band 46. Springer, Berlin (1968)
77. Kuhnigk, K.: On Macaulay duals of Hilbert ideals. J. Pure Appl. Algebra **210**(2), 473–480 (2007). doi:10.1016/j.jpaa.2006.10.014. http://dx.doi.org/10.1016/j.jpaa.2006.10.014
78. Lazarsfeld, R.: Positivity in algebraic geometry, I. Classical setting: line bundles and linear series. In: Ergebnisse der Mathematik und ihrer Grenzgebiete. 3. Folge. A Series of Modern Surveys in Mathematics [Results in Mathematics and Related Areas. 3rd Series. A Series of Modern Surveys in Mathematics], vol. 48. Springer, Berlin (2004).
79. Lefschetz, S.: L'analysis situs et la géométrie algébrique. Gauthier-Villars, Paris (1950)
80. Li, J., Zanello, F.: Monomial complete intersections, the weak Lefschetz property and plane partitions. Discrete Math. **310**(24), 3558–3570 (2010). doi:10.1016/j.disc.2010.09.006. http://dx.doi.org/10.1016/j.disc.2010.09.006
81. Lindsey, M.: A class of Hilbert series and the strong Lefschetz property. Proc. Am. Math. Soc. **139**(1), 79–92 (2011). doi:10.1090/S0002-9939-2010-10498-7. http://dx.doi.org/10.1090/S0002-9939-2010-10498-7

82. Lossen, C.: When does the Hessian determinant vanish identically? (On Gordan and Noether's proof of Hesse's claim). Bull. Braz. Math. Soc. (N.S.) **35**(1), 71–82 (2004). doi:10.1007/s00574-004-0004-0. http://dx.doi.org/10.1007/s00574-004-0004-0

83. Macaulay, F.S.: On the resolution of a given modular system into primary systems including some properties of Hilbert numbers. Math. Ann. **74**(1), 66–121 (1913). doi:10.1007/BF01455345. http://dx.doi.org/10.1007/BF01455345

84. Macaulay, F.S.: The Algebraic Theory of Modular Systems. Cambridge Mathematical Library. Cambridge University Press, Cambridge (1994). Revised reprint of the 1916 original, With an introduction by Paul Roberts

85. Macdonald, I.G.: Symmetric Functions and Hall Polynomials, 2nd edn. Oxford Mathematical Monographs. The Clarendon Press Oxford University Press, New York (1995). With contributions by A. Zelevinsky, Oxford Science Publications

86. MacMahon, P.A.: Combinatory Analysis. Two volumes (bound as one). Chelsea Publishing, New York (1960)

87. Maeno, T.: Lefschetz property, Schur-Weyl duality and a q-deformation of Specht polynomial. Commun. Algebra **35**(4), 1307–1321 (2007). doi:10.1080/00927870601142371. http://dx.doi.org/10.1080/00927870601142371

88. Maeno, T., Numata, Y.: Sperner property and finite-dimensional gorenstein algebras associated to matroids (2011). Preprint, arXiv:1107.5094

89. Maeno, T., Watanabe, J.: Lefschetz elements of Artinian Gorenstein algebras and Hessians of homogeneous polynomials. Ill. J. Math. **53**(2), 591–603 (2009). http://projecteuclid.org/getRecord?id=euclid.ijm/1266934795

90. Maeno, T., Numata, Y., Wachi, A.: Strong Lefschetz elements of the coinvariant rings of finite Coxeter groups. Algebras Represent. Theory **14**(4), 625–638 (2011). doi:10.1007/s10468-010-9207-9. http://dx.doi.org/10.1007/s10468-010-9207-9

91. Martsinkovsky, A., Vlassov, A.: The representation rings of $k[x]$. preprint (2004). http://www.math.neu.edu/~martsinkovsky/GreenExcerpt.pdf

92. Matlis, E.: Injective modules over Noetherian rings. Pac. J. Math. **8**, 511–528 (1958)

93. Matsumura, H.: Commutative ring theory. In: Cambridge Studies in Advanced Mathematics, vol. 8, 2nd edn. Cambridge University Press, Cambridge (1989). Translated from the Japanese by M. Reid

94. McDaniel, C.: The strong lefschetz property for coinvariant rings of finite reflection groups. J. Algebra **331**, 68–95 (2011)

95. McMullen, P.: The maximum numbers of faces of a convex polytope. Mathematika **17**, 179–184 (1970)

96. McMullen, P.: The numbers of faces of simplicial polytopes. Isr. J. Math. **9**, 559–570 (1971)

97. Meyer, D.M., Smith, L.: The Lasker-Noether theorem for unstable modules over the Steenrod algebra. Commun. Algebra **31**(12), 5841–5845 (2003). doi:10.1081/AGB-120024856. http://dx.doi.org/10.1081/AGB-120024856

98. Meyer, D.M., Smith, L.: Realization and nonrealization of Poincaré duality quotients of $\mathbb{F}_2[x, y]$ as topological spaces. Fund. Math. **177**(3), 241–250 (2003). doi:10.4064/fm177-3-4. http://dx.doi.org/10.4064/fm177-3-4

99. Meyer, D.M., Smith, L.: Poincaré duality algebras, Macaulay's dual systems, and Steenrod operations. In: Cambridge Tracts in Mathematics, vol. 167. Cambridge University Press, Cambridge (2005). doi:10.1017/CBO9780511542855. http://dx.doi.org/10.1017/CBO9780511542855

100. Mezzetti, E., Miró-Roig, R.M., Ottaviani, G.: Laplace equations and the weak lefschetz property. Can. J. Math. (2012, online first). doi:10.4153/CJM-2012-033-x. http://dx.doi.org/10.4153/CJM-2012-033-x

101. Migliore, J., Nagel, U.: A tour of the weak and strong lefschetz properties (2011). Preprint, arXiv:1109.5718v2 [math.AC]

102. Migliore, J.C., Miró-Roig, R.M., Nagel, U.: Monomial ideals, almost complete intersections and the weak Lefschetz property. Trans. Am. Math. Soc. **363**(1), 229–257 (2011). doi:10.1090/S0002-9947-2010-05127-X. http://dx.doi.org/10.1090/S0002-9947-2010-05127-X

103. Morita, H., Wachi, A., Watanabe, J.: Zero-dimensional Gorenstein algebras with the action of the symmetric group. Rend. Semin. Mat. Univ. Padova **121**, 45–71 (2009)
104. Motzkin, T.S.: Comonotone curves and polyhedra. In: Bull. Am. Math. Soc. [154], p. 35. doi:10.1090/S0002-9904-1957-10068-8. http://dx.doi.org/10.1090/S0002-9904-1957-10068-8
105. Mumford, D.: Stability of projective varieties. L'Enseignement Mathématique, Geneva (1977). Lectures given at the "Institut des Hautes Études Scientifiques", Bures-sur-Yvette, March-April 1976, Monographie de l'Enseignement Mathématique, No. 24
106. Murthy, M.P.: A note on factorial rings. Arch. Math. (Basel) **15**, 418–420 (1964)
107. Nagata, M.: Local rings. In: Interscience Tracts in Pure and Applied Mathematics, vol. 13. Interscience Publishers a division of Wiley, New York (1962)
108. Numata, Y., Wachi, A.: The strong Lefschetz property of the coinvariant ring of the Coxeter group of type H_4. J. Algebra **318**(2), 1032–1038 (2007). doi:10.1016/j.jalgebra.2007.06.016. http://dx.doi.org/10.1016/j.jalgebra.2007.06.016
109. Oda, T.: Torus embeddings and applications. In: Tata Institute of Fundamental Research Lectures on Mathematics and Physics, vol. 57. Tata Institute of Fundamental Research, Bombay (1978). Based on joint work with Katsuya Miyake
110. Oda, T.: Convex bodies and algebraic geometry. In: Ergebnisse der Mathematik und ihrer Grenzgebiete (3) [Results in Mathematics and Related Areas (3)], vol. 15. Springer, Berlin (1988). An introduction to the theory of toric varieties, Translated from the Japanese
111. Okon, J.S., Vicknair, J.P.: A Gorenstein ring with larger Dilworth number than Sperner number. Can. Math. Bull. **43**(1), 100–104 (2000). doi:10.4153/CMB-2000-015-2. http://dx.doi.org/10.4153/CMB-2000-015-2
112. Okon, J.S., Rush, D.E., Vicknair, J.P.: Numbers of generators of ideals in a group ring of an elementary abelian p-group. J. Algebra **224**(1), 1–22 (2000). doi:10.1006/jabr.1999.8082. http://dx.doi.org/10.1006/jabr.1999.8082
113. Okonek, C., Schneider, M., Spindler, H.: Vector bundles on complex projective spaces. In: Progress in Mathematics, vol. 3. Birkhäuser Boston, Boston (1980)
114. Pandharipande, R.: A compactification over $\overline{M}_g$ of the universal moduli space of slope-semistable vector bundles. J. Am. Math. Soc. **9**(2), 425–471 (1996). doi:10.1090/S0894-0347-96-00173-7. http://dx.doi.org/10.1090/S0894-0347-96-00173-7
115. Perles, M.A.: A proof of Dilworth's decomposition theorem for partially ordered sets. Isr. J. Math. **1**, 105–107 (1963)
116. Popov, V.L.: On the stability of the action of an algebraic group on an algebraic variety. Math. USSR Izvestiya **6**(2), 367 (1972). doi:doi:10.1070/IM1972v006n02ABEH001877. http://dx.doi.org/10.1070/IM1972v006n02ABEH001877
117. Proctor, R.A.: Solution of two difficult combinatorial problems with linear algebra. Am. Math. Mon. **89**(10), 721–734 (1982). doi:10.2307/2975833. http://dx.doi.org/10.2307/2975833
118. Pukhlikov, A.V., Khovanskiĭ, A.G.: The Riemann-Roch theorem for integrals and sums of quasipolynomials on virtual polytopes. Algebra i Analiz **4**(4), 188–216 (1992)
119. Reid, L., Roberts, L.G., Roitman, M.: On complete intersections and their Hilbert functions. Can. Math. Bull. **34**(4), 525–535 (1991). doi:10.4153/CMB-1991-083-9. http://dx.doi.org/10.4153/CMB-1991-083-9
120. Roberts, P.C.: A computation of local cohomology. In: Commutative Algebra: Syzygies, Multiplicities, and Birational Algebra, South Hadley, MA, 1992. Contemporary Mathematics, vol. 159, pp. 351–356. American Mathematical Society, Providence (1994)
121. Sagan, B.E.: The symmetric group: Representations, combinatorial algorithms, and symmetric functions. In: Graduate Texts in Mathematics, vol. 203, 2nd edn. Springer, New York (2001).
122. Sally, J.D.: Numbers of generators of ideals in local rings. Marcel Dekker, New York (1978)
123. Sekiguchi, H.: The upper bound of the Dilworth number and the Rees number of Noetherian local rings with a Hilbert function. Adv. Math. **124**(2), 197–206 (1996). doi:10.1006/aima.1996.0082. http://dx.doi.org/10.1006/aima.1996.0082

124. Shephard, G.C., Todd, J.A.: Finite unitary reflection groups. Can. J. Math. **6**, 274–304 (1954)
125. Shioda, T.: On the graded ring of invariants of binary octavics. Am. J. Math. **89**, 1022–1046 (1967)
126. Smith, L.: Note on the realization of complete intersections algebras as the cohomology of a space. Quart. J. Math. Oxford Ser **33**, 379–384 (1982)
127. Smith, L.: Polynomial invariants of finite groups. In: Research Notes in Mathematics, vol. 6. A K Peters Ltd., Wellesley (1995)
128. Solomon, L.: Invariants of finite reflection groups. Nagoya Math. J. **22**, 57–64 (1963)
129. Solomon, L.: Partition identities and invariants of finite reflection groups. Nagoya Math. J. **22**, 57–64 (1963)
130. Solomon, L.: Invariants of euclidian reflection groups. Trans. Am. Math. Soc. **113**, 274–286 (1964)
131. Specht, W.: Die irreduziblen Darstellungen der symmetrischen Gruppe. Math. Z. **39**(1), 696–711 (1935). doi:10.1007/BF01201387. http://dx.doi.org/10.1007/BF01201387
132. Sperner, E.: Ein Satz über Untermengen einer endlichen Menge. Math. Z. **27**(1), 544–548 (1928). doi:10.1007/BF01171114. http://dx.doi.org/10.1007/BF01171114
133. Stanley, R.P.: Cohen-Macaulay complexes. In: Higher Combinatorics, Proc. NATO Advanced Study Inst., Berlin, 1976, pp. 51–62. NATO Adv. Study Inst. Ser., Ser. C: Math. and Phys. Sci., 31. Reidel, Dordrecht (1977)
134. Stanley, R.P.: Hilbert functions of graded algebras. Adv. Math. **28**(1), 57–83 (1978)
135. Stanley, R.P.: The number of faces of a simplicial convex polytope. Adv. Math. **35**(3), 236–238 (1980). doi:10.1016/0001-8708(80)90050-X. http://dx.doi.org/10.1016/0001-8708(80)90050-X
136. Stanley, R.P.: Weyl groups, the hard Lefschetz theorem, and the Sperner property. SIAM J. Algebr. Discrete Methods **1**(2), 168–184 (1980). doi:10.1137/0601021. http://dx.doi.org/10.1137/0601021
137. Stanley, R.P.: Combinatorics and commutative algebra. In: Progress in Mathematics, vol. 41, 2nd edn. Birkhäuser Boston, Boston (1996)
138. Stanley, R.P.: Enumerative combinatorics, vol. 1. In: Cambridge Studies in Advanced Mathematics, vol. 49. Cambridge University Press, Cambridge (1997). With a foreword by Gian-Carlo Rota, Corrected reprint of the 1986 original
139. Steinberg, R.: Differential equations invariant under finite reflection groups. Trans. Am. Math. Soc. **112**, 392–400 (1964)
140. Stong, R.E.: Poincaré algebras modulo an odd prime. Comment. Math. Helv. **49**, 382–407 (1974)
141. Stong, R.E.: Cup products in Grassmannians. Topol. Appl. **13**(1), 103–113 (1982). doi:10.1016/0166-8641(82)90012-8. http://dx.doi.org/10.1016/0166-8641(82)90012-8
142. Terasoma, T., Yamada, H.: Higher Specht polynomials for the symmetric group. Proc. Jpn. Acad. Ser. A Math. Sci. **69**(2), 41–44 (1993). http://projecteuclid.org/getRecord?id=euclid.pja/1195511538
143. Trung, N.V.: Bounds for the minimum numbers of generators of generalized Cohen-Macaulay ideals. J. Algebra **90**(1), 1–9 (1984). doi:10.1016/0021-8693(84)90193-5. http://dx.doi.org/10.1016/0021-8693(84)90193-5
144. Watanabe, J.: A note on Gorenstein rings of embedding codimension three. Nagoya Math. J. **50**, 227–232 (1973)
145. Watanabe, J.: Some remarks on Cohen-Macaulay rings with many zero divisors and an application. J. Algebra **39**(1), 1–14 (1976)
146. Watanabe, J.: The Dilworth number of Artinian rings and finite posets with rank function. In: Commutative Algebra and Combinatorics, Kyoto, 1985. Advanced Studies in Pure Mathematics, vol. 11, pp. 303–312. North-Holland, Amsterdam (1987)
147. Watanabe, J.: m-full ideals. Nagoya Math. J. **106**, 101–111 (1987). http://projecteuclid.org/getRecord?id=euclid.nmj/1118780704
148. Watanabe, J.: A note on complete intersections of height three. Proc. Am. Math. Soc. **126**(11), 3161–3168 (1998). doi:10.1090/S0002-9939-98-04477-3. http://dx.doi.org/10.1090/S0002-9939-98-04477-3

149. Watanabe, J.: A remark on the hessian of homogeneous polynomials. In: Bogoyavlenskij, O., Coleman, A.J., Geramita, A.V., Ribenboim, P. (eds.) The Curves Seminar at Queen's, vol. XIII. Queen's Papers in Pure and Applied Mathematics, vol. 119, pp. 171–178. Queen's University, Kingston (2000)

150. Watanabe, J.: On the minimal number of the quotient of a complete intersection by a regular sequence. Proc. Sch. Sci. Tokai Univ. **47**, 1–10 (2012)

151. Watanabe, J.: On the theory of Gordan-Noether on the homogeneous forms with zero Hessian. Proc. Sch. Tokai Univ. **48** (2013) (to appear)

152. Weyl, H.: The classical groups. In: Princeton Landmarks in Mathematics. Princeton University Press, Princeton (1997). Their invariants and representations, Fifteenth printing, Princeton Paperbacks

153. Wiebe, A.: The Lefschetz property for componentwise linear ideals and Gotzmann ideals. Commun. Algebra **32**(12), 4601–4611 (2004). doi:10.1081/AGB-200036809. http://dx.doi.org/10.1081/AGB-200036809

154. Youngs, J.W.T.: The November meeting in Evanston. Bull. Am. Math. Soc. **63**(1), 29–38 (1957). doi:10.1090/S0002-9904-1957-10068-8. http://dx.doi.org/10.1090/S0002-9904-1957-10068-8

Index

T. Harima et al., *The Lefschetz Properties*, Lecture Notes in Mathematics 2080,
DOI 10.1007/978-3-642-38206-2, © Springer-Verlag Berlin Heidelberg 2013

LECTURE NOTES IN MATHEMATICS

 Springer

Edited by J.-M. Morel, B. Teissier; P.K. Maini

Editorial Policy (for the publication of monographs)

1. Lecture Notes aim to report new developments in all areas of mathematics and their applications - quickly, informally and at a high level. Mathematical texts analysing new developments in modelling and numerical simulation are welcome.

   Monograph manuscripts should be reasonably self-contained and rounded off. Thus they may, and often will, present not only results of the author but also related work by other people. They may be based on specialised lecture courses. Furthermore, the manuscripts should provide sufficient motivation, examples and applications. This clearly distinguishes Lecture Notes from journal articles or technical reports which normally are very concise. Articles intended for a journal but too long to be accepted by most journals, usually do not have this "lecture notes" character. For similar reasons it is unusual for doctoral theses to be accepted for the Lecture Notes series, though habilitation theses may be appropriate.

2. Manuscripts should be submitted either online at www.editorialmanager.com/lnm to Springer's mathematics editorial in Heidelberg, or to one of the series editors. In general, manuscripts will be sent out to 2 external referees for evaluation. If a decision cannot yet be reached on the basis of the first 2 reports, further referees may be contacted: The author will be informed of this. A final decision to publish can be made only on the basis of the complete manuscript, however a refereeing process leading to a preliminary decision can be based on a pre-final or incomplete manuscript. The strict minimum amount of material that will be considered should include a detailed outline describing the planned contents of each chapter, a bibliography and several sample chapters.

   Authors should be aware that incomplete or insufficiently close to final manuscripts almost always result in longer refereeing times and nevertheless unclear referees' recommendations, making further refereeing of a final draft necessary.

   Authors should also be aware that parallel submission of their manuscript to another publisher while under consideration for LNM will in general lead to immediate rejection.

3. Manuscripts should in general be submitted in English. Final manuscripts should contain at least 100 pages of mathematical text and should always include

   - a table of contents;
   - an informative introduction, with adequate motivation and perhaps some historical remarks: it should be accessible to a reader not intimately familiar with the topic treated;
   - a subject index: as a rule this is genuinely helpful for the reader.

   For evaluation purposes, manuscripts may be submitted in print or electronic form (print form is still preferred by most referees), in the latter case preferably as pdf- or zipped psfiles. Lecture Notes volumes are, as a rule, printed digitally from the authors' files. To ensure best results, authors are asked to use the LaTeX2e style files available from Springer's web-server at:

   ftp://ftp.springer.de/pub/tex/latex/svmonot1/ (for monographs) and
   ftp://ftp.springer.de/pub/tex/latex/svmultt1/ (for summer schools/tutorials).

Additional technical instructions, if necessary, are available on request from lnm@springer.com.

4. Careful preparation of the manuscripts will help keep production time short besides ensuring satisfactory appearance of the finished book in print and online. After acceptance of the manuscript authors will be asked to prepare the final LaTeX source files and also the corresponding dvi-, pdf- or zipped ps-file. The LaTeX source files are essential for producing the full-text online version of the book (see http://www.springerlink.com/openurl.asp?genre=journal&issn=0075-8434 for the existing online volumes of LNM). The actual production of a Lecture Notes volume takes approximately 12 weeks.

5. Authors receive a total of 50 free copies of their volume, but no royalties. They are entitled to a discount of 33.3 % on the price of Springer books purchased for their personal use, if ordering directly from Springer.

6. Commitment to publish is made by letter of intent rather than by signing a formal contract. Springer-Verlag secures the copyright for each volume. Authors are free to reuse material contained in their LNM volumes in later publications: a brief written (or e-mail) request for formal permission is sufficient.

Addresses:
Professor J.-M. Morel, CMLA,
École Normale Supérieure de Cachan,
61 Avenue du Président Wilson, 94235 Cachan Cedex, France
E-mail: morel@cmla.ens-cachan.fr

Professor B. Teissier, Institut Mathématique de Jussieu,
UMR 7586 du CNRS, Équipe "Géométrie et Dynamique",
175 rue du Chevaleret
75013 Paris, France
E-mail: teissier@math.jussieu.fr

For the "Mathematical Biosciences Subseries" of LNM:

Professor P. K. Maini, Center for Mathematical Biology,
Mathematical Institute, 24-29 St Giles,
Oxford OX1 3LP, UK
E-mail : maini@maths.ox.ac.uk

Springer, Mathematics Editorial, Tiergartenstr. 17,
69121 Heidelberg, Germany,
Tel.: +49 (6221) 4876-8259

Fax: +49 (6221) 4876-8259
E-mail: lnm@springer.com